Potts Models and Related Problems in Statistical Mechanics

SERIES ON ADVANCES IN STATISTICAL MECHANICS*

Editor-in-Chief: M. Rasetti *(Politecnico di Torino, Italy)*

Published

Vol. 7: The Hubbard Model – Recent Results
edited by M. Rasetti

Vol. 8: Statistical Thermodynamics and Stochastic Theory of Nonlinear Systems Far From Equilibrium
by W. Ebeling & L. Schimansky-Geier

Vol. 9: Disorder and Competition in Soluble Lattice Models
by W. F. Wreszinski & S. R. A. Salinas

Vol. 10: An Introduction to Stochastic Processes and Nonequilibrium Statistical Physics
by H. S. Wio

Vol. 12: Quantum Many-Body Systems in One Dimension
by Zachary N. C. Ha

Vol. 13: Exactly Soluble Models in Statistical Mechanics: Historical Perspectives and Current Status
edited by C. King & F. Y. Wu

Vol. 14: Statistical Physics on the Eve of the 21st Century: In Honour of J. B. McGuire on the Occasion of his 65th Birthday
edited by M. T. Batchelor & L. T. Wille

Vol. 15: Lattice Statistics and Mathematical Physics: Festschrift Dedicated to Professor Fa-Yueh Wu on the Occasion of his 70th Birthday
edited by J. H. H. Perk & M.-L. Ge

Vol. 16: Non-Equilibrium Thermodynamics of Heterogeneous Systems
by S. Kjelstrup & D. Bedeaux

Vol. 17: Chaos: From Simple Models to Complex Systems
by M. Cencini, F. Cecconi & A. Vulpiani

Vol. 18 Statistical Mechanics of Magnetic Excitations: From Spin Waves to Stripes and Checkerboards
by E. Rastelli

Vol. 19: An Introduction to Stochastic Processes and Nonequilibrium Statistical Physics
by H. S. Wio, R. R. Deza & J. M. López

*For the complete list of titles in this series, please go to
http://www.worldscientific.com/series/sasm

Series on Advances in Statistical Mechanics – Volume 5

Potts Models and Related Problems in Statistical Mechanics

Paul Martin
City University, London

World Scientific

NEW JERSEY • LONDON • SINGAPORE • BEIJING • SHANGHAI • HONG KONG • TAIPEI • CHENNAI

Published by

World Scientific Publishing Co. Pte. Ltd.
5 Toh Tuck Link, Singapore 596224
USA office: 27 Warren Street, Suite 401-402, Hackensack, NJ 07601
UK office: 57 Shelton Street, Covent Garden, London WC2H 9HE

British Library Cataloguing-in-Publication Data
A catalogue record for this book is available from the British Library.

First published 1991
Reprinted 2015

Series on Advances in Statistical Mechanics — Vol. 5
POTTS MODELS AND RELATED PROBLEMS IN STATISTICAL MECHANICS

ISBN-13 978-981-02-0075-6
ISBN-10 981-02-0075-7

Preface

This book is largely a rearrangement of a set of lectures I gave at RIMS, Kyoto in spring 1989. The additional material is either from lectures I *attended* in that period, or arose in the process of rearrangement. The book itself was written at Birmingham University Mathematics Department (summer 89) and City University Mathematics Department (autumn 89 onward).

The central theme of Potts models was chosen with the aim of drawing together, in a systematic way, some key ideas from the continuing explosion of statistical mechanical research. The acid test of relevance to this well established physical picture is one sure way to force a coherent direction on the general explosion of results.

The potential drawback of writing a book on such a vital subject is that one feels more and more pressure each week to incorporate the very latest results. Succumbing to this pressure would obviously result in a rather slowly converging process. In any case, most up to the minute research turns out not, on mature reflection, to be suitable for a textbook. I therefore introduced a cut-off on new material after spring 1989 (except where directly pertinent to the presentation of material already included). At time of going into print no aspect of the subject which has matured since then seems, by its omission, to leave a gap in the presentation.

Of course the subject will soon move on, but the fundamentals will still be fundamental. With this in mind, another aim was to provide a handy work of reference for my own research use, just as Baxter's (1982) book (with a different slant and broader but less recent perspective) has become for so many.

I have tried to make the mathematical development as explicit as is consistent with a finite number of pages. As a physicist I have in particular strived to convey some of the pleasure I have found in learning and applying algebraic techniques to physical problems... and physics is forever trying to teach us algebra.

London, England — Paul Martin

Acknowledgments

Large debts of gratitude are due to B W Westbury, T Miwa, M Jimbo, A Kuniba, M Okado, J B Martin, A J McKane, G P McCauley, M B Green, R A Wilson, G P Launer, M M Martin, C J B Martin and P E Woodley. Thanks are also due to M Wadati, T Deguchi, A J Guttman, E Date, Y Yamada, H N V Temperley and R Curtis.

Contents

Chapter 1

Introduction

Why Potts models? Why statistical mechanics?

I do not think there need be any mathematical model faithfully describing the universe as a whole. There are, however, some very beautiful mathematical models for physical phenomena perceived in restricted kinematic regimes. To me, then, a mathematical model which takes as input a description of a system at one length scale, and gives as output a description of this system at a different scale, is particularly exciting. Equilibrium statistical mechanics concerns models of this type.

In addition to this aesthetic consideration I note three major practical motivations for the application of statistical mechanics. One is the common need to determine properties of a physical system on a macroscopic scale from an initial description of the system on a relatively small (microscopic) but finite scale. Another is the discrete approximation of quantum field theory, in which the microscopic scale becomes infinitessimal. The third is the use of these physically supported pictures of statistical mechanics to provide insight into the many glorious mathematical spin-offs from the subject.

The Potts models are a special and easily defined class of statistical mechanical models, as we will see. Nonetheless, they are richly structured enough to illustrate almost every conceivable nuance of the subject. In particular, they are at the centre of the most recent explosion of interest generated by the confluence of conformal field theory, knot theory, quantum groups and integrable systems.

We are fortunate that *all* problems in statistical mechanics seem to be related to Potts models. Fortunate, because this means that a general discussion of the subject can be couched in Potts model terms. These models are invaluable in that they allow a ready understanding of their own basic physical significance and, compared to many of their more purely mathematically motivated counterparts, exhibit a robust insensitivity to boundary conditions away from the critical region. Any new result can be measured soberly against its implications for the Potts models. At the same time they present a great challenge as the most tantalising of unsolved models....

1.1 On layout and objectives

There exist several splendid books and reviews having some overlap of ambit with the present work. Baxter's (1982) book is a particularly fine example, and we will mention many others as we go along. The material which has been covered has been covered well, and there is no point in going over the same ground again here. On the other hand, there is plenty of scope for progress in the same philosophical territory, without duplicating technical details which are already reported so lucidly. With this in mind an exhaustive survey of work in the field of statistical mechanics has been omitted in preference to one which, while self-contained, covers predominantly new ground. The review material necessary for self-containedness has, as far as possible, been given a novel slant. This book is not intended, then, to be an alternative to, or review of, existing works in the field, but rather a companion to them.

One of the striking features of recent developments in statistical mechanics has been their profound interest to physicists and mathematicians alike. Another feature has been the bewildering proliferation of solvable models and solutions to the star-triangle relations. This Gibbsian/Boltzmannian pulchritude has tended to defy any systematic categorisation. At present, the physical significance of these models is not uniformly clear. However, just as mathematics has been the source of many new models, so the needs and perspectives of physics should provide the main source of organisational criteria.

The first theme introduced here, therefore, will be the basic physical motivation common to these models. This is reviewed in so far as it is needed to understand and organise the models. The basic calculational techniques involved are also introduced, and a framework for understanding and assessing mathematical *results* in physical terms is established. The Potts model is used as an example throughout. This theme might be generally characterised as classical statistical mechanics for mathematicians.

Having established this foundation we then develop an algebraic framework for the statistical mechanical notion of equivalence of models (a key step in categorisation), which leads us to regard the *transfer matrix* as a representative of an element of an associative algebra. It becomes natural, therefore, to characterise and classify such models by the algebra they represent. The development of this characterisation is our second running theme. For example, the most popular such algebra is the Temperley-Lieb algebra, which is the algebraic embodiment of the Potts model. We give a detailed description of this algebra, so that the reader is eventually equipped to go out and construct arbitrary representations, and hence all Temperley-Lieb solutions to the star-triangle relations. This theme can perhaps be characterised as algebra for physicists.

Finally, we discuss some of the ways in which the fruit of the union of ideas coming respectively from the mathematical and physical perspectives may be harvested! This subject is currently the focus of a huge global research drive. The objective here is to get us to the point where we can join in this drive, rather than to attempt to chronicle it fully.

Table of rough interdependence of chapters

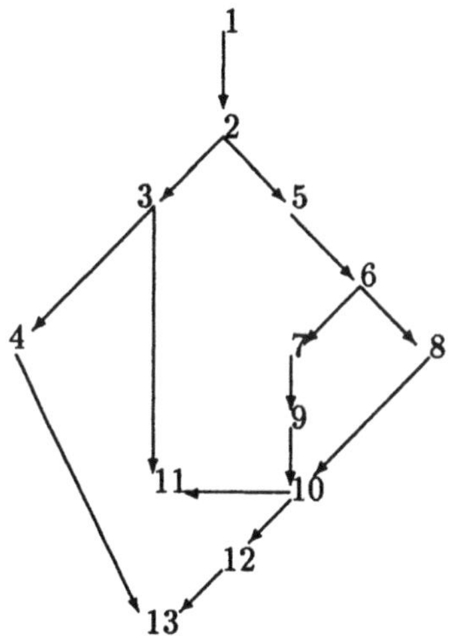

Referencing, in a work with a primarily pedagogical ambit, is often a compromise between fastidious acknowledgement of original sources, however obscure, and recommendation of the clearest pedagogical exposition. Here we will not compromise. I apologise in advance to offended parties! The method of referencing is to give names, and dates where necessary to avoid ambiguity, in the text. The full references are then given in the bibliography.

Notations

In statistical mechanics the singleminded persuit of a universally standardised notation has a tendency to become counter productive. It is sensible, nonetheless, to standardise notation as far as possible. There follows a list of standard symbols and notations. Many of them are sufficiently ubiquitous in the literature as to require no explanation, others will be unfamiliar. Anyway, here they are.

$\mathbf{C}$ Field of complex numbers. Unless otherwise stated we will work over the field of complex numbers throughout;

$\mathbf{R}$ Real numbers;

$\mathbf{Z}$ Integers;

$\mathbf{N}, \mathbf{Z}_+$ Natural numbers, positive integers;

Z_Q cyclic group of order Q (i.e. additive version: elements $p = 0, 1, .., Q-1$, composition given by addition mod Q; multiplicative version: elements $\exp(2\pi ip/Q)$);

$[p]$ integer part of $p \in \mathbf{R}$;

d dimension of a physical system, i.e. $d \in \mathbf{N}$;

1_d d dimensional unit matrix;

$$\text{diagonal}(a,b,c) = \begin{pmatrix} a & 0 & 0 \\ 0 & b & 0 \\ 0 & 0 & c \end{pmatrix};$$

$[A, B]$ $= AB - BA$, commutator of matrices A, B;

$\{A, B\}$ $= AB + BA$, anticommutator of matrices A, B;

$\{U_i\ ;\ i = 1, .., k\}$ a set of k objects U_i, e.g. generators of an algebra;

$\delta_{a,b} \quad = \begin{cases} 1 & \text{if } a = b; \\ 0 & \text{otherwise;} \end{cases}$

V_d d dimensional vector space, or specific set of basis elements;

V vector space of indeterminate dimension, *or* specific set of basis elements for V_d;

$\dim V = d$ order of set V;

$A \otimes B$ tensor or direct product of matrices (or vector spaces) A and B;

$A \oplus B$ direct sum of matrices (or vector spaces) A and B;

$\oplus_i d_i A_i$ $= (1_{d_1} \otimes A_1) \oplus (1_{d_2} \otimes A_2) \oplus \ldots$, $d_i \in \mathbf{N}$, A_i matrices;

$\otimes^n V_d$ $= V_d \otimes V_d \otimes V_d \otimes \ldots \otimes V_d$;

S_d simple module of dimension d;

P_d projective module of dimension d;

$End(V_d)$ endomorphisms of V_d;

$B = End_A(V_d)$ A is an algebra, V_d an A-module, then B is the algebra of linear transformations on V_d which commute with the action of A;

$M_d(\mathbf{C})$ algebra of d-dimensional matrices over the complex numbers $\mathbf{C}$;

N Number of particles in system;

$\{\sigma\}$ a set of variables $\{\sigma_i; i = 1, .., N\}$;

$\hom(\{\sigma\}, V)$ hom-set consisiting of all functions on $\{\sigma\}$ to set V, i.e. $\hom(\{\sigma\}, V) \sim V^N$;

$\{\sigma_i\}$ a possible *configuration* of a set of variables $\sigma_i \in V$, i.e. $\{\sigma_i\} \in hom(\{\sigma\}, V)$;

H Hamiltonian;

Z Partition function;

$< O >$ Expectation value of observable O;

$\langle ij \rangle$ nearest neighbour pair of lattice sites, i and j;

T Transfer matrix;

$\prod'_i$ product over i, the prime is a warning that i is incremented by other than +1 (the increment will be posted in the text adjacent to the first occurence of each variety).

We have omitted some definitions of symbols which have come to have two or more standard meanings. In such cases the appropriate definition will be given where needed.

1.2 Statistical mechanics

In the remainder of this chapter we discuss some fundamental notions of statistical mechanics, define Potts models, and introduce the language of *block spin renormalisation*. These ideas will provide the basis for our subsequent physical discussion.

In this section, before proceeding to technical matters, we offer up a thumbnail sketch of the classical equilibrium statistical mechanical world. Its purpose is to provide an opportunity for the reader to compare his or her intrinsic reference frame on the relevant concepts with that of the author. It will also serve to introduce some technical notation. For a traditional (100 page plus) exposition see, for example, Huang (1963).

There is no need to dwell here on the variety of phenomenological motivations for statistical mechanics, which depend on whether the many body physical system being modelled is a gas, a magnet, a biological system, or the universe in general. The way in which equilibrium statistical mechanics treats all these problems is, in its mathematical essence, the same. We will go into details later. The physical interpretation of statistical mechanics is traditionally in terms of gases and magnets. More recently, the physical justification for mathematical studies of the kind we will describe in this book has come increasingly from the discrete modelling of quantum field theory (see, for example, Kogut 1979). However, the intuitive base for this interpretation seems to reside once again in the traditional magnetic interpretation. We will lose nothing in what follows, therefore, if we concentrate mainly on that.

Consider a physical system of many particles, distributed in space and interacting through known forces dependent on this distribution. Even if it is a deterministic system it will not generally be possible *in practice* to calculate its subsequent behaviour from a given initial state. On the other hand, such a mass of information would probably be overwhelming, and hence useless from a physical point of view anyway.

What would be much more useful would be the calculation of bulk properties of the system. These, if any exist, are the results of measurements on the system as a whole with the property that if the system is split into 2 isolated parts along some boundary, and both parts are large enough so that the number of particles close to the boundaries are negligable compared to the numbers in the bulk of the system, then each part will give the same result for an equivalent measurement. The possibility of such a measurement clearly implies a certain resilience of the system..... a sense in which 2 non-identical systems can still be thought of as the same ***kind*** of system. We know from nature that such systems and reproducible measurements exist (cutting an ice cube in half does not change its melting point).

It is not easy to be very precise about these notions in the physical context. We will be able to be more precise once we have set up a mathematical version. In general, particles close to the boundary are those for which the (supposedly ***known***) interaction forces are affected by cutting at the boundary. Negligable numbers of particles close to the boundary then means that the ratio of boundary to bulk numbers tends to zero as $N \to \infty$. Note that these conditions imply that conservation of energy should not play a direct role in bulk properties (since we are ignoring energy escaping across boundaries). It is replaced by a concept of temperature, as we will see.

The problem of prediction of bulk properties of such a system when carrying finite energy is still not one amenable to direct solution. Considering the particles to be fixed in location (in a crystal lattice, for example), and interacting only with respect to some internal freedom (such as magnetic dipole orientation), does not change the magnitude of the problem. Let us in addition, however, consider the situation in which the bulk properties of interest are essentially constant in time. We can then attempt to model the system by equilibrium statistical mechanics. In this scheme the physical distribution and possible states of the particles appear in a recognisable manner, but the microscopic temporal fluctuations between states are averaged and absorbed within a mathematical construct called the partition function.

1.2.1 Partition functions and Hamiltonians

In general terms, the situation is as follows. Let us introduce statistical mechanical versions of the particles - variables σ_i indexed by their location and taking values from some set representing the possible states (configurations) of each particle. Let us denote the complete set of such variables for the system by $\{\sigma\}$. We must then define the classical Hamiltonian $H(\{\sigma\})$ for the system, which is a certain function from the possible configurations of the system as a whole to the real numbers. That is, the domain of H is the set of configurations of the system, and the codomain the reals. The precise form of $H(\{\sigma\})$ depends on the system. The resiliency property of the system is manifested through the existence of an intensive version of H, that is a formulation which is essentially similar for any 2 large subsystems. We further define the statistical mechanical observable $O(\{\sigma\})$ as another function from configurations to the real numbers.

Let us introduce a notation for the sum over all possible configurations of the system $\{\sigma\}$. If σ_i takes values from some set V for $i = 1, 2, .., N$, say, then

$$\sum_{\{\sigma\}} \equiv \sum_{\sigma_1 \in V} \sum_{\sigma_2 \in V} \cdots\cdot \sum_{\sigma_i \in V} \cdots\cdot \sum_{\sigma_N \in V} .$$

We may then define the partition function as a function of real variables β and J_o in the following way

$$Z(\beta, J_o) = \sum_{\{\sigma\}} \exp(\beta H(\{\sigma\}) + J_o.O(\{\sigma\})). \qquad (1.1)$$

This partition function for a statistical mechanical system, $Z(\beta, J_o)$, is a generating function for statistical mechanical *expectation values*. Defining $Z = Z(\beta, 0)$ we then define the expectation value $< O >$ for a given $O(\{\sigma\})$ as a function of β as follows

$$< O > \; = \; \left.\frac{\partial \ln(Z(\beta, J_o))}{\partial J_o}\right|_{J_o=0} \; = \; \frac{\sum_{\{\sigma\}} O(\{\sigma\}) \exp(\beta H)}{Z}. \qquad (1.2)$$

These expectation values can be thought of as predictions for the results of certain bulk measurements on a corresponding equilibrium physical system (although, having introduced the formal definition of equation 1.2, we are free *mathematically* to use any observable O, which may or may not correspond to a bulk property).

In order to understand how the correspondence between physical system and model works (and determines H) we we must first take a closer look at the kind of physical system, and measurement, we hope to model. The system is generally a many body system. It is one which shows no trends in its configurational fluctuations, and hence no systematic variation in its bulk properties, on time scales large or comparable with those of the measurements. The measurements are, in turn, associated with time scales large compared to those of the *thermal* fluctuations of the system.

The thermal fluctuations of a system may then be thought of simply as a mechanism by which it visits a number of its possible states, with various frequencies, during a given time period. Our stipulation is that during the time of a measurement the integrated number is large (although we will have to be a bit more careful about this later on). The measuring instrument will somehow average the values of the observable in these states to arrive at the measurement.

For the purposes of the measurement the system is thus in a time independent, hence equilibrium, state. The statistical mechanical version of the measurement takes advantage of this. It replaces the unknowable microscopic dynamics by a weighted average over all the possible states of the system. The weight for a state, $W(\{\sigma\})$, called the Boltzmann weight, allows simply for the relative likelihood of the system being in that state, given the known parameters of the system. The standard known parameter is the temperature. A few other bulk parameters like an external magnetic field strength may also appear. The Boltzmann weight decides which states are more popular than others, and how much so. Historically, and for physical reasons, this weight has been expressed as the exponential of the inverse temperature times an 'energy' which is a property of the particular state independent of the temperature. We write

$$W(\{\sigma\}) = e^{\beta H(\{\sigma\})}.$$

so that $\beta \propto 1/T$, plays the role of the inverse temperature (we can omit physical constants and units for the present). The energy assignment, or Hamiltonian, is based on some good physical picture of the system at the molecular level (or at whatever microscopic level for which we *have* a 'good physical picture').

We see from equation 1.1 that a given model is mathematically defined by completely specifying $H(\{\sigma\})$. To do this we must enumerate the configurations of the system, and associate an energy (a positive number) to each one.

This is a crucially important step, and we will come to specific physically illustrative examples shortly. However, a number of models have been introduced for which the energetic interpretation is in any case obscure. The assignment of weights has taken place to suit mathematics rather than physics. This need not detract from the interest of such a model. The underlying battle lines in the conflict between energy (as represented by the weight, in whatever form) and entropy (as represented by the possible configurations of the system) survive in essence, *if they exist at all*, independently of how perversely the classical Hamiltonian is formulated. We will return to this point in the next section, but essentially it means that *any* resilient H with an interesting partition function is a legitimate candidate for study.

The interpretation of $O(\{\sigma\})$ is as the idealised (but physically unmeasurable) instantaneous version of an observable O involving the states of the corresponding particles. The result of an attempted physical measurement on the observable O will be time averaged over the response time of the measuring instrument. In the statistical mechanical representation it is replaced by the expectation value $< O >$, a weighted average over all possible configurations. Hence equation 1.2.

Let us now consider the crucial matter of specifying the Hamiltonian. The study of truly arbitrary choices for H would be a huge and largely physically meaningless task. Physically, and to make sense of other statistical mechanical functions to be defined shortly, we will need to be able to simulataneously define a sequence of Hamiltonians associated with different numbers of particles but with the same 'kind' of physical system. As indicated before, this is akin to requiring that different subsystems of a given system correspond to similar systems. Given the form of the Boltzmann weight, and assuming that each subsystem is large enough for boundary numbers to be negligable, this is equivalent to requiring that the Hamiltonian has an intensive version. That is, that H for the composite system is essentially the sum of the Hamiltonians for the subsystems, in the limit of very large subsystems.

To be more specific about this we need to make some preliminary remarks about the distribution of particles, and the way they contribute to H. In this it will be useful to introduce the notion of a composite Hamiltonian $H(\underline{\beta})$ (the dependence on configurations here is implicit), which consists of the entire argument of the exponential in equation 1.1 or, more generally, any linear combination of Hamiltonian functions multiplied by scalar variables:

$$H(\underline{\beta}) = \sum_{sub-Hamiltonians\ j} \beta_j H_j(\{\sigma\}).$$

Substituting this into the argument of the exponential in equation 1.1 allows us to build more general, multi-parameter, models.

Realistically we will also need to introduce the notion of a lattice. This is an art in itself....

1.2.2 Defining lattices

In this book we will be concerned with *lattice* statistical mechanical models (i.e. with models of countably many particles). As we have seen, the definition of such a model involves a ***composite Hamiltonian***, which may be thought of as a map from the configurations of a discrete set of N

variables σ_i ($i = 1, .., N$ with N typically large), to linear combinations of indeterminates $\{\beta_j\}$ ($j = 1, .., p$ with p typically 1).

This map may be broken up as a linear combination of constituent Hamiltonians. The requirement that boundary numbers be negligable in the large N limit suggests that the constituent maps may each be further broken up as a linear combination of maps called *interactions*, each depending non-trivially on only a small subset of the variables $\{\sigma_i\}$; and that the variables in each subset should be 'close together'. We now see that 'particles close to the boundary' should mean 'particles associated with interactions cut by the boundary'. 'Close together' then means 'such that the boundary numbers remain negligable'. This is still not a well defined condition, since it depends on the shape of the boundary. Assuming a smooth boundary, a sufficient, but not necessary, working rule is to consider subsets involving particles which are nearest neighbours at most a couple of times removed.

If $\{\sigma\}_s$ is such a subset and we have interaction $I(\{\sigma\}) = I_s(\{\sigma\}_s)$ we call such an interaction *local* or *short range*. We have

$$H(\{\sigma\}) = \sum_{subsets\{\sigma\}_s} I_s(\{\sigma\}_s).$$

Imposing these conditions does not guarantee boundary *condition* independence in the limit. Indeed much of the interest in statistical mechanics resides in the 'critical points' of β where this may break down - the short range interactions co-operating to induce long range ones (see section 1.4). We want to keep boundary condition dependence out of the effect of the interactions themselves, so that these co-operative events are easy to spot when they occur. Necessary conditions for this property are difficult to determine. It is in a spirit of pragmatism, then, that we choose to accept the suggested conditions on H as an initial organisational criterion for our mathematical search.

In any case, the way in which these subsets are determined is generally most easily described in terms of a *lattice*.

The resiliency requirement suggests that the interactions translate into one another in some way, i.e. that they look the same in different parts of the system. Under these conditions the interaction I fulfils the role of the intensive version of the Hamiltonian. The lattice will also be a convenient way of describing the relative location of particles.

Part of the usual definition of a lattice model, then, is the definition of an appropriate lattice.

For this purpose a lattice is some set of points bounding lines bounding faces bounding bodies (and so on), which we usually think of as being embedded in some d dimensional space. Consider the convex hull of any set of points on the lattice, and suppose that this has dimension $c \leq d$. If no point on the lattice is an interior point of any segment in the convex hull, then this hull is a c-dimensional (lattice) simplex. We will refer to the points, lines, faces and so on generically as simplices, although in practice we tend to use degenerate cases like hypercubes. Typically statistical mechanical models involve interactions associated with $n \leq d$ dimensional simplices, in that the interactions are dependent on degrees of freedom (variables) associated with the $n-1$ dimensional simplices on their boundaries (see chapter 10).

All, or almost all, of the simplices of a given dimension which carry an interaction should be of the same form. In the case of the Potts model, for example, the variables sit on sites and the interactions on bonds, regardless of the dimension of the embedding space.

Making only the stipulation that we can divine the interactions in such cases, it is not necessary to say anything about the embedding space. We could sufficiently describe the lattice by an incidence matrix - an array indicating which points are adjacent (see chapter 2). This is really too loose a description for our purposes, however. We often wish to regard the lattice model as a discretisation of a continuum model, or at least a model resolved on a much finer length scale, in a systematic way. This, together with our previous considerations, means that we must be able to think of any lattice as part of a sequence of lattices which approach a dense covering of the embedding space. Of course in most contexts in solid state physics our lattice sits on the plain, or in euclidean 3 space. But for some applications we need to be able to consider more interesting topologies, where a denser covering cannot be achieved by simply changing the length scale. One resolution of this problem, at least in 2 dimensions, is to consider only lattices which are constructed *starting* from a closed surface, as described below.

Two dimensions

Consider a set of curves on a closed surface (other than the sphere S^2) constructed as follows (see figure 1.1 for an illustrated example). Chop the surface into 3-punctured spheres (which are often called 'pants', as in trousers!) and empty discs (1-punctured spheres). For each pants draw 2 generating elements of the fundamental group. Draw as many non-touching *circles* as you like around each puncture. These must not touch the generators. Draw as many non-touching *lines* as you like (almost) between each pair of punctures. You must choose the total number of lines at a puncture to match its partner in the original surface. For example, in figure 1.1 the two rightmost punctures, which, in the way we have chopped, are partners

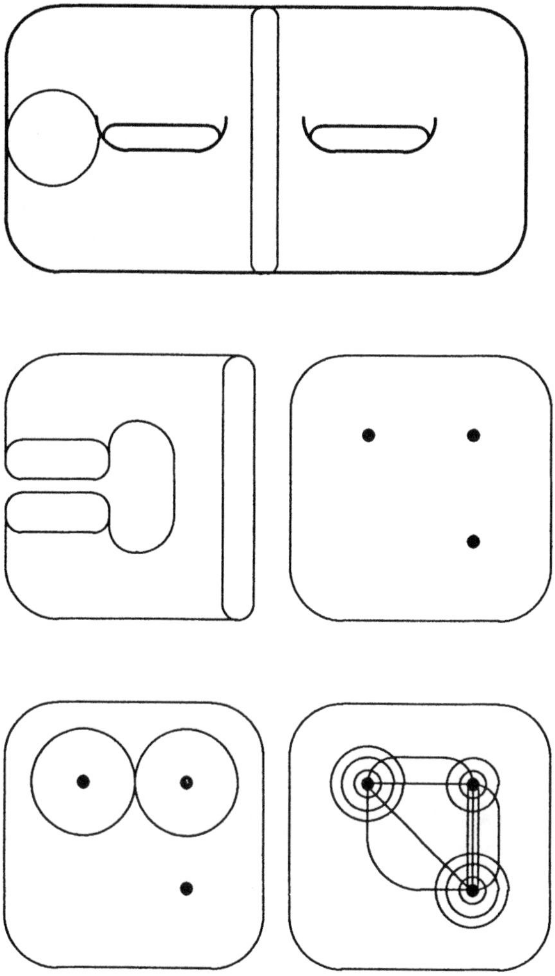

Figure 1.1: The top two pictures indicate one of two possible ways of chopping a double doughnut into two 3-punctured spheres. The bottom picture then illustrates the marking of 2 generating elements of the fundamental group on the left hand sphere, and drawing some possible lattice lines on the other.

in the original surface, must therefore have the same number of incident lines. Sewing up the original surface by identifying lines across boundaries allows some ambiguity (see later).

All crossings in this construction have coordination number 4 and all faces are quadrilateral, except for 2 hexagons per pants (there is a dual construction giving all quadrilateral faces). Arcs of curves between crossings will be called edges.

The mesh of circles and lines we have constructed constitute a lattice. Note the properties of this lattice under the replacement of single lines by multiple parallel lines. The local lattice structure remains unchanged, as does the number of 'anomalous' hexagons. Iterating the replacement generates a sequence of lattices. The local lattice structure remains that of a square lattice throughout.

Once we have established our sequence of lattices in this way, we can take each one individually and make it into a locally regular triangular, or hexagonal, or medial lattice in a systematic way. In each case the number of anomalies remains fixed throughout the sequence, depending only on the original topology.

For triangular lattices we must add a single diagonal edge to each square, in such a way that the diagonals themselves are all (locally) parallel. For hexagonal lattices we take the *dual* of the triangular lattice (i.e. replace $n = 0, 1, 2$ dimensional simplices with $d - n = 2, 1, 0$ dimensional simplices). For medial lattices we take the square lattice and *inscribe* squares at 45° inside each square, so that the new vertices touch the center of each edge. In this case we then discard the original lattice.

To recapitulate: A model is, in principle, associated with a physical space and hence a *sequence* of lattices. Each of these may individually be defined by incidence relations. We have only explained how to derive the sequences in the case of Euclidean spaces and Riemann surfaces. These will be sufficient to cover all the models explicitly discussed in this book.

Armed with a suitable lattice, it then only remains for us to specify the nature of the local interactions I as functions of, say, the variables associated with the boundaries of some class of lattice simplex, in order to arrive at a Hamiltonian. Conversely, changing the lattice whilst holding the nature of the interactions fixed produces an automatic implicit change in the Hamiltonian. By convention we do not regard it as being explicitly changed in this case. In practice, then, a Hamiltonian is specified as a sum of interactions of a fixed form over the appropriate lattice simplices. The sequence of Hamiltonians to which statistical mechanical bulk properties may be attached arises implicitly from the sequence of lattices.

1.2.3 Other statistical mechanical functions

The free energy $f(\beta)$ is defined by

$$f(\beta) = (1/N)\ln(Z)$$

where N is conventionally called the 'size' of the system, since it may typically be expressed as the number of lattice sites. In fact $f(\beta)$ is usually understood to refer specifically to the limit case in which the number of particles is taken to infinity in this expression (by taking the lattice size to infinity while holding the nature of the interactions fixed). This is called the *thermodynamic limit*. The well definedness of such a limit is the mathematical version of our idea of resilience.

The internal energy per site $U(\beta)$ is defined by

$$U(\beta) = -\frac{\partial f(\beta)}{\partial \beta} = \left\langle \frac{-H}{N} \right\rangle .$$

It is also useful to introduce the *expectational probability* for a given configurational feature of a system (eg. for a single variable to take a given fixed value), which is defined to be

$$p(feature) = \frac{1}{Z} \sum_{\{\sigma_i | \, feature\}} \exp(\beta H)$$

where the sum is restricted to configurations which manifest the given feature. We will give examples in and following equation 1.5.

As we have suggested, many of the models which are presently interesting from a mathematical point of view do not correspond particularly closely, in their Hamiltonians, to anything found in nature at the molecular (or smaller) levels. The precise results of statistical mechanical calculations for observables are thus, by and large, unrelated to specific physical problems. The great exception to this rule is the calculation of properties close to phase transitions. Here the sensitivity to the precise details of the Hamiltonian tends to be washed away, and our results may well become physically relevant again. This phenomenon is known as universality. We will give a partial explanation of this effect in section 1.7.

In so far as the Hamiltonians involved *are* appropriate, statistical mechanics provides models of the various wars of order versus disorder in the physical world, together with other conflicts of energy and entropy, and their occasional cataclysmic boundaries of sway, the phase transitions. In order to give an explanation of the way in which it does so, it is necessary to introduce some specific examples. This we will do in the coming sections, where we will discuss some of the most highly idealised and yet richly structured models known to science!

1.3 Potts models

The model we are about to describe was first studied by Potts (1952), following a proposal made by Domb (for a thesis topic). It is a generalisation of a simple model of ferromagnetism introduced by Ising in 1925.

Consider a lattice which, for the sake of definiteness, we will take to be locally square. Associate with each lattice site i a variable or ***spin*** s_i (if the variables in a lattice model are associated with sites we call them spins, and the model a spin model). This variable is to be thought of as somehow representing the local microscopic state of some physical system. In an idealised form, and discarding for the moment the possibility of a continuum of allowed states, we might as well just make it an integer. We will let it take Q possible values: $1, 2, ..., Q$ ($Q = 2$ was the Ising case).

Typically we might think of this number s_i as representing the orientation of some kind of magnetic dipole sitting on the crystal lattice. In this case a crude but effective model of a ferromagnet will arise if we introduce a Hamiltonian which favours, in the sense of large Boltzmann weights, the situation in which nearby variables are similarly oriented. A dipole alignment is something which is realised in external physical space, so we do not need to worry here about the variable being misinterpreted at another site due to a possibly distinct internal identification of states with numbers (c.f. chapters 2 and 10). Physically, we know that dipole interactions tend to have short range, so we will, in the first instance, restrict our Hamiltonian to favour nearest neighbour alignment. The Potts model partition function is thus taken to be

$$Z = \sum_{\{s\}} \exp(\beta \sum_{\langle ij \rangle} \delta_{s_i, s_j}) = \sum_{\{s\}} \prod_{\langle ij \rangle} \exp(\beta \delta_{s_i, s_j}) \tag{1.3}$$

where the product in the latter version is over nearest neighbour pairs of sites $\langle ij \rangle$ (not to be confused with an expectation value $< O >$), and the exponential is the Boltzmann weight for a single interaction.

For each possible value of s_i let us define a distinct unit vector $d(s_i)$ giving the dipole alignment corresponding to the number s_i, and obeying the constraint that

$$\sum_{i=1}^{Q} k_i d(i) = 0 \tag{1.4}$$

if and only if $k_i = k$ is a constant. Then the expectation value $< d(s_i) >$ is called the *spontaneous magnetisation* (non-scalar expectation values are computed component by component).

Now let us take this model apart and see how it might work.

1.4 Phase transitions

At very high temperatures (β small) the exponential weights in equation 1.3 are all close to 1 and the statistical mechanical partition function *overall* gives roughly equal weight to each configuration. We say that high temperature states (with no correlations or co-operative effects) dominate. What this means is that configurations with relatively little order among the spins may be regarded as representative, since large numbers of configurations characterised in this way are present compared to the number of highly ordered configurations. The number of configurations of an essentially similar nature is the *entropy* of such configurations. The notion of similarity of configurations in this context depends in general, therefore, on the properties of a configuration to which the partition function is sensitive. In this case it is clearly just the degree of order, or alignment amongst the spins.

A useful way to visualise this in two dimensions is to draw lines around each region of aligned spins. A highly ordered state has a low density of line elements (their precise distribution is immaterial here). The high entropy situation is a high density of line elements, since there are many arrangements with this property.

This small β picture corresponds to a *physical* system dominated by thermal fluctuations, exploring large regions of phase space almost without prejudice to the energetic cost.

At very low temperatures (large β) the weights depend strongly on the energy of the configurations. The largest weights occur when all the spins are aligned. If the energy dependence is strong enough then the dominant contributions to the partition function will come from these ordered states, despite their low entropy.

This large β picture corresponds to a physical system spending most of its time 'frozen' in an ordered ground state. Since the individual spins all spend most of their time pointing in the same direction anyway, this again is a situation with little long distance correlation of spins. The apparent correlation which does exist is merely a side effect of the spontaneous magnetisation $< d(s_i) >$. It is therefore useful to introduce the notion of subtracted correlation functions

$$< d(s_i)d(s_j) >_s = < d(s_i).d(s_j) > - < d(s_i) > . < d(s_j) > .$$

The subtracted correlation function will then drop off rapidly with the distance between i and j for both high and low temperatures. We will be able to be more specific about this after we have introduced the transfer matrix idea in the next chapter. Only in the crossover region is there any possibility of significant correlations at large distances.

1.4.1 Order parameters

It is clear from equation 1.3 that the spontaneous magnetisation is zero at high temperatures. The partition function as it stands is also clearly symmetrical under re-identification of labels with alignments. That is, for $\mathcal{P}(s_i)$ an arbitrary permutation of $1, 2, .., Q$ then Z is invariant under $d(s_i) \to d(\mathcal{P}(s_i))$. It follows that the spontaneous magnetisation is *always* formally zero unless there are some symmetry breaking boundary conditions. On the other hand we need to be careful about our notion of equilibrium at low temperatures, since it is not very constructive to deny that a finite ferromagnet below its Curie point can be in equilibrium. Of course a physical system will break the symmetry since its state is essentially unique moment by moment. Allowing for some infinitessimal symmetry breaking in the model, then, it is plausible that the spontaneous magnetisation may become non-zero at low temperatures.

Taking a ferromagnet as an example we want to be able to model a physical situation in which, at high temperatures, although it is marginally energetically favourable for adjacent spins to be aligned, thermal fluctuations involving misaligned pairs are not uncommon. Over medium distances there is little net magnetisation, and none over large distances. As the temperature is reduced the range over which spins are aligned increases. It may occur at some point that this range extends to the full size of the system. In this case the local interactions have co-operated to give rise to long range order. This is a co-operative phenomenon. The spontaneous magnetisation is an example of an order parameter, a quantity which signals a phase transition by becoming non-zero.

The Peierls argument

Let us make this idea more specific. The following argument for the existence of a finite spontaneous magnetisation in certain models is due to Peierls.

Define Z_+ as the partition function for a 2 dimensional square lattice $Q = 2$ state (Ising) system with all the spins round the boundary set to $s_i = 1$. Here, for a finite system, the spontaneous magnetisation would be non-zero. On the other hand, since with no symmetry breaking the spontaneous magnetisation would be zero, we might expect that in this (broken symmetry on the boundary) case it would *tend* to zero as the boundaries were taken to infinity, i.e. that we could make the spontaneous magnetisation arbitrarily small by making the system arbitrarily large. This is true at high temperatures. However, we will show that there is a finite temperature below which the spontaneous magnetisation remains larger than some finite positive number however large the system (and hence however far removed the boundary).

The expectational probability that a single spin well away from the boundaries is in state $s_0 = 2$ is

$$p(s_0 = 2) = \frac{1}{Z_+} \sum_{\{s_i | s_0 = 2\}} \exp(\beta H) \tag{1.5}$$

where the sum is over all configurations in which $s_0 = 2$. Following equation 1.4 we put

$$d(s_i) = 3 - 2s_i$$

whereupon

$$< d(s_i) >= 1 - 2\, p(s_0 = 2).$$

The boundary condition ensures that all our lines round regions of aligned spins are closed in each configuration in which $s_0 = 2$. The probability of existence of a specific closed line of length n, $L(n)$, enclosing $s_0 = 2$ *and no other lines which enclose* s_0, is

$$p(L(n)) = \frac{1}{Z_+} \sum_{\{s_i | L(n)\}} \exp(\beta \sum_{(ij) \notin L(n)} s_i s_j).$$

Here the configuration sum is over configurations manifesting the specified line.

Given a configuration in which the line $L(n)$ is present we can associate to it another configuration in which *no* line element from $L(n)$ is present, by flipping the signs of all the spins within $L(n)$ in the original configuration. The Hamiltonian for the new configuration exceeds that for the old configuration by exactly n. This holds for every configuration in the sum above, so

$$p(L(n)) = \frac{1}{Z_+} \exp(-n\beta) \sum_{\{s_i | L'(n)\}} \exp(\beta \sum_{(ij)} s_i s_j).$$

where the configuration sum is over configurations obtained as above. If we replace this sum by a sum over all configurations, the right hand side strictly increases, so

$$p(L(n)) < \exp(-n\beta).$$

Now consider the sum over all possible lines $L(n)$. We have

$$p(s_0 = 2) = \sum_{poss.\ L(n)} p(L(n))$$

and hence

$$p(s_0 = 2) < \sum_{poss.\ L(n)} \exp(-n\beta) = \sum_n N(n)\exp(-n\beta)$$

where $N(n)$ is just the number of shapes of $L(n)$ for a given n.

Since $L(n)$ is of length n and surrounds s_0, it cannot rove further away from s_0 than a distance $n/2$ in a direction parallel to lattice bonds, or $n/2\sqrt{2}$ in a direction at 45°. This range is neatly covered by a square at 45° to the lattice bonds, of diagonal length n. The *total* number of random walks of length n starting in this square is $n^2 4^n/2$, but our line does not have a marked starting point, so $N(n) < n4^n/2$. Altogether we have

$$\begin{aligned} 1/2 \geq \quad p(s_0 = 2) \quad &< \quad \textstyle\sum_{n=4}^{\infty} n4^n \exp(-n\beta)/2 \\ &< \quad (1/2)\textstyle\sum_{n=1}^{\infty} n(4\exp(-\beta))^n \\ &= (1/2)\left(\frac{4\exp(-\beta)}{(1-4\exp(-\beta))^2}\right). \end{aligned}$$

where the left hand inequality arises on symmetry grounds.

Thus, for β sufficiently large, the probability of $s_0 = 2$ is not $1/2$, implying that the spontaneous magnetisation is not zero, however big the lattice. Indeed, however big the lattice the probability can be made arbitrarily small (and hence the spontaneous magnetisation arbitrarily close to 1) simply by making β large. We can thus take the symmetry breaking boundary conditions off to infinity without affecting the result. This completes the argument.

1.4.2 Critical exponents

In the thermodynamic limit then, whilst the magnetisation in an extended region of high temperatures may be strictly zero, in the low temperature region the magnetisation will be finite. This situation can only be modelled by a non-analyticity in the statistical mechanical spontaneous magnetisation. Later we will show *precisely* how statistical mechanics models this phenomenon.

We call the point of crossover a phase transition. The way in which the spontaneous magnetisation tends to zero as a function of $(T - T_c)$ in many systems is in the form

$$M \sim (T - T_c)^{\tilde{\beta}} \quad + \quad \textit{more rapidly vanishing terms}$$

where the positive constant $\tilde{\beta}$ here is called the magnetisation critical exponent (not to be confused with the coupling parameter β).

A phase transition will also manifest itself in the divergence of some β-derivative of the free energy (we will see this explicitly in chapters 3 and 11). The *order* of the transition is the minimum number of differentiations required before a divergence appears. Thus a second order transition, for example, has its first divergence at the level of the specific heat $S(\beta)$ defined by

$$S(\beta) = \frac{\partial U(\beta)}{\partial \beta}.$$

This enables us to define another independent critical exponent α by

$$S \sim (\beta - \beta_c)^{-\alpha}. \tag{1.6}$$

In fact all of the above phenomena are successfully modelled in statistical mechanics! A more explicit version of this discussion will be provided in chapters 2 and 11, when we have developed the appropriate technical hardware and language.

1.5 Dichromatic polynomials

Instead of describing the configurations of the Potts partition function in terms of lines bounding areas (or more generally surfaces bounding regions) of aligned spins, we could equivalently have described bond-connected clusters of aligned spins. In two dimensions there is no combinatorial disitinction between the two pictures but, as we will repeatedly see, they serve to illuminate different areas of our discussion.

Defining $v = e^\beta - 1$ we can rewrite equation 1.3 as

$$Z = \sum_{\{s\}} \prod_{\langle ij \rangle} (1 + v\delta_{s_i,s_j}) \tag{1.7}$$

and expand the product as a sum of terms corresponding to the possible choices of summand from each factor. We can think of the terms in the sum as being represented by the elements of the power set of the set of bonds. That is, possible Z_2 coverings of the bonds of the lattice, which are all possible ways of assigning either 0 or 1 to each bond. Each bond corresponds to a factor in the product, and the covering 0 or 1 at each bond to taking the summand 1 or $(e^\beta - 1)\delta_{s_i,s_j}$ for that factor respectively.

For a covering with l bonds 'filled' (i.e. set to 1) we have a factor v^l together with a product of delta functions which forces to zero every configuration summand in which all the spins connected by filled bonds are not aligned. The total number of configurations which contribute, in a situation in which the filled bonds form c connected clusters, is Q^c. Thus

$$Z = \sum_{bond\ coverings} Q^c v^l.$$

This formulation of the partition function is called a dichromatic polynomial (Baxter 1982), since Q may now be formally regarded as a continuous variable.

Note that if we fix some of the boundary spins to some given set of values, then the number of connected clusters must not count the clusters connected to the fixed spins. We will study the uses of this representation of the Potts partition function in some detail later on.

For the $Q = 2$ state (Ising) model we can reformulate the partition function in yet another way. The argument of the product in equation 1.7 may be trivially rewritten as

$$(1 + (e^\beta - 1)\delta_{s_i,s_j}) = \left(\frac{(e^\beta + 1)}{2} + \frac{(e^\beta - 1)}{2}(2\delta_{s_i,s_j} - 1)\right)$$

and the bond coverings of the lattice may be re-interpretted as corresponding to the possible choices of new summand in the expansion of the resultant product. In this case the sum over configurations kills off any bond coverings in which *odd* numbers of filled bonds touch a site, and gives a factor of 2 per site otherwise. Once again writing N for the number of sites and l for the number of filled bonds; and with M the total number of bonds on the lattice (i.e. $M \sim dN$ in d dimensions), we are left with

$$Z = \left(\frac{e^\beta + 1}{2}\right)^M 2^N \sum_{even\ coverings} (\tanh(\beta/2))^l. \tag{1.8}$$

Here the sum is over bond coverings for which there are no sites touched by an odd number of filled bonds.

Statistical mechanical duality

For the $Q = 2$ model, the set of even coverings in equation 1.8 is, up to boundary effects, the same combinatorial object as the set of possible distributions of lines bounding areas of aligned spins. To see this consider the closedness of these boundary lines. There is a bijection between even coverings and line distributions which takes shapes onto identical, but shifted, shapes. We can write the partition function explicitly in terms of the lines as

$$Z = 2\exp(M\beta) \sum_{lines\ L} \exp(-n(L)\beta)$$

where $n(L)$ is the length of the line L. The correspondence takes $n(L) = l$ where l is the number of filled bonds.

Note that, up to boundary effects, the partition function is thus invariant under the transformations

$$\frac{e^{\beta} - 1}{e^{\beta} + 1} \leftrightarrow e^{-\beta}.$$

This invariance property of the model is called self-duality. It is straightforward to generalise to anisotropic couplings. It is also possible to generalise to all Potts models, and to many other models. We will return to this point when we have enough algebraic hardware to make a straightforward presentation of the general transformation (in chapter 2).

1.6 High and low temperature series

It follows from the foregoing discussion that we can obtain low temperature perturbation expansions for Potts model partition functions and expectation values (and hence estimate critical exponents) by directly enumerating the spin configurations with energies within some region of the maximum energy. The scope for *indirect* enumerations is vast.

As we will see later, most of the series data obtainable for the critical exponents of two dimensional models can be compared indirectly with exact or conjectured exact results. After making the observation that they are not inconsistent, it is impossible to resist dropping the series results in favour of the exact ones. In higher dimensions, convincing series results would have a much more central role to play, but the calculations are much harder.

A tremendous amount of ingenuity (and computer power) has been used in the development of series expansions to the present high orders (see e.g. Guttman and Enting 1988). Unfortunately this approach is beset by a very powerful version of the law of diminishing returns. It is becoming an increasingly difficult area in which to do research. On the other hand, the existing series expansions and their concommitant critical point and exponent estimates still constitute one of the principle sources of immutable data against which more exotic techniques must be proven.

Results

It is not necessary to give an exhaustive list of results to illustrate the pros and cons of this approach. Ignoring two dimensions, as suggested above, and restricting to simple cubic lattices we have the following low temperature expansion for the N site Q-state Potts model partition function:

$$Z = Qe^{3N\beta}\ (1\ +\ (Q-1)Ne^{-6\beta}\ +\ (Q-1)3Ne^{-10\beta} + (Q-1)(Q-2)3Ne^{-11\beta} + (Q-1)N((Q-2)\ +\ (Q-1)/2)(N-7)e^{-12\beta}\ + ...).$$

The series can be extended by a few terms by direct enumeration, then more sophisticated techniques are required. It is perhaps less confusing to present it as a series for the free energy. Below we give the first few terms for the free energy in the case $Q = 3$, using the notation $x = e^{-\beta}$:

$$f = 3\beta + 2x^6 + 6x^{10} + 6x^{11} - 14x^{12} + 30x^{14} + 60x^{15} - 108x^{16} - 144x^{17} + ...$$

In the case $Q = 3$ this and related series have been extended to about 40 terms by Guttman and Enting! For higher Q the known series are somewhat shorter.

Such series contain a lot of information. However, we can already see from the few terms above that they are by no means *immediately* revealing away from small values of the expansion parameter (a friendlier way of presenting data of this type is given in chapter 11). This means that we have to work very hard to unearth critical properties, for example. The series are also strongly dependent on the fine details of the model involved, and frequently give ambiguous indications as to the order of the phase transition. So let us now go in search of a more robust, physically motivated, and pedagogically illuminating approach to the critical region.

1.7 Block spin renormalisation

The starting point for all our machinations was a physical system together with a length scale. The length scale was one on which we supposed that we had a good physical understanding of the degrees of freedom of the system, and their energetics. It does not really matter if these degrees of freedom are fundamental, or even if they exhaustively describe the states of the system at the length scale in question, so long as their use results in a description adequate to model *the properties in which we are interested.*

Let us suppose that we have no preconceived notions about the energetics of a system, but a good idea about the kind of length scale a, and the N degrees of freedom relevant at that scale, with which we want to describe it.

We could, in principle, formulate a general composite Hamiltonian depending on every conceivable type of interaction. This would include a trivial 'interaction', independent of any spins. Each interaction has an associated unknown coupling constant. Most of them, the very long range ones, would hopefully have very small or zero couplings in any practical situation. The Hamiltonian would take the form

$$H(\underline{\beta}) = c + \beta \sum_{\langle ij \rangle} s_i s_j + \beta' \sum_{ij'} s_i s_{j'} + \beta'' \sum_{ijk} s_i s_j s_k + \dots$$

where $\underline{\beta}$ represents all the couplings collectively, j' is a site which is a next-nearest neighbour to i, and so on.

Now we can always re-express the N original degrees of freedom, the spins, as a group of N/L^d 'block' spins and a group of 'internal' spins with a constraint. That is, we can think of grouping the spins into localised blocks of L^d spins, and then arranging the set of configurations of such a block into subsets each of which has, for example, a prevailing (e.g. modal) spin configuration among the L^d spins. We can then label a configuration by the prevailing spin configuration - now called the block spin variable b_i - and the associated internal spins $s_{j(i)}$ (which will take values consistent with the prevailing spin being as required). Then define a blocked spin Hamiltonian $H(\{b_i\})$ by

$$\exp(H(\{b_i\})) = \sum_{\{s_j|b\}} \exp(H) \tag{1.9}$$

where the sum is over allowed values of the internal spins for a given configuration of the block spins. The idea of this is that it plays the role of H in a partition function in which the sum over configurations involves only blocked spins, but which is identical to the original Z, i.e.

$$Z_{blocked} = \sum_{\{b_i\}} e^{H(\{b_i\})} = \sum_{\{b_i\}} \sum_{\{s_j|b\}} e^H = \sum_{\{s\}} e^H = Z.$$

The block spins approximately represent the same system, configuration by configuration, by a cruder picture with fundamental length scale La. The legitimacy of what follows depends on the extent to which this representation distorts, while simplifying, the original information. This in turn depends on the extent to which a block spin represents its block, i.e. the way the block spin is abstracted from the original spin configurations in its block. We have not given a definite procedure, and there may not necessarily *be* a prevailing spin configuration within the block. A random association of block spin configurations with subsets of original spin configurations, for example, would certainly give rise to a meaningless representation of the original system.

For credible results in specific cases some discretion is clearly required here. For the moment we are not in a position to apply, let alone systematise, any discretion. It is sufficient to know that sensible blockings can be found (an example follows shortly). The blocked spins need not be of the same form as the original spins, and indeed forcing them to be so may increase the distortion. However, if they ***can*** reasonably be made so, then a powerful simplification occurs. It is this circumstance, which for the moment we must regard as fortuitous, which we will consider now.

Since the original Hamiltonian includes all possible interactions of this type of variable, then the resultant Hamiltonian for the block spins will necessarily be of the same form. It will just involve new, ***renormalised***, values of the coupling constants. These will refer to the same kind of interactions as before, only on a different length scale. Their values will depend on the values of the original couplings via the equations 1.9. This raises the possibility of iterating the resultant coupling renormalisation transformation.

An example

Before exploring the general consequences of such a procedure let us illustrate the basic idea of the coupling transformation. There is a particularly neat example, due to Niemeyer and vanLeeuwen 1973, with a suitably convincing blocking algorithm. The calculational technique itself has not proved to be very versatile, but it is a beautiful pedagogical tool.

Consider a triangular lattice such as the one shown in figure 1.2. We are going to combine spins into blocks of 3 (as indicated), i.e. $L = \sqrt{3}$.

We will take $Q = 2$ state (Ising) spins. In this case there is a simple and relatively natural candidate for the blocking procedure. The 2 state block spins b_i are given by a majority rule from their three contributions. The internal spins $s_{j(i)}$ associated to block spin b_i can then take just 4 different values for each value of the block spin.

To be completely specific, consider the case in which the lattice consists just of the spins shown in figure 1.3. Here the only possible interactions on the (2 site) block lattice L are the trivial one c_L and the nearest neighbour one β_L. Including only these types of interactions at the original level then, we have, by direct computation,

$$\exp(H(b_1 = b_2 = 1)) = \exp(\beta_L + c_L)$$

$$= (e^{8\beta} + 3e^{6\beta} + 2e^{5\beta} + 3e^{4\beta} + 6e^{3\beta} + e^{2\beta})e^c$$

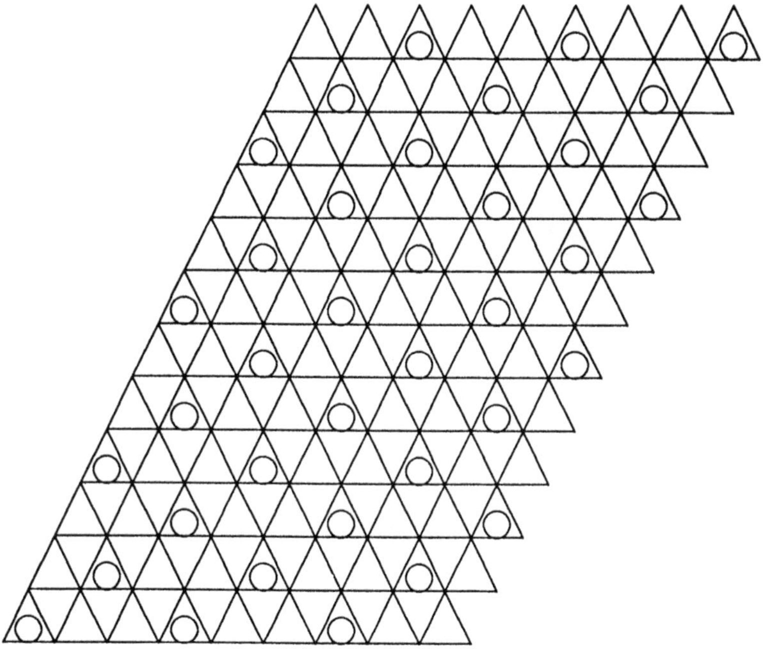

Figure 1.2: A triangular lattice. Block spin triangles are indicated by a circle.

Figure 1.3: A small triangular lattice. Block spin triangles are indicated by a circle.

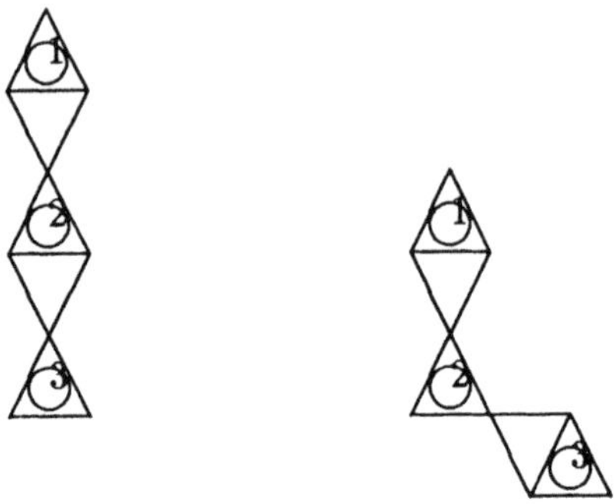

Figure 1.4: Block spins are built out of the spins on ringed triangles.

$$\exp(H(b_1 \neq b_2 = 2)) = \exp(c_L)$$
$$= (2e^{6\beta} + 2e^{5\beta} + 4e^{4\beta} + 6e^{3\beta} + 2e^{2\beta})e^c$$

and an identical pair of equations (by symmetry). Hence we can eliminate c and determine β_L as a function of β.

In general, for a finite initial system, if we get n independent equations we need n coupling parameters at the blocked level. In effect we need a coupling parameter for every possible interaction on the block system. For form invariance of the Hamiltonian we should therefore include just these couplings in the original system.

Consider the two lattices shown in figure 1.4. In either case we may write the various possibilities in the form

$$\exp(H(b_1 = b_2 = b_3)) = \exp(\beta_L^{12} + \beta_L^{23} + \beta'_L + c_L)$$

$$\exp(H(b_1 \neq b_2 = b_3)) = \exp(\beta_L^{(23)} + c_L)$$
$$\exp(H(b_1 = b_2 \neq b_3)) = \exp(\beta_L^{(12)} + c_L)$$
$$\exp(H(b_2 \neq b_1 = b_3)) = \exp(\beta'_L + c_L)$$

where β' is the next-to-next nearest (resp. next nearest) neighbour coupling in the first (resp. second) case. In the second case the two nearest neighbour interaction coupling parameters ($\beta_L^{(12)}$ and $\beta_L^{(23)}$) are manifestly identified, because the (reflection) symmetry of the original system survives in the block system. The third equation above thus becomes the same as the second one. In the *first* case the calculation is slightly more complicated, because the original system does not have an exact reflection symmetry. In such cases $\beta_L^{(12)}$ and $\beta_L^{(23)}$ may have to be treated separately. Either way, after some algebra we can determine the block spin couplings as functions of the original couplings.

Note that in all cases, and regardless of the blocking procedure, the programme amounts to *exactly* re-expressing the partition function. However, the information we will discard at the next stage of this scheme makes results depend on the blocking procedure in practice.

The information we will discard is the renormalisation of the constant term $c \to c_L$. Recall that for the moment we are mainly interested in critical phenomena. Then we are interested in non-analytic behaviour of the statistical mechanical functions at the critical point, and we assume that the constant term does not contribute to this behaviour. Since in general we throw away different bits of information in the constant term depending on how we make the blocking, different blockings produce computationally different results. The test is that the eventual *critical properties* should not be sensitive to these changes. Note that we do not expect to obtain quantitative results for systems far away from criticality by this method.

We will discuss how critical properties are to be extracted shortly. First, for comparison let us consider an alternative procedure for obtaining coupling transformations based on the same idea as above. The calculational method outlined above will clearly become messy for large lattices (although truncated approximations are possible). The following alternative deals with an infinite lattice and attendant approximation from the outset. We write the original triangular lattice Hamiltonian H as a sum of two terms: the interactions between spins within blocks H_0, and those between spins in different blocks H_1. We are still talking about the original spins at this stage.

For any quantity A, say, depending on all the original spins let us define a corresponding quantity $< A >_b$ depending only on the block spins by

$$< A >_b = \frac{\sum_{\{s_j|b\}} e^{H_0} A}{\sum_{\{s_j|b\}} e^{H_0}}.$$

Then trivially

$$\exp(H(\{b_i\}) = \left(\sum_{\{s_j|b\}} \exp(H_0) \right) < \exp(H_1) >_b$$

where, in the general N site triangular lattice case, for example

$$\left(\sum_{\{s_j|b\}} \exp(H_0)\right) = (e^{3\beta} + 3e^{\beta})^{N/3},$$

since the H_0 interactions couple only within blocks. We may thus develop an expansion for $H(\{b_i\})$. In the triangular lattice case we have

$$\begin{aligned} H(\{b_i\}) &= (N/3)\ln(e^{3\beta} + 3e^{\beta}) + < H_1 >_b \\ &+ (1/2)(< H_1^2 >_b - < H_1 >_b^2) + \textit{higher orders}. \end{aligned}$$

In this case H_1 is obtained from

$$H = H_0 + \beta \sum_{ij}{}' \delta_{s_i,s_j}$$

where the sum is over those nearest neighbour interactions which are *between* blocks. Let us compute

$$< H_1 >_b = 2\beta \sum_{(ij)} < \delta_{s_i,s_j} >_b$$

where the sum is now over pairs of adjacent blocks; s_i and s_j are spins in the corresponding adjacent blocks; and we have used the symmetries of the original system. Writing b_i and b_j for the appropriate *block* spins we obtain, by direct calculation,

$$< \delta_{s_i,s_j} >_b = \frac{(4e^{2\beta} + 2e^{4\beta}) + (e^{3\beta} + e^{\beta})^2 \delta_{b_i,b_j}}{(e^{3\beta} + 3e^{\beta})^2}$$

so that

$$< H_1 >_b = \beta_L \sum_{(ij)} \delta_{b_i,b_j} + \dots .$$

The supressed terms do not depend on the block spins, and thus contribute only to the renormalisation of the constant term. The higher order terms in the expansion of $H(\{b_i\})$ give rise to interactions between next-nearest neighbour blocks and so on.

The finite lattice method previously espoused corresponds roughly, but not exactly, to truncating the expansion for an infinite lattice at a given order. Higher orders here correspond to larger lattices, and indeed generate more possible couplings at the renormalised level. Once again, then, correspondingly more couplings are *required*, to maintain form invariance, at the original level.

1.7.1 Fixed points

In general, the output of any such procedure is thus a coupled set of coupling constant recursion relations (depending crucially on the block size L and, hopefully less significantly, on the details of the blocking procedure). This recursion may be written

$$\underline{\beta} \to \underline{\beta_L(\beta)}$$

so that

$$Z_N(\underline{\beta}) = Z_{N/L^d}(\underline{\beta_L}).$$

The free energy per site of the original system and the new system are hence related by

$$f(\underline{\beta}) = L^{-d} f(\underline{\beta_L}).$$

We can think of the free energy as being composed of two different parts:

$$f = \frac{1}{N}\ln(\sum e^c) + \frac{1}{N}\ln(Z|_{c=0}) = f_0 + f_i.$$

The 'constant' part, f_0, comes from the spin independent term $c \to c_L$. As we have said, it is reasonable to suppose that this is not responsible for critical properties of the system, singularities of the specific heat and so on. We attribute these critical properties to the other part, the 'interacting' part f_i. Henceforward we will restrict our attention to f_i.

Consider a thermodynamic limit system with 2 couplings, β and β' say. Suppose that on the $\beta' = 0$ line we have a critical point $\beta = \beta*$ at which the spin-spin correlation range diverges. If we turn on β' infinitessimally we will disturb the critical point a little, gradually exploring a critical line in the (β, β') plane.

If the spin-spin correlation length of the system is X, say, (see chapter 2 for details) then after a block spin transformation it should be $X' = X/L$. At the critical point $X = \infty$, so $X' = \infty$. In other words the blocking takes a critical system to a critical system. On the other hand any finite X will be reduced, i.e. the blocking will move the system further away from criticality.

More generally, suppose there is a fixed point $\underline{\beta*}$ in multi-coupling parameter space, and that we can linearise the block transformations about the fixed point. Regarding the coupling set $\underline{\beta}$ as a vector in parameter space we can introduce a matrix A such that

$$\underline{\beta_L} - \underline{\beta*} \sim A(\underline{\beta} - \underline{\beta*})$$

where the matrix elements of A are given by

$$A_{ij} = \left.\frac{\partial \beta_L^{(i)}}{\partial \beta^{(j)}}\right|_{\underline{\beta}=\underline{\beta*}}$$

with eigenvectors and eigenvalues given by

$$Av_k = l_k v_k$$

say. Let us resolve the displacement of the coupling constants from the fixed point into components in the eigenvector directions

$$(\underline{\beta} - \underline{\beta *}) = \sum_k p_k v_k.$$

Then after many iterations of the transformation the various non-zero components blow up or die away depending on whether l_k is bigger than or less than 1 (respectively 'relevant' or irrelevant). If a system is displaced away from a fixed point along a critical line, or more generally a critical surface, then all the components with $l_k > 1$ continue to vanish, and the remainder tend to zero as the transformation is iterated. If the system is displaced *off* a critical line then some component with $l_k > 1$ becomes nonvanishing, and as the transformation is iterated the discrepancy from criticality will grow. Physically we identify one such component with $l_k > 1$ as a temperature like variable. In general (where not excluded by symmetry) we associate another one with the magnetic field. Diverse investigations indicate that this typically exhausts the set of relevant components (Ravndal 1976).

We will see shortly that the set of eigenvalues obeying $l_k > 1$ are related to critical exponents. Any model on the critical line is transformed into one closer to the fixed point by the blocking transformation, and eventually reaches it after an infinite number of iterations. All the models on the critical line are thus expected to have the same exponents as the fixed point. This is the idea of universality of critical exponents. In order to establish the exponents of a model it is only necesary to show that it is on the same critical line as a model with known exponents. Since our initial Hamiltonian involved both very simple and not so simple interactions it is regarded as reasonable to assume that if a mathematically tractable model and a realistic but intractable model are compatible on symmetry and dimensional grounds, then they will have the same exponents.

Note in particular that if we do not break a symmetry such as spin reversal symmetry with our original Hamiltonian, as we could by putting in a coupling of single spins to an external 'magnetic' field, for example, then it will not be broken after a blocking.

In our $Q = 2$ state triangular lattice example we started with only one non-trivial coupling parameter. The finite lattice method (from figure 1.3) gives, by direct calculation (with $x = \exp(\beta)$)

$$x \longrightarrow x_L = \frac{x^5 - x^4 + 4x^3 - 2x^2 + 5x + 1}{2(x^3 + 2x + 1)}.$$

The fixed points of this transformation are given by

$$(x-1)(x^4-2x^3+2x^2-4x-1)=0$$

so in the physical regime (i.e. the region of ferromagnetic coupling where our choice of blocking makes sense) we have the trivial fixed point $\beta = 0$, and a unique non-trivial fixed point $x = 2.07631...$ (or $\beta = .73...$). The non-trivial fixed point should be compared with the exact result, $\beta = .55..$, due to Onsager (1944). The exponent from our calculation is $l = 1.54...$ (cf. the exact result, which is 1.73..., as we will see in chapters 3 and 4).

In iterating the transformation obtained from the finite lattice we are assuming that the finite lattice transformation somehow approximates the infinite lattice transformation. We have no real control over this approximation, and at present we make no claims for it. The calculation does, however, serve to illustrate the *idea* of block spin transformations, and under the circumstances the results shown above exhibit encouragingly good agreement.

The expansion method gives, at first order, $\beta = .68...$ and $l = 1.62...$, and at second order (three coupling parameters) the unique 'relevant' exponent $1 < l_k = 1.77....$ As we might expect, the errors are reduced, although once again we have no real control over the approximation.

On critical exponents

In general, ignoring the trivial coupling we can represent the position of a system in coupling parameter space by the components p_k. Then in the critical region

$$f_i(p_1,p_2,..) = L^{-d} f_i(l_1 p_1, l_2 p_2,..)$$

Let us look at the 'temperature' parameter p_1 (uniquely identified by symmetry and the requirement that $l_1 > 1$). We have

$$f_i(l_1 p_1,..) = L^d f_i(p_1,..)$$

so

$$f_i(p_1,..) = |p_1|^{-\frac{d\ln(L)}{\ln(l_1)}} f_i(1,..)$$

and the specific heat critical exponent from equation 1.6 is

$$\alpha = 2 - d\ln(L)/\ln(l_1).$$

Note that the sustained absence of magnetic field contributions in our examples, because of the sustained Z_2 symmetry, illustrates the point that the range of universality will be governed in general by the symmetries of the system.

The examples which we have considered here are too small to be of practical use. However, they have facilitated the development of a language which serves for tackling problems in the statistical mechanical modelling of phase transitions in general. We are now ready to tool up for some more realistic calculations.

Chapter 2

Transfer matrices

In this chapter we will begin to address the technical problem of computing partition functions and other statistical mechanical functions. We will introduce the general notion of a *partition vector*, and hence the notion of a *transfer matrix*, which is a certain arrangement of a partition vector.

The basic idea here is that we can compute the partition function for a large system by *composing* partition functions for subsystems whose overlap is at most a common boundary. If the subsystems are disjoint then the composition is simply a product of partition functions (and a direct product of boundary information). If the subsytems have a common boundary, composition is achieved by integrating over all possible configurations of the internalised boundary information.

If the subsystems are indistinguishable then it is possible to systematise their iterated composition. We will show that this can lead to a reduction in the effective dimension of the problem.

The general computational advantages of this transfer matrix approach stem from the reduction in dimension, taking maximum advantage of external symmetries such as translation invariance, and also from couching the problem in an algebraic framework, which allows us to fully exploit any internal symmetries of the system. The transfer matrix approach will also allow us to make mathematically precise more of the general notions of statistical mechanics introduced in the previous chapter. Indeed the principal objective of this chapter is really to introduce a more precise statement of the problems to be solved in statistical mechanics. Specifically we show the importance of the spectrum of the transfer matrix. In the next chapter we will address the computation of this spectrum.

2.1 Partition vectors

As discussed in chapter 1, statistical mechanical models in general embrace variations, among the available simplices, on the roles of carrier of interactions and carrier of variables. For the sake of definiteness and simplicity we begin here with a discussion couched in terms of the Potts model. This serves well to illustrate the basic ideas of the partition vector and transfer matrix, and can be straightforwardly generalised later on.

Consider, then, a system of spins s_x, each taking values from some set V, residing on the crossings (labelled x), and interactions H_i, coupling between spins, residing on the edges (labelled i) of a lattice.

For the moment we will also assume that the lattice is either as defined in section 1.2.2, i.e. on a surface, or is a sublattice thereof (again, a degree of generalisation is straightforward). Recall in particular that all lattice crossings are then intersections of *circles* and *lines*.

Draw a closed curve or curves $\{c\}$ on the surface passing through crossings but not edges of the lattice. Consider the spins inside and on $\{c\}$ (you may have to decide which submanifold is inside). Call these interior and exterior spins respectively. We define $|c|$ as the number of exterior spins of $\{c\}$.

The partition vector $Z[c]$ is a vector in the space of possible configurations of the exterior spins. We will call this space $[c]$ in general, so that for a Q-state model

$$Z[c] \in [c] \sim \otimes^{|c|} V_Q .$$

The components of $Z[c]$ with respect to the configuration space basis each correspond to a particular configuration of the exterior spins. Specifically, each component is defined to be the partition function for the appropriately bounded interior system. That is, $Z[c]_i$ is given by the sum over the *interior* spin configurations of the Boltzmann weights for the interior interactions, with the exterior spin configuration held fixed to configuration i.

For example, if the interior and boundary of $\{c\}$ is the patch of lattice shown in figure 2.1 then the vector has $(\dim V)^{10}$ entries, corresponding to the possible configurations of the $|c| = 10$ boundary sites in the lattice. Each vector entry is a partition function summed over the $(\dim V)^{20}$ configurations of the internal sites, with the external sites fixed to a given choice from the $(\dim V)^{10}$ possibilities. Figure 2.2 shows a similar patch from a different perspective. In this case the vector has $(\dim V)^{16}$ entries.

Now suppose we have two subsystems, bounded by $\{c\}$ and $\{c'\}$ respectively, which touch only along an arc of $\{c\}$. This means that they have $|c \cap c'|$ exterior spins, but no interior spins, in common. Then we can compose $Z[c]$ and $Z[c']$ by a dot product over the $[c \cap c']$ subspace, giving a new

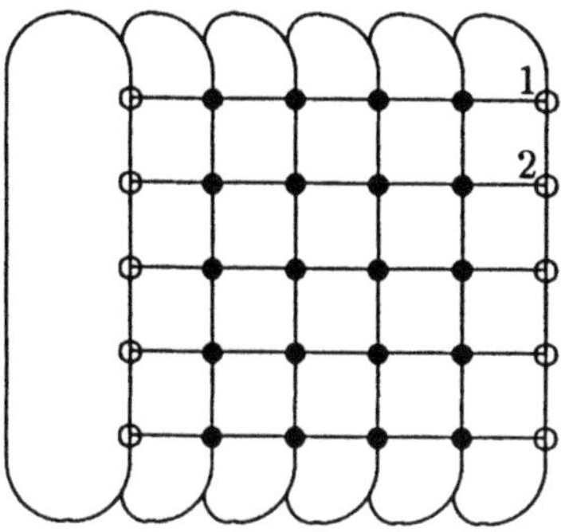

Figure 2.1: A patch of lattice, specifically a 2-punctured sphere containing 6 circles and 5 lines. Boundary (i.e. exterior) sites are marked with empty circles, interior sites with full ones.

partition vector. The boundary for the new partition vector is given by the disjoint union of $\{c\}$ and $\{c'\}$.

Note that, in the case of more general interactions, it would be necessary to regard the boundary as having a *thickness*, corresponding to the range of interactions. That is, more than just the spins on $\{c\}$ would have to be regarded as exterior. In general, a spin in our bounded subsystem must be treated as exterior if it would affect any interaction in the system as a whole which is not determined by the spins in the subsytem. For example, with interactions over three lattice spacings *all* the spins in figure 2.1 would have to be taken as boundary information. The partition vector approach is thus useless for dealing with infinite range interactions, and most powerful when dealing with nearest neighbour interactions.

The transfer matrix

The partition vector is a *vector* because we arranged the partition information as a vector in the space of all configurations of the exterior spins. We could make it into a matrix by collecting the exterior spins into two sets (called *incoming* and *outgoing*) and letting the two indices determine the vector in each subspace (as we will do for the *transfer matrix* shortly). Obviously we could arrange the information as a tensor in various ways instead. A square matrix arrangement is most useful in systems with translation invariance, as we will see.

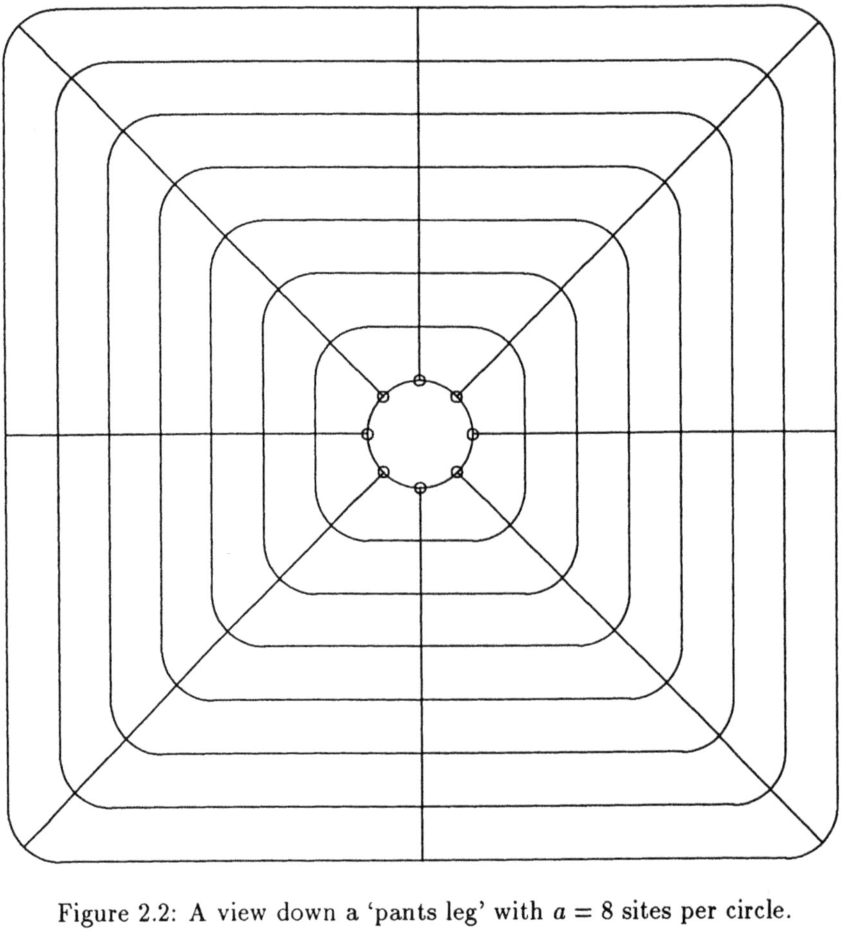

Figure 2.2: A view down a 'pants leg' with $a = 8$ sites per circle.

The three punctured sphere or *pants* and the empty patch are our basic building blocks for surfaces (section 1.2.2). Let us label the three punctures on the boundary of a pants by $\{c : i_1, i_2, i_3\}$. A simple lattice on the pants has only one circle per puncture; and $\underline{a} = (a_1, a_2, a_3)$, say, as the number of lines per puncture, so that $|c| = a_1 + a_2 + a_3$. Note that if a_{ij} is the number of lines between punctures i and j then

$$\underline{a} = (a_{12} + a_{31}, a_{23} + a_{12}, a_{31} + a_{23}).$$

Let us denote the configuration of spins at the boundary at puncture $\{i_j\}$ by i_j, i.e.

$$i_j \in \otimes^{a_j} V.$$

It is fairly natural to arrange the partition vector for this pants in the form $Z[c : i_1 i_2 i_3] = (P_{(\underline{a})})_{i_1 i_2 i_3}$ where the indices i_1, i_2, i_3 give the exterior configuration for each puncture. In this notation the special case $\underline{a} = (a, a, 0)$, i.e. the partition vector for a 2-punctured sphere with only one circle, is denoted $Z[c : l, m, \emptyset] = (T_{(a)})_{lm}$.

The general partition vector for a *pants* with punctures designated $\{c : l_1, l_2, l_3\}$ may then be written

$$Z[c : l_1, l_2, l_3] = \sum_{i_1 i_2 i_3} (P_{(\underline{a})})_{i_1 i_2 i_3} ({T_{(a_1)}}^{b_1})_{i_1 l_1} ({T_{(a_2)}}^{b_2})_{i_2 l_2} ({T_{(a_3)}}^{b_3})_{i_3 l_3} \quad (2.1)$$

where b_1, b_2, b_3 are the numbers of extra circles dressing each puncture.

The object $T_{(a)}$ is called the a-site cylindrical layer transfer matrix. It now acts as a linear transformation on $\otimes^a V$.

2.1.1 On internalising a common boundary

Matching punctures may be sewn together by matrix multiplication as we see in equation 2.1. In general some intermediate matrix is needed to give the required identification of spins (and hence lines) at the common boundary. Even if we adopt the convention of organising single spin subspaces in the order in which the spins appear round the boundary, as we will, there are still rotational and reflectional ambiguities.

Let us define a translation matrix $G^2_{(a)}$ which cycles the site subspaces. With $e_i \in V$, a configuration of a single spin i, we have

$$G^2_{(a)}(e_1 \otimes e_2 \otimes ... \otimes e_a) = (e_a \otimes e_1 \otimes e_2 \otimes ... \otimes e_{a-1}).$$

Meanwhile the reflection matrix $M_{(a)}$ is given by

$$M_{(a)}(e_1 \otimes e_2 \otimes ... \otimes e_a) = (e_a \otimes e_{a-1} \otimes e_{a-2} \otimes ... \otimes e_1).$$

Punctures are then joined by

$$..T_{(a)}(M_{(a)})^m(G_{(a)})^{2p}T_{(a)}.. \qquad (m = 0, 1;\ p = 0, 1, .., a-1),$$

where the matrices $M^m_{(a)}$ and $G^{2p}_{(a)}$ cover all possible identifications of lines at the boundary. We call the trailing boundary spins in the left hand transfer matrix $(T_{(a)})_{ij}$ above (configuration determined by the right hand index j) the outgoing set, and the leading boundary spins in the right hand transfer matrix the incoming set.

More generally, the puncture repair kit is

$$M^m_{(a)}G^{2p}_{(a)}P^j_{(a)} \qquad (j = 0, 1, .., \dim V - 1)$$

with the parallel transporter $P_{(a)}$ defined by

$$P_{(a)}(e_1 \otimes ... \otimes e_a) = (p(e_1) \otimes ... \otimes p(e_a))$$

where $p(e_i) \in V$ cycles the single spin configurations in some order.

With open boundary conditions in the circle direction, the circle and line construction gives an obvious correspondence between each spin in the outgoing set and one in the incoming. That is to say, the one on the same line. It is natural to organise the two configuration space bases in the same way. With periodic boundary conditions the correspondence is less absolutely compelling, and we could introduce a shift. Note that a translation by a sites does not affect the composite partition vector.

We note the following points. Here

$$G^{2a}_{(a)} = M^2_{(a)} = P^Q_{(a)} = 1.$$

If the interaction strengths in $T_{(a)}$ are the same on each line edge and the same on each circle edge then $T_{(a)}$ is translation invariant:

$$[T_{(a)}, G^2_{(a)}] = 0.$$

The transfer matrix may be used to associate a partition function with the torus in various ways. For example

$$Z = Tr(T^M_{(a)})$$

has a lines, while

$$Z' = Tr(T^M_{(a)}G^2_{(a)})$$

has one line, wound round the torus a times, and so on.

2.1.2 Local transfer matrices

The *local partition vector* $z[i]$ for a model is a partition vector whose boundary encloses a single interaction i, with Hamiltonian contribution H_i. It is a vector in the space of configurations *only* of the variables which are non-trivially involved in H_i. Thus each component in the vector is labelled by a configuration of these variables, and is given by $\exp(\beta H_i)$, which is completely determined by that configuration.

If we can arrange things so that incorporating a new interaction into a partition vector preserves the number of variables in the boundary, then we can necessarily arrange the corresponding local partition vector as a matrix.

Consider, then, the transfer matrix for adding a single interaction i to the partition vector $Z[c]$. This matrix will be given by the matrix arrangement of $z[i]$ direct producted with a copy of the unit matrix for each spin on the boundary not involved in interaction i. We write

$$t_i = 1_Q \otimes 1_Q \otimes ... \otimes z[i] \otimes ... \otimes 1_Q$$

for this local transfer matrix.

There are various ways to proceed with this, depending on the relative orientation we choose for the direction of bonds and the direction of evolution of the lattice as we add complete transfer layers. It is sometimes helpful to think of the latter as a Euclidean time direction, in the field theory sense. There is no *a priori* reason why the lattice bonds should be oriented parallel and perpendicular to this direction.

For the moment, however, let us suppose that sites on a given circle correspond to points at fixed time, and those on a line to points at fixed spatial position. We then want the transfer matrix for adding an interaction which is either timelike or spacelike. That is to say, for including an interaction between s_x and s_{x+1} adjacent on $\{c\}$; or between s_x and $s_{x'}$ internalising x and thus modifying $\{c\}$ to $\{c'\}$. Some examples of each type are illustrated in figure 2.3, where again internal spins are marked with solid circles and external ones with empty circles.

The Potts model

In the Potts model the interaction is, for example,

$$H_i = \beta\delta_{s_x,s_{x'}}$$

where x and x' are the sites at either end of the edge i. To be explicit we will consider $Q = 2$. Let us ignore, for the moment, the trailing boundary

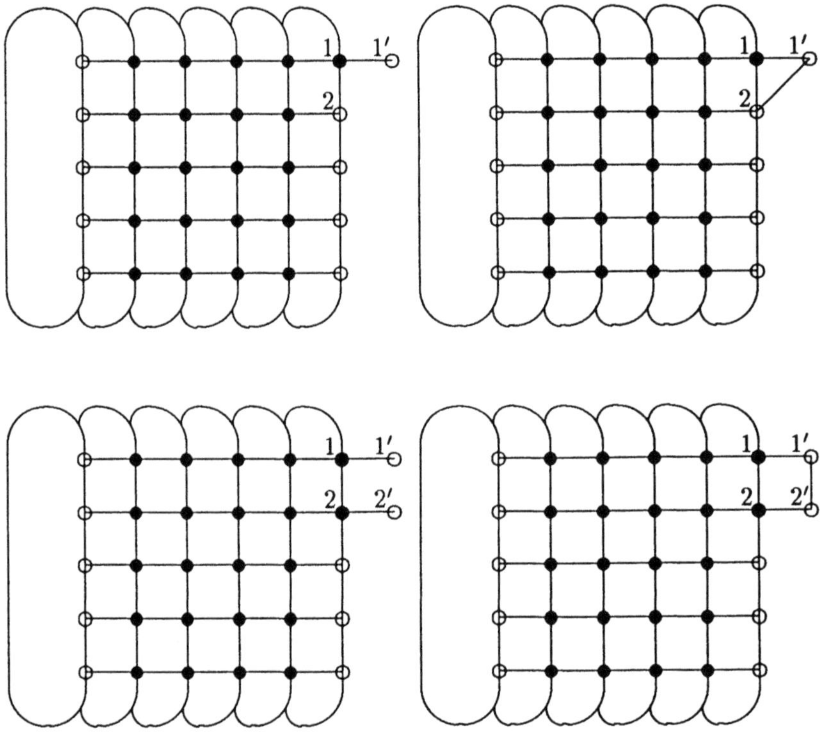

Figure 2.3: Building lattice interactions onto the partition vector $Z[c]$ associated with the lattice shown in figure 2.1, using local transfer matrices.

information for the lattice shown in figure 2.1, and the leading boundary information for all but the two sites labelled 1 and 2. For each configuration of the ignored boundary spins we thus have a vector of partition functions

$$Z[c] = (Z_{++}, Z_{+-}, Z_{-+}, Z_{--})$$

where the indices refer to the (dipole) configuration of the two spins $s_1, s_2 \in V = \{+1, -1\}$ (see equation 1.4). In other words, for a given configuration of the ignored spins these last two pieces of information completely determine the boundary configuration, and we can compute a partition function. If we include the effect of the interaction on the bond between sites 1 and 1′ (see figure 2.3) then, with $y = \exp(\beta)$, the new vector of partition functions will be

$$Z[c'] = (yZ_{++} + Z_{-+} \ , \ yZ_{+-} + Z_{--} \ , \ Z_{++} + yZ_{-+} \ , \ Z_{+-} + yZ_{--})$$

where the positions in the vector now correspond to the configuration of spins at sites 1′ and 2 (which form the new boundary, site 1 having been internalised). The single interaction transfer matrix for this interaction, $t_{1.}(y)$ say, is thus given by

$$Z[c]t_{1.}(y) = Z[c']$$

that is

$$t_{1.}(y) = \begin{pmatrix} y & 0 & 1 & 0 \\ 0 & y & 0 & 1 \\ 1 & 0 & y & 0 \\ 0 & 1 & 0 & y \end{pmatrix}. \tag{2.2}$$

We can similarly compute the matrix for the interaction on the edge between 2 and 2′

$$t_{2.}(y) = \begin{pmatrix} y & 1 & 0 & 0 \\ 1 & y & 0 & 0 \\ 0 & 0 & y & 1 \\ 0 & 0 & 1 & y \end{pmatrix}; \tag{2.3}$$

and then that for the interaction between 1′ and 2′

$$t_{12}(y) = \begin{pmatrix} y & 0 & 0 & 0 \\ 0 & 1 & 0 & 0 \\ 0 & 0 & 1 & 0 \\ 0 & 0 & 0 & y \end{pmatrix}; \tag{2.4}$$

and so on.

Here we have used one of various workable notations available for labelling interactions, i.e. labelling by the variables involved (1. is taken to signify 11′, and so on). In this case the problem of labelling interactions reduces to one of labelling the variables themselves, but of course the variables themselves may simply be numbered sequentially as they occur on the boundary.

Another popular notation for two dimensional models is just to number all the *interactions* in the layer sequentially. In the bulk of the system, in this alternative scheme, we usually label edges on circles with even numbers and those on lines with odd (see section 2.8). It is useful to be able to use either notation. Note that we have distinguished them by dotting the single number index in the variable notation.

Applying the matrices in different *orders* builds different shaped networks of lattice interactions. The sequence from top left to bottom left to bottom right in figure 2.3 (the order $Z[c]t_1t_2t_{12}$) builds a square; while that from top left to top right ($Z[c]t_1t_{12}$) builds a triangle.

Building a layer transfer matrix

Altogether, a complete layer of square lattice interactions is incorporated, in our 5 site layer example, by

$$T_{(5)} = t_{1.}t_{2.}t_{3.}t_{4.}t_{5.}t_{12}t_{23}t_{34}t_{45}t_{51}.$$

and more generally by

$$T_{(N)} = \left(\prod_{i=1}^{N} t_{i.}\right)\left(\prod_{i=1}^{N} t_{ii+1}\right) \tag{2.5}$$

where $t_{NN+1} = t_{N1}$. After this, the boundary *shape* is the same as it was at the begining (one complete layer has been internalised) and we may simply apply the same object again to iterate the procedure. In such a transfer matrix the states which become internalised when it is applied to a partition vector are called the incoming ones, and those which form the new boundary of the system are called the outgoing ones.

By omitting t_{N1} in $T_{(N)}$ we build a layer with the *circle* of interactions left open. This open layer transfer matrix is denoted $T_{(N')}$. The periodic closure is lost, but note that in this particular construction we have not kept the state of spins on the resultant boundary as boundary information. In fact we have simply summed over all the possibilities at this boundary. This means that we cannot compose with another partition vector at this boundary, and so it must now be regarded as a true boundary of the system as a whole.

It is worth noting that certain multispin interactions can be incorporated into a partition vector by using appropriate combinations of the *same* local transfer matrices (see Baxter, Temperley and Ashley 1978, for example).

2.2 Algebraic formulation

In this section we will develop a coupling parameter independent formal algebraic framework for building transfer matrices. The general procedure is to resolve the transfer matrix into sums of constant matrices, with coefficients which are functions of the coupling parameters. We will illustrate the merits of this approach.

In general, the single interaction transfer matrix t_i (where i is some appropriate interaction label) may be written in the form:

$$t_i(y) = a(y)1 + b(y)g_i + c(y)h_i + .. \tag{2.6}$$

where $a(y), b(y), c(y)$,.. are scalar functions of the coupling strengths (here generically represented by y) and $g_i, h_i, ...$ are constant matrices.

Spin models

Although each particle in a system usually has the same kind of configuration space, it is useful to be able to formally distinguish the single particle subspaces of the space of configurations of the whole system. Thus we introduce $V^{(x)}$ to denote the space of configurations of particle x.

For spin models, if a spin s_x is internalised by $t_i(y)$ the matrices $1, g_i, h_i$,.. map $V^{(x)} \rightarrow V^{(x')}$; if t_i includes an interaction between s_x and s_{x+1} adjacent on $\{c\}$ then they are diagonal on $V^{(x)} \otimes V^{(x+1)}$. In any case, for a given i they span, at most, some $End(V)$.

In many cases the symmetries of the system simplify this. The number of summands in equation 2.6 is a lower bound on the number of distinct possibilities for the interaction energy (i.e. the order of the image of the configuration space under H_i). Thus, for example, the Potts model has simply

$$t_i(y) = a(y)1 + b(y)g_i \ . \tag{2.7}$$

Any $t_i(y)$ that does not change the boundary $\{c\}$ only changes the coupling. To see this, consider the effect of applying $t_{12}(y)$ *again* to the bottom right picture in figure 2.3. A double bond between the same two spins effectively just counts the same interaction twice or, equivalently, doubles the interaction strength. In this sense, each local transfer matrix acts like a kind of rotation of the local coupling parameter, an observation which we will exploit in chapter 4, when solving the Ising model.

It follows that there exists a function $f(y)$ such that $t_i^2(y) = t_i(f(y))$ and, using equation 2.7, we have

$$g_i^2 + \beta g_i + \alpha 1 = 0 \tag{2.8}$$

where α and β are constants, that is, they are independent of the couplings. For the

$$Q = 4\cos^2(\pi/r)$$

state Potts model with

$$q = -\exp(\pi i/r) \tag{2.9}$$

we usually take $\alpha = -q^2$, $\beta = -1 + q^2$, for reasons which will become clear shortly.

It also follows, provided the local transfer matrices are symmetric, that we can form a symmetric layer transfer matrix. For example, defining

$$S_{(N)} = \left(\prod_{i=1}^{N} t_{ii+1}\right)^{1/2}, \tag{2.10}$$

for example

$$S_{(5)} = (t_{12}t_{23}t_{34}t_{45}t_{51})^{1/2},$$

we have the symmetric Potts transfer matrix:

$$S_{(N)}T_{(N)}S_{(N)}^{-1} = \left(\prod_{i=1}^{N} t_{ii+1}\right)^{1/2} \left(\prod_{i=1}^{N} t_{i.}\right) \left(\prod_{i=1}^{N} t_{ii+1}\right)^{1/2}. \tag{2.11}$$

2.2.1 General local interactions

What about more general models (with local interactions)?

Suppose there are s p-state variables in each incoming and outgoing edge of an N site layer. In the Potts case $s = N$ and $p = Q$; in the 6-vertex model case, for example (see chapter 12), $s = 2N$ and $p = 2$. Suppose further that the interaction at position i depends on the configuration of m nearby incoming and m nearby outgoing spins, and temperature variable y. Then the *interaction matrix* $R(y) \in End(\otimes^m V_p)$ is the matrix arrangement of the local partition vector $z[i](y)$, and:

(i) The matrix $t_i(y) \in End(\otimes^s V_p)$ acts trivially on all the subspaces of configurations of spins not involved in interaction i, that is

$$t_i(y) = 1_p \otimes 1_p \otimes .. \otimes R(y) \otimes 1_p .. \otimes 1_p,$$

where the product contains $(s-m)$ unit matrices; and

(ii) The matrix $R(y)$ *could*, at its most general, assign a different weight to each of the p^m possible configurations of the variables involved in interaction i. To obtain an expression of the form of equation 2.6 it would then be necessary to build $R(y)$ out of representations of a p^m dimensional matrix algebra. We call this the 'local' algebra for the model, because it covers only a single interaction. In this case the matrices $g_i, h_i, ..$ in equation 2.6 span $End(\otimes^m V_p)$.

As we will see, their are plenty of models of all shapes and sizes characterised by the fact that $R(y)$ is built out of representations of a 2 dimensional subalgebra (i.e. equations 2.7 and 2.8 apply). This is the simplest non-trivial special case. Later we will furnish examples involving 3,4,5,... dimensional subalgebras.

2.2.2 Transfer matrix algebras

Since the relation 2.8 is independent of coupling parameters the operator g_i can then be regarded as a fundamental building block for an algebraic formulation of the transfer matrix. The full transfer matrix will be expressible as a sum of products of such objects, with all coupling dependent information occuring in scalar factors. We call equation 2.8 (or any other element of a set of relations determining the local algebra) a *local* relation.

Returning to figure 2.3, we note that, as we build another transfer matrix layer onto the lattice (i.e. take into account the effect of another layer of interactions) the order of inclusion of interactions which are not dependent on a common variable is unimportant. This means that the operators g_i, g_j commute if they do not act on any single variable configuration subspace in common (they are not *adjacent*). We call

$$g_i g_j = g_j g_i \qquad (i, j \text{ not adjacent}) \tag{2.12}$$

the *commuting* relation.

It remains, of course, to describe the non-trivial relations between products of adjacent operators, and more generally, non-commuting composites. In the two dimensional Potts model (and many other two dimensional models, see later) each operator is adjacent to two others. Let us consider the case in which the quadratic local relation 2.8 applies. If we suppose that there exist relations between products involving only adjacent pairs, then on symmetry grounds these should take the form

$$g_i g_j g_i g_j \ldots = g_j g_i g_j g_i \ldots \qquad (i, j \text{ adjacent}) \tag{2.13}$$

where both products contain the same number of factors. If they contain one or two factors the model is trivial. In almost all the cases we will consider, including the Potts model, the products contain three factors. We then have the *braid* relation:

$$g_i g_j g_i = g_j g_i g_j \qquad (i, j \text{ adjacent}). \tag{2.14}$$

Later we will give a direct physical explanation for the source of this relation in the case of the Potts model. This relation is also common among models for which the quadratic local relation 2.8 does not apply.

In general, the total set of matrices $g_i, h_i, ..$ required to construct a transfer matrix in this way generate a *transfer matrix algebra*. We will see shortly that such algebras can play an important role, both in statistical mechanical computation, and in characterising models. To this latter end it is also useful to determine a complete set of relations obeyed by the matrices, and hence to regard them as a representation of a set of abstract generators for the algebra. The relations 2.8, 2.12 and 2.14 define a *Hecke* algebra (see chapter 9).

2.3 Automorphisms of the lattice

Definition

Having constructed our lattice as part of a sequence (see chapter 1) we can then safely describe it *individually* in an algebraic way, with an incidence matrix I. Let L be the space of all sites in the lattice. Then $I \in End(L)$ and the matrix entries I_{ij} are 1 if sites i and j are nearest neighbours and zero otherwise. For our purposes the automorphisms of a lattice are the subgroup of permutations P of the set of sites which preserve the incidence matrix, i.e., if $P(i)$ is the image of i under permutation P, then those permutations for which $P(i) = k, P(j) = l$ implies $I_{ij} = I_{kl}$.

Realisation

In the formalism we have described, the group of such permutations depends very much on the global topology. However, since the single bond transfer matrices are local they are not necessarily sensitive to the global topology (it depends on the model). The Potts model, for instance, suggests a natural topology of its own (as we will see). For the moment we might as well work with our favourite topology.

On the torus, we have two directions worth of periodic translations and some rotations. Considering the single circle sublattice corresponding to the transfer matrix we have periodic translations and a reflection. A translation permutes the *configurations* by cycling them within certain subsets. It cycles the single spin subspaces.

Conjugation by the translation matrix $G^2_{(a)}$, or any invertible multiple of it $AG^2_{(a)}$ where the matrix A commutes with g_i, takes $g_i \to g_{i+2}$. It follows that the translation matrix is in the algebra generated by the g_i matrices. By the same argument so is $M_{(a)}$.

Translation in the perpendicular direction may be achieved by conjugation with the transfer matrix itself. There are only a very limited number of rotations on the lattice. These points will be discussed in chapter 11.

2.4 Perron Frobenius theorem

In the physical coupling regime (i.e. β real) the transfer matrix has all positive entries. A matrix or vector with this property is called positive. The Perron Frobenius theorem states that a finite dimensional positive matrix has a unique largest magnitude eigenvalue, which is itself positive, and the associated eigenvector is also positive. This result is used heavily in statistical mechanics. Its worth justifies the effort required to give a proof....

Proof:

All our matrices and vectors have dimension D. We will take all vectors x with entries x_i to be normalised by

$$\sum_{i=1}^{D} x_i = 1.$$

We write $K > 0$ for K a positive matrix, and $M \geq 0$ ($x \geq 0$) for a matrix (vector) with non-negative entries. For a matrix $K > 0$ let S_K be the set of non-negative real numbers λ for which there exist vectors $x \geq 0$ such that

$$Kx \geq \lambda x.$$

Then

$$\sum_{i,j} K_{ij} x_j \geq \lambda$$

and

$$\sum_{ij} K_{ij} \geq \sum_{ij} K_{ij} x_j.$$

If λ_0 is the largest $\lambda \in S_K$ then for some $x^{(0)}$ we have

$$Kx^{(0)} \geq \lambda_0 x^{(0)}.$$

If λ_0 is not an eigenvalue of K then without loss of generality we have

$$\sum_j K_{1j} x_j^{(0)} - \lambda_0 x_1^{(0)} = d > 0.$$

However, for $y \geq 0$ defined by

$$y_i = x_i^{(0)} + (d/2\lambda_0)\delta_{1,i}$$

we then have $Ky > \lambda_0 y$, which contradicts the maximum property of λ_0. Thus λ_0 is an eigenvalue. Furthermore $x^{(0)}$ is a strictly positive eigenvector, since

$$\sum_j K_{kj} x_j^{(0)} = \lambda_0 x_k^{(0)}.$$

Now suppose there is an eigenvalue c with eigenvector z for which $|c| \geq \lambda_0$. Then with $|z|$ the vector given by $|z|_i = |z_i|$ we have

$$K|z| \geq |c||z|$$

(since the absolute value of the sum of two complex numbers of given magnitude is maximised if they are colinear). Thus $|c| = \lambda_0$ and

$$K|z| = |c||z| = |cz| = |Kz|$$

Since K is positive all components of z must be colinear in the complex plane, so $c = \lambda_0$.

Finally, suppose there is a vector z linearly independent of $x^{(0)}$ and not necessarily positive, for which

$$Kz = \lambda_0 z.$$

Then we may construct $w = x^{(0)} + ez$ where e is the smallest scalar for which one or more of the components of w is zero. But then $w \geq 0$ is an eigenvector with eigenvalue λ_0, so $w > 0$ and we have a contradiction. Thus there can be no such z, and λ_0 is non-degenerate.

This completes the proof of the Perron-Frobenius theorem.

2.5 The free energy

Note from equation 2.1 that the bulk of the calculation of the partition function can be reduced to repeated multiplication of a vector by transfer matrices, or to multiplying many transfer matrices together. For example, introducing T_I as the transfer matrix appropriate for the Ith circle (in general the circle transfer matrices *may* be distinguishable by their couplings), the partition function associated with a cylindrical lattice built out of M single circle transfer matrix layers (i.e. a 2 punctured sphere with M circles) and with free boundary conditions (i.e. summing over all configurations of the boundary spins) is simply

$$Z = \sum_{j,k} \left(\prod_{I=1}^{M} T_I \right)_{jk} . \tag{2.15}$$

If $T_I = T$ then

$$Z = \sum_{j,k} (T^M)_{jk} . \tag{2.16}$$

This latter expression is valid supposing that the lattice of consists of *identical* circle layers. If in addition the couplings within a layer are all the same, or distinguished only between horizontal and vertical bonds, then the partition function above may alternatively be computed as a trace of a product of open ended *line* transfer matrices.

It is helpful to proceed formally in terms of the left and right eigenvectors and the eigenvalues of the transfer matrix. Suppose that we have a transfer matrix of dimension D associated with a real Hamiltonian and real coupling parameter. Suppose further that this matrix is diagonalisable by some similarity transformation

$$T \to STS^{-1} . \tag{2.17}$$

(a non-diagonal Jordan form merely obfuscates, rather than invalidates, the following argument).

Then at a given value of the coupling parameter we may define D ket vectors v_i, with a partial order, by

$$Tv_i = \lambda_i v_i \tag{2.18}$$

for $i = 0, .., D-1$, where $\lambda_i > \lambda_j$ implies $j > i$. Note that λ_0 is the unique first in the partial order for all real couplings, by the Perron Frobenius theorem.

Similarly we have D bra vectors w_i,

$$w_i T = \lambda_i w_i \tag{2.19}$$

where

$$w_j v_i = \delta_{i,j} \qquad\qquad \sum_i v_i w_i = 1_D \tag{2.20}$$

The boundary conditions involve summing over the matrix elements with some weight $\omega(i)$ associated to each leading and trailing edge configuration i. They can therefore be represented by bra and ket vectors themselves. Let $|A(i)>$ be the vector which picks out a specific configuration i, that is $(|A(i)>)_j = \delta_{i,j}$ in the configuration space basis. Then we have the boundary ket vector

$$|\omega>=\sum_i \omega(i)|A(i)> . \tag{2.21}$$

Alternatively, these boundary vectors can be decomposed as a linear combination of eigenvectors,

$$|\omega>=\sum_i \omega_i v_i \tag{2.22}$$

where ω_i is a scalar, and similarly for $<\omega'|$. We will adopt the local convention that indices in brackets refer to the configuration space basis, while indices in subscript refer to the Jordan basis. We obtain the N layer partition function

$$Z_{(N)} =<\omega'|T^M|\omega>=\sum_i \omega'_i\omega_i(\lambda_i)^M . \tag{2.23}$$

The free energy in the limit $M \to \infty$ is thus

$$f = \lim_{M\to\infty}(1/M)\ln(Z_M) = \lambda_0 \tag{2.24}$$

provided that $\omega'_0, \omega_0 \neq 0$. This latter condition is clearly satisfied provided all the weights are non-negative and at least one is positive. So, apart from this physicality condition, we observe boundary condition independence of the free energy in the large M limit.

The free energy of a system on a lattice with an infinite number of circle layers may thus be computed as the largest eigenvalue of the *single* circle transfer matrix. It is in this sense that the transfer matrix approach leads to a reduction in the effective dimension of the problem.

2.6 Correlation functions

Consider the subtracted correlation function for two separated local measurements on observables $\theta_{(0,0)}$ and $\theta_{(x,y)}$

$$<\theta_{(0,0)}\theta_{(x,y)}>_s \; = \; <\theta_{(0,0)}\theta_{(x,y)}> - <\theta_{(0,0)}><\theta_{(x,y)}> .$$

We will take the former observable to be a function of the spin variables within a layer and close to some nominal origin $(0,0)$; and the latter to be an identical function translated to (x,y). We suppose that both observables are entirely within a part of the lattice which can be constructed by composing a sequence of identical transfer matrices, so that coordinates refer to a site on a circle which is x transfer matrix layers away, and a line which is y sites away in the layer, measured from the nominal origin.

Define $D_y(j)$ as the result of a measurement θ made on a lattice layer when in configuration j, with the origin of coordinates within the lattice layer taken to be at $-y$. We then define matrices θ_y by:

$$\theta_y = \sum_j |A(j)> \; D_y(j) \; < A(j)|. \tag{2.25}$$

We define the scalar $d_{ik}(y)$ by

$$\theta_y = \sum_{i,k} d_{ik}(y) \; v_i w_k \tag{2.26}$$

It will not be necessary to compute the *value* of $d_{ik}(y)$.

With $< a'|, |a >$ denoting our prefered boundary vectors, we can now write the subtracted correlation function as

$$\begin{aligned} < \theta_{(0,0)}\theta_{(x,y)} >_s \;&=\; \frac{< a'|T^M \theta_0 T^x \theta_y T^{N-M-x}|a >}{Z_{(N)}} \\ &\quad - \frac{< a'|T^M \theta_0 T^{N-M}|a >< a'|T^{M+x}\theta_y T^{N-M-x}|a >}{Z^2_{(N)}} \\ &= (1/Z_{(N)}) \left(\sum_{i,j,k} a'_i \lambda_i^M d_{ij}(0) \lambda_j^x d_{jk}(y) \lambda_k^{N-M-x} a_k \right) \\ &\quad -(1/Z^2_{(N)}) \left(\sum_{i,j} a'_i \lambda_i^M d_{ij}(0) \lambda_j^{N-M} a_j \right) \\ &\quad \cdot \left(\sum_{i,j} a'_i \lambda_i^{M+x} d_{ij}(y) \lambda_j^{N-M-x} a_j \right) . \end{aligned} \tag{2.27}$$

Let

$$\chi = 1/(\ln(\lambda_0) - \ln(\lambda_{1'}))$$

where $\lambda_{1'}$ is the largest eigenvalue after λ_0 for which $d_{0i}(0), d_{i0}(y) \neq 0$. Then in the limit of $N >> M >> x >> 1, y$ we obtain, for fixed y,

$$\begin{aligned} < \theta_{(0,0)}\theta_{(x,y)} >_s \;&\rightsquigarrow\; \textstyle\sum_{i\neq 0} d_{0i}(0) d_{i0}(y) \left(\frac{\lambda_i}{\lambda_0}\right)^x \\ &\sim\; x^{-c} \exp(-x/\chi) \end{aligned}$$

where c is some number, and χ is called the correlation length.

We see that the long distance properties of the model are determined by the spectral gaps of the transfer matrix.

Recall that one signal of a second order phase transition is a divergent correlation length. This occurs when a correlation eigenvalue $\lambda_{1'}$ becomes degenerate with the free energy eigenvalue λ_0. By the Perron Frobenius theorem this can only happen for real β in the thermodynamic limit.

2.7 Spectrum of the transfer matrix

In any case it is clear that the spectrum of the transfer matrix contains crucial information. At least for finite dimension D this spectrum is given by the roots of the characteristic polynomial of T, that is, for λ a scalar variable, by the roots of

$$S(T, \lambda) = \det(T - \lambda\, 1_D).$$

If we only want the spectrum, and not the eigenvectors, then it is sufficient to solve the polynomial. In principle, the calculation of the eigenvalues is simplified by, as a first stage, the factorisation of the characteristic polynomial into its irreducible components (Ahlfors 1979) with coefficients polynomial in $y = \exp(\beta)$ (denoted $\in C[y]$).

The characteristic polynomial is a Dth order polynomial in λ, with coefficients which are polynomials in $y = \exp(\beta)$ (and, if required, $\sqrt{Q}$), since the matrix elements of T have these properties, at least for the Potts model. The irreducible factorisation of the polynomial over $\exp(\beta)$ is the (unique) factorisation which preserves polynomial coefficients in $\exp(\beta)$. It corresponds to the complete block diagonalisation of T using only similarity transformations with entries in the same field.

Once this has been achieved, the eigenvalues within a block form an algebraic function in y. In particular they are analytic continuations of one another in y.

Part of the block diagonalisation is achieved by using the external symmetries (translation and reflection) of the layer, and part by using the internal symmetries. This latter procedure corresponds to exhibiting the irreducible content of the matrices g_i as a representation of the appropriate algebra. Thus the algebraic functions in the eigenvalue spectrum are labelled by the representation and the external symmetry properties.

Note that the spectrum of the transfer matrix is insensitive to the choice of basis for a representation of the transfer matrix algebra involved. Thus if two models have transfer matrices containing the same representation as an irreducible component, then the corresponding part of each spectrum will be identical. If the largest eigenvalues in the region of β real coincide (i.e. the representation carrying the largest eigenvalue is common to both transfer matrices) then the models are said to be *equivalent.* Baxter's working definition of equivalence (Baxter 1985) also requires the order parameters to coincide. We will return to this point later.

In principle we could ignore the models, and simply obtain the spectrum of the transfer matrix, regarded as an element of the abstract algebra, in each irreducible representation. It is certainly an important first move to investigate the algebra associated with a model.

It is important to know *all* the relations among the generators associated with a transfer matrix algebra A, say, since any representation of an algebra B defined by a *subset* of the relations, which does not obey the full set, will have multiplicity zero in the transfer matrix representation of A.

2.8 Potts models

We will now construct the Potts transfer matrix explicitly. The precise construction can depend on the boundary conditions. We will discuss these in the next section.

In general it is sufficient to construct the local transfer matrices and then use equation 2.5, or some other product of local transfer matrices as appropriate. Let $x = (e^{\beta} - 1)/\sqrt{Q}$ and

$$U_{i.} = \left(1/\sqrt{Q}\right) \, (1_Q \otimes 1_Q .. \otimes M \otimes .. \otimes 1_Q).$$

with $M \in End(V_Q)$ given by $M_{ab} = 1$ $(a, b \in V)$ and appearing in the ith position in the product. Then generalising from equation 2.2 we can write the local transfer matrix $t_{i.}(x) \in End(\otimes^N V_Q)$ as

$$t_{i.}(x) = \sqrt{Q}\, x \, 1_{Q^N} + \sqrt{Q}\, U_{i.} = \sqrt{Q}x \left(1_{Q^N} + \frac{1}{x} U_{i.}\right).$$

With $D \in End(V_Q \otimes V_Q)$ given by

$$D_{(a\otimes b, c\otimes d)} = \delta_{a,c}\delta_{b,d}\delta_{a,b}$$

$(a, b, c, d \in V)$, and

$$U_{ii+1} = U_{i+1i} = \sqrt{Q} \, (1_Q \otimes 1_Q .. \otimes D \otimes .. \otimes 1_Q),$$

it follows that

$$t_{ii+1}(x) = 1_{Q^N} + xU_{ii+1}. \tag{2.28}$$

In interaction notation we have the alternative labelling for the matrices $U_{i.}$ and U_{ii+1}:

$$U_{i.} = U_{2i-1}$$

$$U_{ii+1} = U_{2i}.$$

The matrices U_i (i dotted integer or double integer; or equivalently $i = 1, 2, ..., 2N-1$) generate the same algebra as the matrices

$$g_i = 1 - qU_i$$

(from equation 2.8 and below) in the Potts model case. The matrices U_i obey the Temperley-Lieb relations (Temperley and Lieb 1971)

$$\begin{array}{rcll} U_iU_i & = & \sqrt{Q}U_i & \\ U_iU_jU_i & = & U_i & (i,j \ \mathit{adjacent}) \\ U_iU_j & = & U_jU_i & (i,j \ \mathit{not\ adjacent}) \end{array}$$

together with some additional (generically trivial) relations which we will discuss in chapter 6. These matrices constitute the Potts representation of the Temperley-Lieb algebra.

Note that, although $U_{i.}, U_{ij}$ are of equivalent standing in the algebra, $t_{i.}(x)$ and $t_{ij}(x)$ are *different* functions of x. In fact $(x/\sqrt{Q})t_{i.}(1/x)$ and $t_{ij}(x)$ are formally the same functions of x.

We will use the notation $t_I(x)$ when we do not want to specify an interaction type, but refer to an abstract interaction.

2.9 Standard transfer matrices and duality

Here we collect some remarks about the different ways of building a transfer layer, and about boundary conditions.

For a given model in 2 dimensions, and the given transfer matrix layering direction, a closed layer transfer matrix is any ordering of the local transfer matrices in which the cylindrical completion is present. Any such matrix will be similar, i.e. conjugate, to the one defined by $T_{(N)}$ (equation 2.5). It will thus have the same spectrum.

With $t_{NN+1} = t_{N1}$ we define a translation invariant 2 parameter special case of $T_{(N)}$ by

(i) *'Closed stepladder'*

$$T_{(N)}(x_1, x_2) = \left(\prod_{j=1}^{N} t_{j.}(x_1)\right)\left(\prod_{j=1}^{N} t_{jj+1}(x_2)\right).$$

The standard alternative

(ii) *'Closed stairway'*

$$T'_{(N)}(x_1, x_2) = \prod_{j=1}^{N} (t_{j.}(x_1)\, t_{j\,j+1}(x_2))$$

is not, in itself, translation invariant (but see chapter 12 for an alternative formulation).

It is also possible to similarity transform the *open* layer transfer matrices (cylindrical completion absent) for a large class of 2 parameter lattice models, including the above models, into either of the *standard forms*

(i) *'Open stepladder'*

$$T_{(N')}(x_1, x_2) \quad = \quad \left(\prod_{j=1}^{N} t_{j.}(x_1)\right) \left(\prod_{j=1}^{N-1} t_{jj+1}(x_2)\right)$$

(ii) *'Open stairway'*

$$T'_{(N')}(x_1, x_2) \quad = \quad t_{1.}(x_1)t_{12}(x_2)t_{2.}(x_1)t_{23}(x_2)...t_{N-1N}(x_2)t_{N.}(x_1)$$

or into any other product in which each local transfer matrix (except $t_{N1}(x_2)$) appears exactly once. We will return to the wedge or rotation transfer matrix $S(x_1, x_2)$ defined by

$$T'_{(N')}(x_1, x_2) = S(x_1, x_2)\left(T_{(N')}(x_1, x_2)\right)(S(x_1, x_2))^{-1}$$

later.

Recall that, for the Potts model, the factors $t_i(x)$ each introduce the effect of one horizontal (i dotted integer) or vertical (i double integer) lattice bond interaction into the Hamiltonian. The matrices T and T' just differ, therefore, in the order in which the effect of various interactions is incorporated into the Hamiltonian, and hence in the effective layer orientation. For the Potts model T builds layers parallel to the horizontal bonds, while T' builds stairway layers at 45°. The direction of *propagation* of the layers is the same in each case however, i.e. perpendicular to horizontal bonds.

Duality

Note that the spectrum of the transfer matrix is thus invariant, up to boundary effects and overall factors, under the statistical mechanical duality transformation defined by

$$\begin{aligned} U_{i.} &\rightarrow U_{ii+1}; \\ U_{ii+1} &\rightarrow U_{i+1.}; \\ x_1 &\rightarrow (x_2)^{-1}; \\ x_2 &\rightarrow (x_1)^{-1}. \end{aligned}$$

Specifically, for example, the spectrum of the matrix

$$T_{(N'')}(x_1, x_2) = \left(\prod_{j=1}^{N-1} t_{j.}(x_1)\right) \left(\prod_{j=1}^{N-1} t_{jj+1}(x_2)\right)$$

is invariant up to a factor $(x_1 x_2)^{N-1}$.

In the sense that performing this transformation twice produces a translation it acts like 'half' a translation. It is called a lattice model duality transformation (c.f. section 1.5). For an alternative, topological explanation of duality see Savit 1980 (and c.f. chapter 6).

Hintermann, Kunz and Wu (1978) have proved that the isotropic (i.e. $x_1 = x_2$) $Q > 4$ state Potts models each have a unique phase transition for $\beta > 0$. The result also holds for $Q = 2, 3, 4$. It follows that a critical point in the chosen 2 parameter sub-manifold of general coupling parameter space is any point on the line

$$x_1 x_2 = 1,$$

since these are the fixed points of the duality coupling transformation (see also Baxter 1982). We will write the one parameter family of *critical* Potts transfer matrices as

$$T(1/x, x) = T(x).$$

More generally we can define arbitrary one parameter families of transfer matrices by

$$T_f(x) = T(f(x), x)$$

for some single valued function $f(x)$. So, for example, $f(x) = 1/x$ is the critical line and $f(x) = x$ is the isotropic line. Once given $f(x)$ we use the convention that the local transfer matrix when written in interaction labeling notation, i.e. $t_i(x)$ $(i = 1, 2, .., 2N-1$, or $2N)$, has the same formal functional dependence on x for all i. That is

$$t_i(x) = 1 + xU_i.$$

This helps to manifest the duality symmetry, and will streamline the abstract discussion in the next chapter.

In general then, up to an overall factor of the form $(\sqrt{Q}x)^N$, we have

$$\begin{aligned} T'_f(x) &= t_1(1/f(x))\ t_2(x)\ t_3(1/f(x))\ t_4(x) \ldots. \\ &= (1 + U_1/f(x))(1 + xU_2)(1 + U_3/f(x))\ldots \end{aligned} \tag{2.29}$$

where

$$t_{ii+1}(x) = t_{2i}(x)$$

and

$$\frac{1}{\sqrt{Q}f(x)}t_{i.}(f(x)) = t_{2i-1}(1/f(x)).$$

We will now also redefine the 2 parameter transfer matrix $T'(x_1, x_2)$ so as to keep the simple form

$$T'(x_1,x_2) = t_1(x_1)t_2(x_2)t_3(x_1)t_4(x_2)....$$

$$= (1 + x_1U_1)(1 + x_2U_2)(1 + x_1U_3)(1 + x_2U_4)....$$

This just means that the arguments x_1 and x_2 here are no longer both related to the Potts couplings β_1, β_2 in the same way. We have

$$x_2 = \frac{\exp(\beta_2) - 1}{\sqrt{Q}}$$

$$x_1 = \frac{\sqrt{Q}}{\exp(\beta_1) - 1}.$$

Note that $T'(x_1, x_2)$ gives the Potts transfer matrix up to an overall scalar factor, $(\sqrt{Q}x_1)^N$. Since such a factor only gives a regular additive contribution to the free energy it will not be necessary to follow its adventures too closely.

2.9.1 Alternative layering directions

Although $T'_{(N)}$ is not translation invariant we can use it to compute a periodic lattice partition function. For example with $N = 2$,

$$(T'_{(2)})^M = \underbrace{t_0t_1t_2t_3}_{T'_{(2)}}\ \underbrace{t_0t_1t_2t_3}_{T'_{(2)}}\ \underbrace{t_0t_1t_2t_3}_{T'_{(2)}}\ \ldots\ \underbrace{t_0t_1t_2t_3}_{M^{th}\ copy}\ .$$

However, a brief examination of this example will show that the periodic identification of boundary sites is staggered. This is annoying, because $T'_{(N)}$ has useful properties from other points of view which the stagger obfuscates. One resolution of this problem is to use a different transfer matrix layer propagation direction. In general, we can propagate parallel to an arbitrary line on the lattice, but more boundary information will have to be kept.

Consider layering at 45° to the bond directions, with periodic identification at the ends of the layer. Let $N = 4M$ (with $M \in Z_+$, for later computational convenience) be the number of sites in a zigzag layer, as exemplified in figure 2.4. The layer transfer matrix here, Y_N, may be thought of as a product of two sublayer transfer matrices X and W, as shown.

Figure 2.4: Building a square lattice by adding transfer layers at 45° in the case $N = 8$. The left and right endpoints are identified in each layer, giving a skew periodicity compared to the natural periodic completion.

As soon as we have decided how to arrange the boundary information the matrices for X and Y become well defined and can thus be written down *directly*, i.e. all in one go. However, if we add layers in this way, and insist on the 'no redundancy' condition that $X, W \in End(\otimes^{N/2} V_2)$ (i.e. that X maps from the space of configurations of odd numbered sites in the zigzag layer to configurations of even numbered sites in the zigzag layer, and W vice versa) then X and W cannot be built as products of local operators $t_i(x)$. For example, with $N = 4$

$$X = \begin{pmatrix} y_1^2 y_2^2 & y_1 y_2 & y_1 y_2 & 1 \\ y_1 y_2 & y_1^2 & y_2^2 & y_1 y_2 \\ y_1 y_2 & y_2^2 & y_1^2 & y_1 y_2 \\ 1 & y_1 y_2 & y_1 y_2 & y_1^2 y_2^2 \end{pmatrix}. \qquad (2.30)$$

Let us define the following elements of the Potts representation of the Temperley-Lieb algebra with $2N - 1$ generators (i.e. twice as many generators and twice as many spins as we should strictly need in the present situation):

Define matrices A_j for $j = 1, 2, 3, 4$ by

$$A_j = Q^{(-1)^{j+1} M} \prod_{i=1}^{N/2} U_{4i-j},$$

and, recalling that

$$t_{2i-1}(x) = 1 + xU_{2i-1}$$

so that

$$\sqrt{Q}xt_{2i-1}(1/x) = \sqrt{Q}x + \sqrt{Q}U_{2i-1},$$

define matrix $B(y_1, y_2)$ by

$$B(1+\sqrt{Q}\,x_1, 1+\sqrt{Q}x_2)$$

$$= (Qx_1x_2)^{N/2}\; t_1(1/x_1)\, t_3(1/x_2)\, t_5(1/x_1)\, t_7(1/x_2)....t_{2N-1}(1/x_2)$$

$$= (y_1 - 1 + \sqrt{Q}\;U_1)(y_2 - 1 + \sqrt{Q}\;U_3)(y_1 - 1 + \sqrt{Q}\;U_5)(y_2 - 1 + \sqrt{Q}\;U_7)...$$

Note that

$$B(1,1) = A_1A_3$$

$$\lim_{y\to\infty} \frac{B(1,y)}{y^{N/2}} = A_3$$

$$\lim_{y\to\infty} \frac{B(y,1)}{y^{N/2}} = A_1$$

and so on.

Then relaxing the no redundancy condition we can write, for example,

$$X(y_1, y_2) \rightsquigarrow A_4\; B(y_1, y_2)\;\; A_2A_1A_4\; \in End(\otimes^N V_2)$$

$$W(y_1, y_2) \rightsquigarrow A_4A_1A_2\; B(y_2, y_1)\;\; A_4$$

where $\rightsquigarrow$ signifies the use of a basis with redundancy. This just means that we are only interested in the subspace projected out by the idempotent A_4. We have then effectively split each spin in the original problem into 2 separate spins and then recovered the original problem by freezing them together with infinite ferromagnetic interactions.

Note that we have arranged the parameters in the matrix B so that the arguments of X and W are the original Potts exponentiated couplings $y_i = \exp(\beta_i)$.

The whole construction is illustrated in figure 2.5.

The translation matrix G here is illustrated in figure 2.6. It is given by

$$G = \lim_{y\to\infty} \frac{X(y,1)W(y,1)}{y^N} = A_4A_1A_2A_3A_4,$$

with

$$G^{-1} = \lim_{y\to\infty} \frac{X(1,y)W(1,y)}{y^N} = A_4A_3A_2A_1A_4.$$

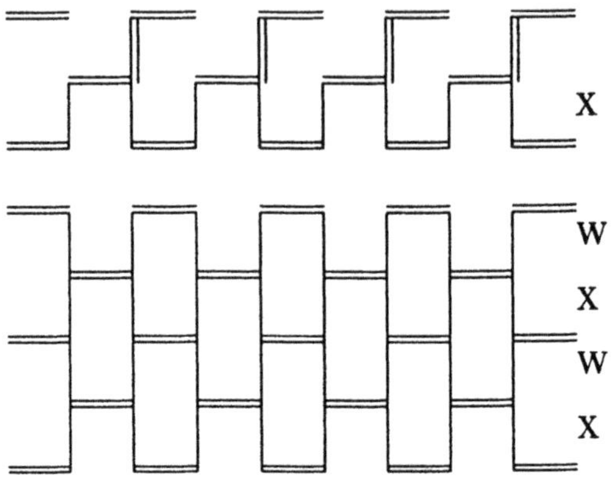

Figure 2.5: Building layers at 45° with standard local transfer matrices. Double lines represent frozen interactions, so the spins at each end may be identified (contracting these lines to points, then, we recover the previous figure). Absent lines correspond to decoupled spins. The left and right endpoints are identified in each layer.

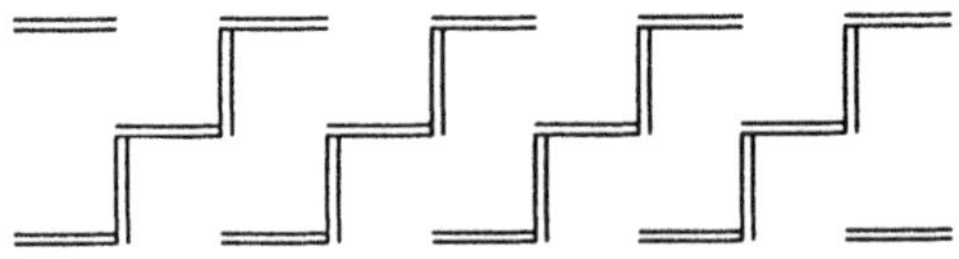

Figure 2.6: The translation G.

2.10 Spectrum inversion transformations

It is a widespread property of transfer matrices that the inverse matrix may also be written as some form of transfer matrix. We will see in the next chapter that this can lead to some very useful hints about the spectrum. Let us first illustrate the property with a familiar example....

Note that for each local transfer matrix $t_i(x)$ built using an algebra with a quadratic local relation we can necessarily define a function $x'(x)$ such that

$$t_i(x)\, t_i(x'(x)) = 1. \tag{2.31}$$

For example, in the Hecke algebra (and hence Temperley-Lieb algebra) case we have

$$(1 + xU_i)(1 + x'(x)U_i) = 1$$

giving

$$x'(x) = \frac{-x}{1 + \sqrt{Q}x}.$$

Let us define a *reverse* stairway transfer matrix, using the reflection matrix $M_{(a)}$, by

$$\tilde{T}'(x_1, x_2) = M_{(a)} T'(x_1, x_2)\, M_{(a)}^{-1}.$$

Since from equation 2.31

$$T'(x_1, x_2)\tilde{T}'(x'(x_1), x'(x_2)) = 1$$

it follows that the layer transfer matrix $T'(x'(x_1), x'(x_2))$ has spectrum inverse to that of $T'(x_1, x_2)$.

Note that, since T' and $\tilde{T}'$ here commute, they have the same eigenvectors. The corresponding eigenvalues are thus mutually inverse. In general the similarity transformation obtained by conjugating with $M_{(a)}$ may permute the eigenvalues (regarded as functions of x_1, x_2) in pairs. This means that the analytic continuation of an eigenvalue from $x \to x'$ does not necessarily give the inverse. It would be useful to have a transformation which strictly inverted each eigenvalue, so we need some way of avoiding the permutation effect.

Let us define the involutive inversion transformation

$$x \to x'(x).$$

Recalling that the Potts coupling parameters are given by $x_2 = (y_2 - 1)/\sqrt{Q}$ and $x_1 = \sqrt{Q}/(y_1 - 1)$ the inversion transformation may be written:

$$y_1 \to 2 - Q - y_1$$

$$y_2 \to 1/y_2.$$

We can get another more structured transformation if we build a transfer matrix which is unchanged with horizontal and vertical (odd and even) couplings interchanged. Alternatively, note that the free energy, at least, would be unchanged with odd and even couplings interchanged on the assumption of 90° rotation invariance of the lattice. The inversion transformation could then be augmented to

$$y_1 \to 2 - Q - y_2$$
$$y_2 \to 1/y_1$$

(see section 3.3). This transformation has a rich Q dependent structure. For $Q = 2$ it has order 4, for $Q = 3$ it has order 6, and every even order is realised by some Q value.

The case of Y_N

The inversion relation is slightly complicated by the change of transfer matrix formulation in section 2.9.1. Let the matrix P be such as to take a configuration to the configuration with all spins permuted by the transformation

$$s_x \to p(s_x) = (s_x)_{mod\ 2} + 1$$

(c.f. $Q = 2$ in equation 2.1.1). In our representation this is

$$P = A_4 B(0,0) A_4.$$

Now consider a small section of the 4 parameter transfer matrix layer $X(y_1, y_2)\ W(y_1', y_2')$ of the form

$$t_i(x_1)\ t_{i+2}(x_2)\ U_{i+1}\ t_i(x_1')\ t_{i+2}(x_2').$$

Can we arrange it so that W is the inverse of X? With the benefit of our discussion in section 2.10 we try $y_1' = 1/y_1$ and $y_2' = -y_2$. From section 2.8 and section 2.9 this is then, supressing tensor products with unit matrices,

$$\begin{pmatrix} y_1 & 1 & 0 & 0 \\ 1 & y_1 & 0 & 0 \\ 0 & 0 & y_1 & 1 \\ 0 & 0 & 1 & y_1 \end{pmatrix} \begin{pmatrix} y_2 & 0 & 1 & 0 \\ 0 & y_2 & 0 & 1 \\ 1 & 0 & y_2 & 0 \\ 0 & 1 & 0 & y_2 \end{pmatrix} \begin{pmatrix} 1 & 0 & 0 & 0 \\ 0 & 0 & 0 & 0 \\ 0 & 0 & 0 & 0 \\ 0 & 0 & 0 & 1 \end{pmatrix}$$

$$\cdot \begin{pmatrix} 1/y_1 & 1 & 0 & 0 \\ 1 & 1/y_1 & 0 & 0 \\ 0 & 0 & 1/y_1 & 1 \\ 0 & 0 & 1 & 1/y_1 \end{pmatrix} \begin{pmatrix} -y_2 & 0 & 1 & 0 \\ 0 & -y_2 & 0 & 1 \\ 1 & 0 & -y_2 & 0 \\ 0 & 1 & 0 & -y_2 \end{pmatrix}$$

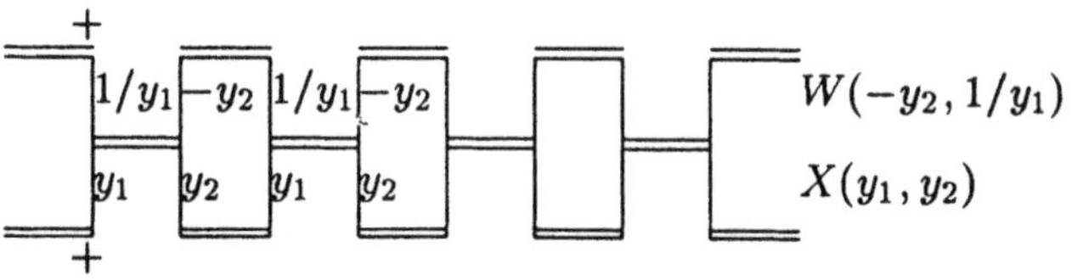

Figure 2.7: Checking an identity from the text. Double lines represent frozen interactions, so the spins at each end may be identified. Absent lines correspond to decoupled spins. The left and right endpoints are identified in each layer.

$$= \begin{pmatrix} 1-y_2^2 & \cdot & 0 & y_2(y_1-1/y_1) \\ \cdot & 1-y_2^2 & y_2(y_1-1/y_1) & 0 \\ 0 & y_2(y_1-1/y_1) & 1-y_2^2 & \cdot \\ y_2(y_1-1/y_1) & 0 & & 1-y_2^2 \end{pmatrix}$$

where we have only written the important (as it will turn out) matrix elements out explicitly. We see from the position of the vanishing elements that if the two spins marked + in figure 2.7 both take the value +1 (or both −1) then the next two spins on their immediate right must take the same value as each other, or else the associated matrix element of XW above is zero. Iterating, it then follows immediately from the periodicity of the lattice that either no 2 spins as marked are the same anywhere along the layer

$$X(y_1,1)W(-1,1/y_1) = (y_1 - 1/y_1)^{N/2} P$$

or all such pairs of spins are either $(+,+)$ or $(-,-)$

$$X(1,y_2)\, W(-y_2,1) = y_2^{N/2}(1/y_2 - y_2)^{N/2}\, A_4$$

and altogether

$$X(y_1,y_2)\, W(-y_2,1/y_1) = y_2^{N/2} X(y_1,1)\, W(-1,1/y_1) + X(1,y_2)\, W(-y_2,1).$$

Using the translation matrix G we can write

$$W(y_1,y_2) = X(y_1,y_2)\, G$$

and hence rewrite the modified inversion identity as

$$y_2^{N/2}\, X(y_1, y_2)\, X(-y_2, 1/y_1)\, G = (y_1 - 1/y_1)^{N/2}\, P + (1/y_2 - y_2)^{N/2}\, A_4. \tag{2.32}$$

In forthcoming chapters we will apply this technology to the problem of computing the transfer matrix spectrum.

Special points of the inversion transformation

In the Temperley-Lieb and Hecke algebra cases the inversion transformation itself has a non-trivial fixed point, $x = -2/\sqrt{Q}$, i.e.

$$t_i^2(-2/\sqrt{Q}) = 1.$$

Since the eigenvalues are permuted this does not imply that the transfer matrix is the identity matrix at this point, but rather that the spectrum of the fixed point matrix contains eigenvalues in reciprocal pairs.

The Potts model representation of the Temperley-Lieb algebra is unitary (in fact every U_i is real and symmetric). This means that the hermitian conjugate of a local transfer matrix is the same matrix with the coupling x complex conjugated. Thus

$$T'(x)^\dagger = M_{(a)} T'(x^*) M_{(a)}^{-1}$$

and the solutions of

$$x'(x) = x^*$$

i.e. $|x|^2 = -2Re(x)/\sqrt{Q}$, obey

$$T'(x)T'(x)^\dagger = 1$$

i.e. each eigenvalue has unit magnitude. This holds for any unitarisable representation (see chapter 6). Note that the solutions for x here include, when $|q| = 1$, the 'braid point' $x = -q$.

All these properties hold separately in each appropriate irreducible representation of the transfer matrix algebra.

Chapter 3

On commuting transfer matrices

We saw in chapter 2 that many problems in statistical mechanics can be related to the problem of computing the spectrum of the transfer matrix T, or part thereof. Fortunately for those of us seeking a challenge, this spectrum has not been calculated in general!

One property which turns out to be closely linked to the computability of this spectrum (the 'solvability' of the model) is commutativity, up to boundary terms, of all the different layer transfer matrices obtained by moving within an extended submanifold of coupling parameter space. This is essentially because commuting matrices have the same set of eigenvectors. That is, the eigenvectors must be constant under the variation of coupling parameters within the submanifold.

In this chapter we will look at conditions for commutativity for two dimensional models. We will then look at other properties of the transfer matrix which, if present, may be taken in conjunction with commutativity and used to solve a model. For example, double periodicity of eigenvalues with respect to some complex intra-submanifold coupling parameter.

Methods based more or less on the idea of commuting transfer matrices have been used in the solution of many two dimensional statistical mechanical models (several, unfortunately, of questionable physical significance). As for the Potts models, however, only the $Q = 2$ case has been solved away from criticality. For this reason, and given our ambit in this book, when illustrating the method we will be forced to lean heavily on the Ising model example. This case has also been solved by many other means, some of the most illuminating of which we discuss in chapter 4. We will persevere with commuting transfer matrices in spite of this apparent redundancy, however. Because of its implications for algebra, and the wider field of statistical mechanics, the idea of the commuting transfer matrix is worth studying in its own right (see also Baxter 1982,1985).

3.1 Yang-Baxter equations

A sufficient condition for the commutativity of distinct layer transfer matrices of $T'_{(N')}$ type is the existence of a corresponding crypto-commutativity property of *local* transfer matrices $t_I(x)$ at different values of x. In general x is to be understood here as a generic label representing values for the possibly manifold coupling parameters of the model. However, we will mainly consider examples in which it is a single parameter, as in equation 2.28. Specifically then, the commutativity condition is that there exist matrices $R_I(x)$ for which the following equations hold:

With N the number of sites in the transfer matrix layer, and for $i = 1, 2, .., N-1$, then in Potts model site labelling notation

$$t_{i.}(x)\ t_{ii\pm1}(x')\ R_{i.}(x''(x,x')) = R_{ii\pm1}(x''(x,x'))\ t_{i.}(x')\ t_{ii\pm1}(x) \quad (3.1)$$

$$t_{ii\pm1}(x)\ t_{i.}(x')\ R_{ii\pm1}(x''(x,x')) = R_{i.}(x''(x,x'))\ t_{ii\pm1}(x')\ t_{i.}(x) \quad (3.2)$$

These are called the Yang-Baxter (YB) equations. If such a set of matrices $R_I(x)$ exists, then they are said to give a solution to the YB equations for the local transfer matrix $t_I(x)$.

Suppose there exist matrices $R_I(x)$ for which the YB equations 3.1 and 3.2 hold. Consider repeated applications of these equations to the product

$$\begin{aligned} p(x,x',x'') &= t_{1.}(x)\, t_{12}(x) t_{2.}(x) ... t_{N-1N}(x)\, t_{N.}(x) \\ &\qquad .t_{1.}(x')\, t_{12}(x')\, t_{2.}(x') \ ...\ t_{N-1N}(x')\, t_{N.}(x') \\ &\qquad .(t_{N.}(x'))^{-1}\, R_{N.}(x'')\, t_{N.}(x) \\ &= t_{1.}(x)\, t_{12}(x)\, t_{2.}(x) \ ...\ t_{N-1N}(x) \\ &\qquad .t_{1.}(x')\, t_{12}(x')\, t_{2.}(x') \ ...\ t_{N-1.}(x') \\ &\qquad .\underbrace{[t_{N.}(x)\, t_{N-1N}(x')\, R_{N.}(x'')]}_{cf.\ LHS\ of\ equation\ 3.1}\, t_{N.}(x) \end{aligned} \quad (3.3)$$

where we also repeatedly use the commuting relation equation 2.12, so as to move some functions of x to the right and functions of x' to the left. These manipulations eventually yield

$$p(x,x',x'') =$$
$$t_{1.}(x)\, R_{1.}(x'')\; t_{12}(x')\, t_{2.}(x') \dots t_{N.}(x')\; t_{1.}(x) t_{12}(x) t_{2.}(x) \dots t_{N.}(x).$$

This process can be seen diagrammatically in figure 3.1.

3.1.1 Commuting layer transfer matrices

If we restrict attention to the one parameter submanifold of general coupling parameter space defined, in the Potts model case, by the anisotropic critical surface

$$x = 1/x_1 = x_2,$$

i.e. $f(x) = 1/x$ in equation 2.29, then $p(x,x',x'')$ from equation 3.3 may be written, up to boundary factors, as the product of two layer transfer matrices

$$p(x,x',x'') = T'_{(N')}(x)\, T'_{(N')}(x')\quad (t_{N.}(x'))^{-1}\, R_{N.}(x'')\, t_{N.}(x).$$

The last three factors all pertain to the last interaction at the end of the layer, which is why we refer to them as boundary factors.

Using the YB equations we have thus shown that

$$T'_{(N')}(x)\, T'_{(N')}(x')\quad (t_{N.}(x'))^{-1}\, R_{N.}(x'')\, t_{N.}(x)$$
$$= t_{1.}(x)\, R_{1.}(x'')\, (t_{1.}(x'))^{-1}\quad T'_{(N')}(x')\, T'_{(N')}(x).$$

This is the commutativity condition for layer transfer matrices analogous to the YB equation for local transfer matrices (cf. equation 3.1).

We call $R_I(x)$ the R matrix for local transfer matrix $t_I(x)$. A corollary of the YB equations, arising when three different transfer matrices are commuted through each other in different orders, is that the $R_I(x)$ matrices obey:

$$R_{i.}(x)\, R_{ii\pm1}(x')\, R_{i.}(x''(x,x')) = C\, R_{ii\pm1}(x''(x,x'))\, R_{i.}(x')\, R_{ii\pm1}(x).$$

where C is a central factor in the transfer matrix algebra. This means that if an $R_I(x)$ matrix exists for some set $\{t_I(x)\}$, and itself constitutes a plausible local transfer matrix, then it automatically satisfies the YB equations with itself as R matrix. This is called an adjoint solution to the YB equations, by analogy with the representation theory of Lie groups (in which the structure constants may be arranged to form the adjoint representation).

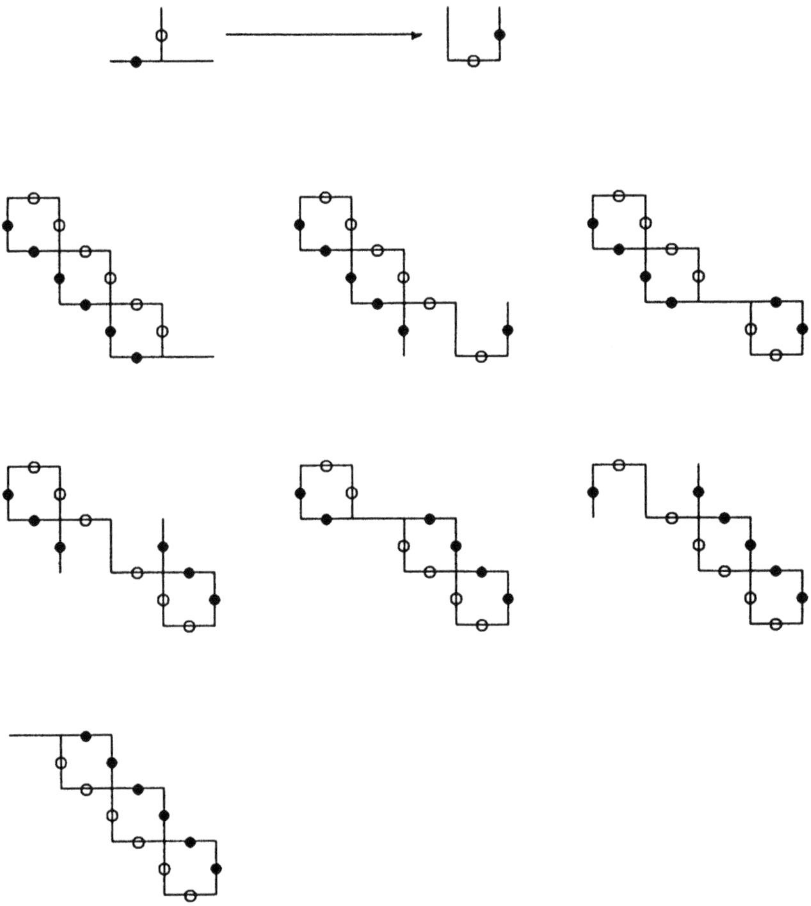

Figure 3.1: The Yang Baxter equation and its application to commuting transfer matrices in the Potts formulation. Bonds marked with solid circles are $t_i(x)$; empty circles are $t_i(x')$, and no circles $R_i(x'')$.

Star-triangle relations

Conversely, direct computation will confirm that a quadratic local relation (equation 2.8), together with the braid and commuting relations (equations 2.14 and 2.12), are sufficient to ensure that a local transfer matrix itself gives a solution to the YB equations via $R_I(x) = t_I(x)$. In general if $R_I(x) = t_I(x)$ the YB equations will be called *star-triangle* relations. This name is derived from a particular graphic representation, see Baxter 1982, and is sometimes reserved for a more powerful set of relations, see section 3.4.

Assuming that interactions can be labelled along the layer, so that interaction $i+1$ has a variable in common with interaction i, we have simply:

$$t_i(x)\ t_{i+1}(x')\ t_i(x'') \quad = \quad t_{i+1}(x'')\ t_i(x')\ t_{i+1}(x).$$

In fact we will only give examples of solutions to the Yang-Baxter equation which are also star-triangle relations. It is as well to realise, however, that it ain't necessarily so!

Since their transfer matrix algebra satisfies equation 2.8, equation 2.14 and equation 2.12, the 2 dimensional Potts models are (part of a large class of) 2 dimensional models which satisfy the star-triangle relations in *some* one dimensional sub-manifold of coupling parameter space.

This then implies that layer transfer matrices differing only in having different values of a coupling parameter which moves *within* this sub-manifold commute with each other (up to boundary factors).

Let us assume for the moment that the boundary factors can be eliminated by an appropriate choice of boundary conditions, so the transfer matrices strictly commute. They all thus have the same eigenvectors. This means that the eigenvectors for the model considered over the subset of general parameter space covered by all such manifolds can only depend on some complementary inter-manifold parameters. If this subset was a significant proportion of general parameter space then re-expressing the model in terms of the intra and inter commuting submanifold parameters would clearly be likely to produce a simplification in the eigenvalue problem. Unfortunately the *only* commuting one parameter submanifold for the $Q > 2$ state Potts model is the one we have already exhibitted, the manifold containing all horizontal to vertical bond strength anisotropies at criticality. As a consequence the models cannot be simplified and solved in general by this method. A notable exception is $Q = 2$, which we will return to shortly.

On the other hand, the critical region is of great interest, so the star-triangle relations are of great importance despite their limitations. Applying the Temperley-Lieb relations to the Potts local transfer matrix on the critical line, that is (supressing only some overall factors) to

$$t_i(x) = 1 + xU_i,$$

we find that the YB equations as written above are satisfied when

$$x''(x, x') = (x' - x)/(1 + x\sqrt{Q} + xx')$$

or, with ϵ, δ non-vanishing complex numbers, and q defined as in equation 2.9,

$$x''(\underbrace{-q - \epsilon}_{x}, -q - \delta) = \frac{(\epsilon - \delta)}{(-q\epsilon + q\delta)}. \tag{3.4}$$

The *critical* Potts models have indeed been solved in some sense using this result (see chapter 12).

3.2 Algebraic consequences of the relations

Note the 'trivial' case of the star triangle relations in the Temperley-Lieb case, from 3.4,

$$x = x' \neq -q^{\pm 1}$$

implies $x'' = 0$. That is, at the layer transfer matrix level, identical matrices commute trivially. Analogous trivial limits occur quite generally.

3.2.1 The braid point

Note also the special (fixed point) case

$$x = x' = x'' = -q.$$

After a little algebra we can confirm that the matrices $t_i(-q)$ satisfy the relations for generators of the $2N$ string braid group B_{2N} (Birman 1974, see also chapter 13). That is with

$$g_i^{\pm 1} = t_i(-q^{\pm 1}) = (1 - q^{\pm 1}U_i) \tag{3.5}$$

the star-triangle relation at the fixed point becomes the braid relation

$$g_i g_{i+1} g_i = g_{i+1} g_i g_{i+1}.$$

At this special (and unphysical, i.e. complex β) value of the coupling parameter, then, each interaction matrix may be thought of as representing a braid crossing. This suggests quite a helpful graphical way of thinking of

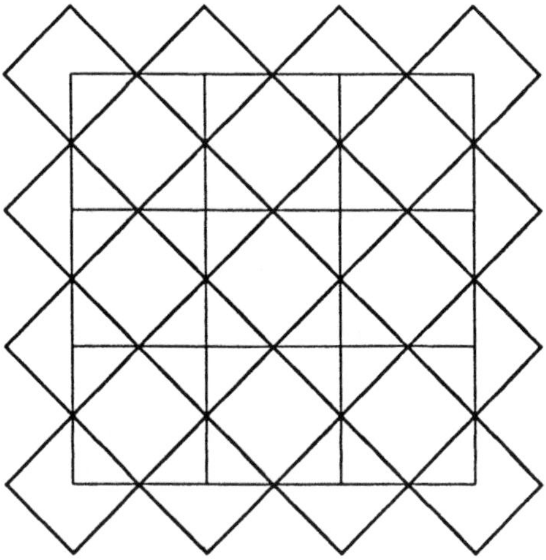

Figure 3.2: The construction of the medial lattice (thick lines) associated to the square lattice (thin lines).

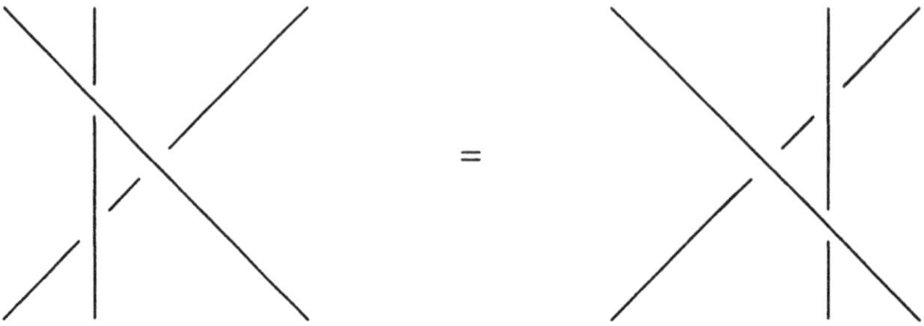

Figure 3.3: Diagrammatic representation of the braid relation.

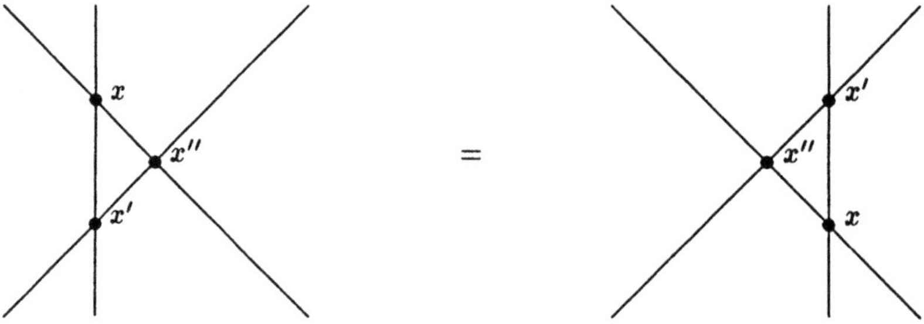

Figure 3.4: Generalised braid representation of the Star-Triangle relation.

the star-triangle relations in general. Suppose we recall the medial lattice for the square lattice as defined in chapter 1. Each crossing of the medial lattice corresponds to an interaction (figure 3.2). We now think of each interaction as a kind of generalised braid crossing, so the edges of the medial lattice become arcs of braid strands. The braid relation (figure 3.3) is deformed into the full star-triangle relation by replacing each crossing with a 'blob' carrying the appropriate interaction parameter (figure 3.4). Note, for example, that drawing the medial lattice for the YB equation in figure 3.2 we recover this picture up to irrelevant deformations of the braid strands.

Recall that the quadratic local relation

$$g_i^2 = (1 - q^2)g_i + q^2$$

together with the braid and commuting relations are a weaker set of relations than the Temperley-Lieb relations. They are the defining relations for an A_n-type Hecke algebra (see chapter 9).

To recapitulate: the Temperley Lieb or Hecke relations imply solutions of the YB equations, and the YB equations have fixed points which imply realisations of the braid relations. However realisations of the braid relations do not necessarily imply realisations of the Hecke relations, since the quadratic relation need not be satisfied (see later).

Up to degeneracy the algebra then determines the spectrum, and hence the correlation functions, of the model in question.

3.2.2 Translations revisited

For *any* model providing solutions to the star triangle relations with a braid limit $t_i \to g_i$, a viable alternative to the translation matrix $G^2_{(a)}$ in the puncture repair kit of section 2.1.1 is the square of

$$G_{(a)} = \prod_{m=1}^{2a-1} g_m$$

or alternatively of

$$\tilde{G}_{(a)} = \prod_{m=1}^{2a-1} g_m^{-1}.$$

This works because

$$G^{\pm 1}_{(a)}\ g_i\ (G_{(a)})^{\mp 1} = g_{i\pm 1},$$

by repeated application of the braid relations, and similarly for $\tilde{G}$. In general there is a distinction between the two periodic completions

$$G_{(a)}\, g_{2a-1}\, (G_{(a)})^{-1} = G^{-1}_{(a)}\, g_1\, G_{(a)} = g_0$$

and

$$\tilde{G}_{(a)}\, g_{2a-1}\, (\tilde{G}_{(a)})^{-1} = \tilde{G}^{-1}_{(a)}\, g_1\, \tilde{G}_{(a)} = \tilde{g}_0,$$

but both g_0 and $\tilde{g}_0$ obey the braid relations with both g_1 and g_{2a-1}. We call the local transfer matrices obtained from these derived braid crossings the *natural* completions of the open bounded transfer matrix to periodic boundaries.

Also,

$$M'_{(a)} = \prod_{l=1}^{2a-1} \left(\prod_{m=1}^{2a-l} g_m \right)$$

gives the reflection

$$M'_{(a)}\, g_i\, (M'_{(a)})^{-1} = g_{2a-i}.$$

Finally,

$$M'_{(a)} g_0 (M'_{(a)})^{-1} = \tilde{g}_0.$$

We give a diagrammatical derivation of these results in chapter 13.

Note that

$$M'^{\,2}_{(a)} = (G_{(a)})^{2a}$$

is only necessarily central here, while the corresponding matrix is equal to unity in the definitions of $G^2_{(a)}$ and $M_{(a)}$ (section 2.1.1). Explicit calculations reveal that this central element is indeed *not* unity in this construction, for some model based representations. The new translator $(G_{(a)})^2$ is itself quite a different looking matrix from the original $G^2_{(a)}$, even in the Potts case.

Discrete translation invariance is thus implicitly realised in the operator formalism associated with the YB equations, as the automorphism realised by conjugation by $(G_{(a)})^2$. The duality transformation (which is like a 90° rotation for the Potts model at criticality, in the sense that horizontal and vertical couplings are interchanged) is realised as conjugation by $G_{(a)}$.

Note that $G_{(a)}$ is itself a critical Potts transfer matrix, although the coupling parameter is required to take a non-physical value, $x = -q$. This means that under the special unitarity conditions of section 2.10 each eigenvalue of $G_{(a)}$ has unit magnitude.

3.3 Alternative layering directions

Recall that the star-triangle relation implies commutativity of $T'_{(N')}$, i.e. open stairway, type (but not $T_{(N')}$, i.e. open stepladder, type) transfer matrices, up to a boundary effect. In order to illustrate the solution of a model it will be helpful to eliminate this boundary effect. At the same time we will be able to invoke the augmented inversion transformation discussed in section 2.10.

It is possible to eliminate the boundary effect, and thus realise strict commutativity, by using periodic boundary conditions. However, the periodic matrix required (denoted $T'_{(N)}$) *cannot* be written as a product of local transfer matrices in the usual way. In Potts language this is because the local transfer matrix $t_i(x)$ transfers attention from one spin i to a spin i' at a later 'time' (in the Euclidean field theory sense). This means that we *discard* the information about the earlier spin (internalise it). However, a spin on the periodic seam should not be internalised until the layer is completed, as its state is required to determine the final interaction energy (for the bond which closes the layer).

We can compute a periodic lattice partition function using $T'_{(N)}$, but the staggered periodic identification of boundary sites obscures the usefulness of the star-triangle relation. Because now every factor t_i meets 2 others in the layer with which it does not commute, a pair of layers with different interactions cannot be rearranged so that the relation can be used. For example, adopting the shorthand i' for $t_i(x')$ (and so on), in

$$0'0''0^{-1} \;\; 0123 \;\; 0'1'2'3'$$

we cannot rearrange to obtain the factor $0''10'$ (that is, as in the star-triangle relation) because $0'$ does not commute with 3.

One resolution of this problem is to use the 45° transfer matrix layering direction discussed in section 2.9.1. Recall that the layer transfer matrix here, Y_N, consists of a product of two sublayer transfer matrices X and W, as shown in figure 2.4. The two sublayer transfer matrices can be made to commute on certain one parameter submanifolds of coupling parameter space. For the Ising model these manifolds include off-critical couplings.

3.4 The two dimensional Ising model

Let us restrict attention to the Ising model. In this case we now have enough technical hardware to illustrate the usefulness of the YB equations in determining transfer matrix eigenvalues. In this section we will be deviating only in perspective from the presentation in Baxter (1982).

With $Q = 2$, the star triangle relation implies that

$$X(y_1, y_2)W(y_1', y_2') = X(y_1', y_2')W(y_1, y_2) \tag{3.6}$$

provided that

$$(y_1 - 1/y_1)(y_2 - 1/y_2) = (y_1' - 1/y_1')(y_2' - 1/y_2') = k^{-1}. \tag{3.7}$$

The one parameter submanifolds of commuting transfer matrices here are thus characterised by the value of the alternative parameter k. The critical case is $k = 1$.

To see this result first note that the star-triangle relation may be deformed into the relation

$$x''x'U_i\,[t_{i-1}(1/x'')\,t_{i+1}(x)\,U_{i+2}\,t_{i+1}(1/x')\,]\,U_i$$
$$= U_i\,[U_{i-1}\,t_{i+1}(x')\,t_{i+2}(x'')\,t_{i+1}(x)]\,U_i.$$

This relation is readily confirmed by using the Temperley-Lieb relations. The sense in which it is a deformation of the star-triangle relation may be seen from figure 3.5. The justification for the diagrammatic representation of U_i used in this picture is given in chapter 6.

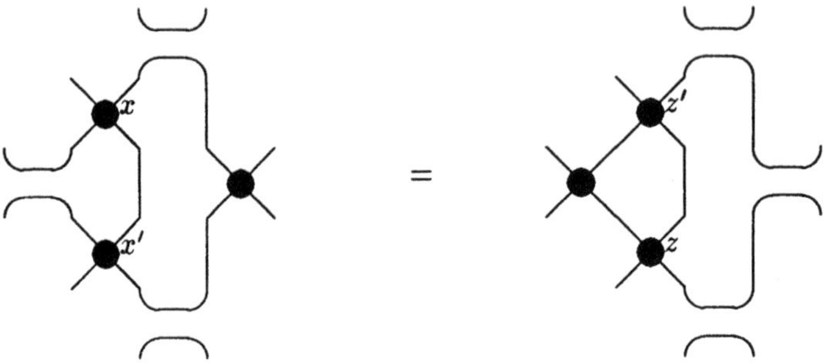

Figure 3.5: The deformed star-triangle relation. In general $z = 1/x$ and $z' = 1/x'$, but for the Q = 2-state Potts model a more flexible parameterisation arises.

When $Q = 2$ the Potts representation of the TL algebra obeys the additional (unitarity) condition

$$1 - \sqrt{Q}(U_i + U_{i+1}) + U_iU_{i+1} + U_{i+1}U_i = 0.$$

This reduces the number of independent terms in the deformed star-triangle relation above to the point where a more flexible parameterisation arises:

$$U_i\ [t_{i-1}(x'')\ t_{i+1}(x)\ U_{i+2}\ t_{i+1}(x')\ ]\ U_i$$
$$= K\ U_i\ [U_{i-1}\ t_{i+1}(z')\ t_{i+2}(z'')\ t_{i+1}(z)]\ U_i. \qquad (3.8)$$

where K is a scalar parameter, and four relations relate the 7 parameters. Eliminating K, x'' and z'', then one remaining relation relates 4 parameters. This is equation 3.7.

Repeated application of this improved relation to

$$X(y_1, y_2)\ Y(y_1', y_2') = A_4 \prod_{i=1}^{N/2} \left(t_{4i-3}t_{4i-1}U_{4i-2}t'_{4i-1}t'_{4i-3}\right)\ A_4$$

$$= A_4\left((1 + xU_1)\ t_3U_2t_3't_1'\right) \prod_{i=2}^{N/2} \left(t_{4i-3}t_{4i-1}U_{4i-2}t'_{4i-1}t'_{4i-3}\right) A_4$$

yields equation 3.6. The process is illustated by applying figure 3.5 repeatedly to figure 3.6.

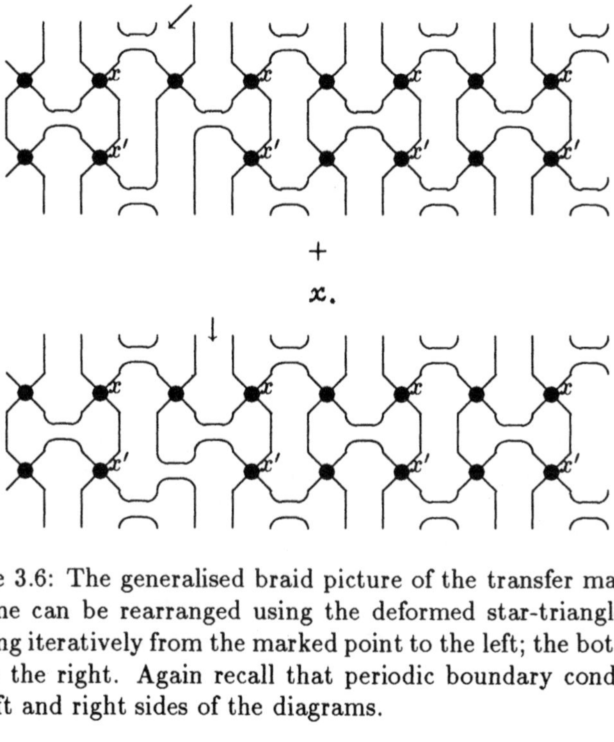

Figure 3.6: The generalised braid picture of the transfer matrix XY. The top line can be rearranged using the deformed star-triangle relation and working iteratively from the marked point to the left; the bottom line working to the right. Again recall that periodic boundary conditions connect the left and right sides of the diagrams.

Note that the new parameter k, from equation 3.7, is invariant under (amongst other things) the following transformations of the old parameters:

$$(y_1, y_2) \to (y_2, y_1)$$

$$(y_1, y_2) \to (-y_1, -y_2)$$

$$(y_1, y_2) \to (1/y_1, -y_2).$$

The translation matrix G of section 2.9.1 has undefined k ($= 0.\infty$) and in fact commutes with all X and Y.

3.4.1 Useful transfer matrix identities for $Q = 2$

Note that both $X(y_1, y_2)$ and $W(y_1, y_2)$ have entries of the form $y_1^n \; y_2^m$, where $n + m$ is even, so

$$X(y_1, y_2) = X(-y_1, -y_2), \tag{3.9}$$

$$W(y_1, y_2) = W(-y_1, -y_2). \tag{3.10}$$

Also, for example from equation 2.30,

$$(y_1 y_2)^{N/2} \; X(1/y_1, 1/y_2) = X(y_1, y_2) \; P \tag{3.11}$$

and

$$(y_1 y_2)^{N/2} \; W(1/y_1, 1/y_2) = W(y_1, y_2) \; P. \tag{3.12}$$

The transformations involved here all preserve k. They can be regarded simply as transformations of a complementary variable, u say. Let us suppose that we can define u so that the transformation of the arguments of X in equation 3.9 is always given by

$$u \to u + 2iK$$

where K is a positive constant. Then regarded as a function of u every eigenvalue of X and hence Y will be periodic of period $2iK$. The transformation in equation 3.11 has the same effect as $2iK$ in only 2 places ($y = \pm i$). It is thus possible to suppose that we can arrange u so that this transformation has the effect

$$u \to u - 2iK + 2K'$$

where K' is another positive constant. The common points of the transformations would be at $u = \pm\infty$. We will be even more presumptuous, and assume that the transformation

$$(y_1, y_2) \to (-y_2, 1/y_1)$$

which also preserves k, can be written

$$u \to u + K'.$$

Note that there is no symmetry of the transfer matrix associated with this transformation, but that it is consistent with the previous ones!

In any case we see that the eigenvalues of $(y_1 y_2)^{-N/2} Y$ will be doubly periodic functions of u at each fixed value of k. This has a profound baring on their form, as we will see shortly.

Following Hancock (1958) (and hence Hermite) let us introduce

$$q = \exp(-\pi K'/K)$$

i.e. $0 < q < 1$ (the use of q here is a universal standard elliptic function notation - it should not be confused with q in, say, equation 3.5, which is unfortunately also used as standard in this entirely different context). Further let

$$\phi(u) = 1 + \exp(\pi i u/K)$$

(period $2K$, and $\phi(K) = 0$). Then for

$$\Phi(u) = \prod_{n=1}^{\infty} \phi(u + (2n-1)iK')\, \phi(-u + (2n-1)iK'),$$

that is

$$\Phi(u) = \prod_{n=1}^{\infty} (1 + 2q^{2n-1}\cos(\pi u/K) + q^{4n-2})$$

(convergent for $|q| < 1$), we have

$$\Phi(u + 2iK') = \Phi(u)\frac{\phi(-u - iK')}{\phi(u + iK')} = \exp(-\pi i(u + iK')/K)\, \Phi(u).$$

We say that $\Phi(u)$ is *quasi-periodic* with period $2iK'$. From the definition it has exactly one zero per period rectangle. Introducing the constant

$$A = (1 - q^2)(1 - q^4)(1 - q^6)...$$

we define

$$H(u) = A\Phi(u - u')$$

where $u' = K + iK'$ is the zero of Φ.

Now consider a doubly periodic function $f(u)$, such as one of our eigenvalues. If we can arrange u so that $f(u)$ is analytic up to n simple poles per period rectangle, at $u_1, u_2, .., u_n$ say, then

$$f(u) = Ce^{i\lambda u} \prod_{j=1}^{n} \frac{H(u - v_j)}{H(u - u_j)}$$

where C, λ and v_j are constants obeying a couple of constraints (but not enough to determine them).

To see this note that the integral of $f'(u)/f(u)$ round a period rectangle is zero because of the double periodicity. From Cauchy's theorem then, the number of poles equals the number of zeros. Suppose we could locate all the zeros and form the trial function

$$g(u) = C' e^{i\lambda' u} \prod_{j=1}^{n} \frac{H(u - v_j)}{H(u - u_j)}$$

then $f'/f - g'/g$ is doubly periodic and strictly analytic. By Liouville's theorem it is thus a constant, c say. Integrating it we then have

$$\ln(f/g) = cu + d$$

where d is another constant. QED.

In the next section we will show how to obtain enough constraints to locate the zeros of the eigenvalues of Y by using the inversion relation.

3.4.2 Applying the inversion relation

Remarkably, the augmented inversion transformation of section 2.10 also leaves k unchanged, so it transforms between commuting transfer matrices. Given their periodic properties, this will in fact lead to a solvable equation for the eigenvalues!

Recall the inversion relation 2.32. Since all the matrices involved commute *at a given value of k*, they have a common set of eigenvectors. Regarding the equations 3.9, 3.11, 2.32 as acting on such an eigenvector $v(k)$, with

$$A_4\, v(k) = v(k)$$

$$P\, v(k) = +v(k)$$

(i.e., since A_4 is a projection, restricting consideration to half of the complete set of eigenvectors), and

$$G\, v(k) = v(k)$$

(i.e. restricting to translation invariant states); then defining the scalar $E(x,t)$ by

$$(xt)^{-M}\, X(x,t)\;\; v(k) = E(x,t)\, v(k),$$

(where the arguments are now understood to be related in such a way as to maintain a fixed value of k) we obtain, for example, the scalar equation

$$E(x,t)E(-t,1/x) = (t - 1/t)^{N/2} + (1/x - x)^{N/2}.$$

Strictly speaking, we do not assume that $E(-t, 1/x)$ is the continuation of the eigenvalue $E(x,t)$. We mean here that it is the eigenvalue of the transformed Y with with the same eigenvector.

We will now use the periodicity properties of E to locate its zeros. First we must find a variable u with the required properties.

The point $u = \infty$ corresponds to the braid point of the local transfer matrix (remember $y = i$ here, so $x = (i-1)/\sqrt{2} = -q$ when $Q = 2$) .

3.4.3 Reparameterised YB equation

The Temperley-Lieb solutions to the YB equation coming from the critical (i.e. $f(x) = 1/x$) $Q = 2$ state Potts model, for example, may be reparameterised to get the YB equation in the form

$$t_i(x(u))\, t_{i+1}(x(u'))\, t_i(x(u'')) \;=\; t_{i+1}(x(u''))\, t_i(x(u'))\, t_{i+1}(x(u))$$

where $u' = u + u''$, as follows. Put $x(u) = x, x(u') = x', x(u'') = x''$ and

$$\tan(u') = x' \frac{\sqrt{2} + x'}{1 + \sqrt{2}x'}$$

then with

$$x' = \frac{x + x'' + \sqrt{2}xx''}{1 - xx''}$$

we have

$$x(u) = \frac{1 + sin(u) - cos(u)}{\sqrt{2}cos(u)} \tag{3.13}$$

and similarly for x', x'', with $u' = u + u''$.

We introduce an additive parameterisation for other fixed values of k by generalising the critical additive parameterisation . First we need some notation.

On elliptic functions

For each constant $0 < k < 1$ it is possible to choose a constant $0 < q < 1$ such that

$$k = 4q^{1/2} \prod_{n=1}^{\infty} \left(\frac{(1 + q^{2n})}{(1 + q^{2n-1})} \right)^4 .$$

Then define

$$k' = \prod_{n=1}^{\infty} \left(\frac{(1 - q^{2n-1})}{(1 + q^{2n-1})} \right)^4 = +\sqrt{1 - k^2}$$

and

$$K = \int_0^{\pi/2} \frac{da}{\sqrt{1 - k^2 \sin^2(a)}}$$

or equivalently

$$K = \pi/2 \prod_{n=1}^{\infty} \left(\frac{(1 + q^{2n-1})(1 - q^{2n})}{(1 - q^{2n-1})(1 + q^{2n})} \right)^2$$

and $K' = K \ln(1/q)/\pi$.

For fixed k define entire (that is, everywhere analytic) functions of a variable u by

$$H(u) = 2q^{1/4} \sin(\frac{\pi u}{2K}) \prod_{n=1}^{\infty} \left(1 - 2q^{2n} \cos(\pi u/K) + q^{4n}\right) \left(1 - q^{2n}\right)$$

(note that this is identical to the previous definition) and

$$\Theta(u) = \prod_{n=1}^{\infty} \left(1 - 2q^{2n-1} \cos(\pi u/K) + q^{4n-2}\right) \left(1 - q^{2n}\right)$$

and $H_1(u) = H(u + K)$ and $\Theta_1(u) = \Theta(u + K)$.

The *Jacobian elliptic functions* are then the meromorphic functions of u defined by

$$sn(u) = \frac{H(u)}{\sqrt{k}\Theta(u)}$$

$$cn(u) = \frac{\sqrt{k'}H(u + K)}{\sqrt{k}\Theta(u)}$$

$$dn(u) = \frac{\sqrt{k'}\Theta(u + K)}{\Theta(u)}.$$

We will need the following identities:

$$F_1(u) = \frac{H(u)\Theta(u)H_1(u)\Theta_1(u) - H(v)\Theta(v)H_1(u)\Theta_1(u)}{H(u - v)\Theta(u + v)H_1(0)\Theta_1(0)} = 1 \qquad (3.14)$$

Proof:

Defining

$$\mu(u) = q^{-1}e^{-\pi i u/K}$$

then

$$H(u+2iK') = -\mu(u)H(u)$$
$$H_1(u+2iK') = \mu(u)H_1(u)$$
$$\Theta(u+2iK') = -\mu(u)\Theta(u)$$
$$\Theta_1(u+2iK') = \mu(u)\Theta_1(u)$$

so $F_1(u)$ is doubly periodic. Furthermore, the zeros of the denominator occur when $u = v$ and $u = -iK' - v$. These are also the points where the terms in the numerator cancel each other, so $F_1(u)$ is entire. QED.

Similarly

$$\Theta^2(u)\Theta^2(v) - H^2(u)H^2(v) = \Theta(u-v)\Theta(u+v)\Theta^2(0)$$

and hence

$$\frac{H(u-v)}{\Theta(u-v)} = \frac{H_1(0)\Theta_1(0)}{\Theta^2(0)} \frac{\frac{H_1(v)\Theta_1(v)}{\Theta(v)H(v)} - \frac{H_1(u)\Theta_1(u)}{\Theta(u)H(u)}}{\frac{\Theta(u)\Theta(v)}{H(u)H(v)} - \frac{H(u)H(v)}{\Theta(u)\Theta(v)}}. \tag{3.15}$$

Elliptic function parameterisation

Noting equation 3.7, we can generalise the critical additive parameterisation of section 3.4.3 to other fixed values of k in the range by introducing

$$y_1^{\pm 1} = cn(iu) \mp sn(iu)$$
$$y_2^{\pm 1} = \frac{i(dn(iu) \pm 1)}{k\ sn(iu)}.$$

This, then, is how we parameterise each fixed k manifold of commuting transfer matrices by the single variable u.

In the new parameterisation the coupling transformations in the modified inversion relation may be simplified using the following fixed k identities:

$$y_1 = cn(iu) - isn(iu) \stackrel{u \to u+K'}{\longrightarrow} -i\frac{dn(iu)}{k\ sn(iu)} - i\frac{1}{k\ sn(iu)} = -y_2$$

$$y_2 = \frac{i(dn(iu)+1)}{k\ sn(iu)} \stackrel{u \to u+K'}{\longrightarrow} cn(iu) + isn(iu) = 1/y_1.$$

Introducing the new parameterisation we can discard x and t in favour of u (and implicitly k). We have, from equations 3.9, 3.11 and 2.32

$$E(u) = E(u+2K')$$

$$E(u) = E(u - 2iK)$$

$$E(u)E(u + K') = \left(\frac{-2}{ksn(iu)}\right)^{N/2} + (-2sn(iu))^{N/2} \tag{3.16}$$

The first two of these, together with the property that every element in $(y_1 y_2)^{-M} X(u)$, and hence $E(u)$, is entire up to (at worst) an overall factor of $(H(iu)\Theta(iu))^{-M}$, imply by our general theorem that

$$E(u) = Ce^{\lambda u} \, (H(iu)\Theta(iu))^{-M} \prod_{j=1}^{2M} H(iu - iu_j)$$

where C and λ are constants . The zeros u_j can then be located using equation 3.16:

The zeros of the right hand side of 3.16 occur when

$$\frac{H(iu)}{\Theta(iu)} = k^{1/2} sn(iu) = e^{i(2j+1)\pi/N} \tag{3.17}$$

$(0 \leq j < N)$. Suppose $iu = iu_j$ is the solution to this equation for some j (altogether N possibilities, occuring in complex conjugate pairs). If iu_j is a solution then so is $iu_j + iK'$. There are thus too many solutions unless $(iu_j)^* = iu_j + iK'$, hence $iu_j = v_j \pm iK'/2$ where v_j is real (it will not be necessary to determine v_j). We then have various possibilities for $E(u)$ depending on how we distribute the elements of these pairs between $E(u)$ and $E(u + K')$. For definiteness let us consider the case in which all the zeros in $E(u)$ are of the form $v_j - iK'/2$. Then

$$E(u) = Ce^{\lambda u} \, (H(iu)\Theta(iu))^{-M} \prod_{j=1}^{N/2} H(iu - v_j + iK'/2).$$

The corresponding eigenvalue of the full transfer matrix Y is $E^2(u)$, which may now be rearranged as

$$E^2(u) = C^2 e^{2\lambda u}$$

$$\cdot \prod_{j=1}^{N/2} \frac{H(iu - v_j + iK'/2)\Theta(iu - v_j - iK'/2) \; iq^{1/4} \; e^{\pi(-u - iv_j + K'/2)/(2K)}}{(H(iu)\Theta(iu))}.$$

From the periodicity properties λ must be such that all the exponential arguments cancel, and we have

$$E^2(u) = C' \prod_{j=1}^{N/2} \left(\frac{H(iu - v_j + iK'/2)}{\Theta(iu - v_j + iK'/2)}\right.$$

$$\left. \frac{\Theta(iu - v_j - iK'/2)\,\Theta(iu - v_j + iK'/2)}{(H(iu)\Theta(iu))} \right) .$$

where C' is another constant. Using equation 3.15 and defining $V_j = v_j - iK'/2$ we can rewrite this as

$$E^2(u) = C' \prod_{j=1}^{N/2} \left(\frac{H_1(0)\Theta_1(0)}{\Theta^2(0)} \frac{\frac{H_1(V_j)\Theta_1(V_j)}{\Theta(V_j)H(V_j)} - \frac{H_1(iu)\Theta_1(iu)}{\Theta(iu)H(iu)}}{\frac{\Theta(iu)\Theta(V_j)}{H(iu)H(V_j)} - \frac{H(iu)H(V_j)}{\Theta(iu)\Theta(V_j)}} \right.$$

$$\left. \cdot \frac{\Theta(iu - v_j - iK'/2)\,\Theta(iu - v_j + iK'/2)}{(H(iu)\Theta(iu))} \right) .$$

The overall factor coming from taking the last factor out of the product is a doubly periodic function with the same set of zeros and poles as the right hand side of the reparameterised inversion identity, equation 3.16. Recalling equation 3.17 we see that taking the denominator of the first factor out of the product gives the inverse of the same object. Cancelling these, then, with the appearance of another overall constant C'', gives

$$E^2(u) = C'' \prod_{j=1}^{N/2} \left(\frac{H_1(0)\Theta_1(0)}{\Theta^2(0)} \left(\frac{H_1(V_j)\Theta_1(V_j)}{\Theta(V_j)H(V_j)} - \frac{H_1(iu)\Theta_1(iu)}{\Theta(iu)H(iu)} \right) \right) .$$

This is valid for all $0 < k < 1$, so we can eliminate u (and k) in favour of y_1 and y_2 again. Using

$$cn^2(u) + sn^2(u) = 1$$

and

$$dn^2(u) + k^2 sn^2(u) = 1$$

(proof similar to that of equation 3.15) we find that

$$k' \frac{H_1(V_j)\Theta_1(V_j)}{H(V_j)\Theta(V_j)} = k^{1/2} \left(\frac{1}{k\ sn^2(V_j)} - k - 1/k + k\ sn^2(V_j) \right)^{1/2} .$$

We can thus eliminate V_j using equation 3.17. Altogether we obtain

$$E^2(y_1, y_2) = C'' \prod_{j=1}^{N/2} \left[\frac{(y_1 + 1/y_1)(y_2 + 1/y_2)}{4} \right.$$

$$\left. + \left((1 + k^2)/k^2 - 2k^{-1} \cos(2(2j+1)\pi/N) \right)^{1/2} \right] .$$

This is the result we have been pursuing! Since the duality transformation maps $0 < k < 1$ into $\infty > k > 1$, and this result is invariant under the duality transformation, it holds everywhere except possibly at $k = 1$. For finite N it clearly holds everywhere on analytical grounds.

In fact this eigenvalue turns out to be the largest one in the physical region, the one responsible for the free energy. It may seem like an anti-climax, but it crucially 'reveals' the presence of a sensible thermodynamic limit with a second order phase transition at $k = 1$ in the $Q = 2$-state model! In chapters 4 and 11 we look at the *way* in which it reveals these things....

Chapter 4

On exactly solved cases

Among the thermodynamic limit Potts models in two or more dimensions only $Q = 2, d = 2$ has been solved for arbitrary couplings (c.f. chapter 11). Beside some results for two dimensional models at criticality discussed in chapter 12 some other non-trivial special cases, equivalent to colouring problems, have been solved (see, for example, Baxter 1987). Three dimensions remains uncharted territory (although plenty of exact finite lattice results are available)!

The eight vertex model, equivalent to a general 2-valued site variable model of interactions round a square face, and the hard hexagon model, have been solved by Baxter (see his book, 1982). Other solved models appear in Andrews, Baxter and Forrester 1984, Date *et al* 1987, Akutsu *et al* 1986, and several other of the references at the end of this book. Although these solutions do not pertain directly to the Potts models away from criticality, they are interesting in their own right.

The two dimensional Ising model has been solved in many different ways. A comparison between some of the more elegant methods of solution provides an exellent way of reviewing the mathematical tools potentially at our disposal for addressing other Potts models.

4.1 The two dimensional Ising model

The next few sections describe various very disitinctive methods of solution. We work with the Ising model in the $Q = 2$ state Potts model formulation.

The first method discussed below is essentially Kaufman's (1949) method. The second is that of Schultz, Mattis and Lieb (1964), which is related to Onsager's (1944) method, and the third is due to Kac and Ward (1952) and Vdovichenko (1965). After looking at those it is worth having another look at Baxter's commuting transfer matrix method, which was discussed in chapter 3.

4.1.1 Solution by rotations

The single bond transfer matrices for Potts models may be thought of as giving representatives of rotations in a plane, in the sense that the effect of applying such a matrix twice is to count the same bond interaction again, or equivalently to double the coupling constant on that bond. In other words the coupling constant corresponds to some multiple of the rotation angle.

In the case of the Ising model ($Q = 2$, so $r = 4$), for example, we note that the fourth power of the braid generator

$$g_i \mapsto t_i(-q) = 1 - qU_i$$

is unity, so we associate this matrix with the rotation through the real angle $\pi/2$.

This is not as facile an observation as it might seem. In 3 dimensions a rotation through $\pi/2$ about the x-axis followed by a rotation through $\pi/2$ about the y-axis followed by a rotation through $\pi/2$ about the x-axis is equivalent to the same operation with the x and y axes interchanged. In other words, extending to D dimensions, and (arbitrarily) assigning a total order to D orthogonal axes, we have a realisation of the braid group as follows: For a, b axes adjacent in the order let w_{ab} signify a rotation in the ab plane, then

$$w_{ab}w_{bc}w_{ab} = w_{bc}w_{ab}w_{bc}. \tag{4.1}$$

In fact we have much more than this, since *any* pair of planes give a braid relation, but we will not need this property (see the chapter on graph Temperley-Lieb algebras).

It is easy to see that equation 4.1 extends to a solution of the Yang-Baxter equations, although the relationships between the rotation angles and the spectral parameters u of chapter 3 are rather complicated (see later). This means that we can think of the strings of our braid as corresponding to orthogonal axes, and the braid generators as rotations in the plane of their associated pairs of axes.

The representation of the braid group here just comes from the transfer matrix in the usual way. Only the interpretation is new. However, associated with this interpretation comes another representation, in terms of the D dimensional orthogonal matrices. Instead of our bond transfer matrix $t_i(y)$ we can write the corresponding rotation matrix $w_{ab}(\theta) \in End(V_D)$, which acts trivially except in the 2 dimensional subspace associate with the axes a and b, where it acts as

$$\begin{pmatrix} \cos(\theta) & \sin(\theta) \\ -\sin(\theta) & \cos(\theta) \end{pmatrix} \tag{4.2}$$

The axes a and b are determined by the interaction i (i and $i+1$ have an axis in common, i and $i+2$ do not), and the relationship between $y = \frac{e^{\beta}-1}{\sqrt{Q}}$ and θ is determined by noting that $t_i^4(-q) = 1$ implies that

$$\beta = i\pi/2$$

corresponds to

$$\theta = \pi/2.$$

The easiest way to see that we have a solution of the Yang-Baxter equations at the level of rotations is to note that it is sufficient to check the relation

$$w_{ab}(\theta)w_{bc}(\psi)w_{ab}(\phi) = w_{bc}(\phi)w_{ab}(\psi)w_{bc}(\theta)$$

for all θ, ϕ (with ψ determined) for any triple of axes a, b, c. In this case we can use the quaternionic realisation of 3 dimensional rotations. Defining the quaternionic generators $1, i, j, k$ by the relations

$$ii = jj = kk = -1 \quad ij = -ji \quad ik = -ki \quad jk = -jk$$

we must check that

$$(\cos(\theta/2) + i\sin(\theta/2))(\cos(\psi/2) + j\sin(\psi/2))(\cos(\phi/2) + i\sin(\phi/2))$$

$$= (\cos(\phi/2) + j\sin(\phi/2))(\cos(\psi/2) + i\sin(\psi/2))(\cos(\theta/2) + j\sin(\theta/2))$$

where each factor corresponds to a rotation through twice the anglular argument shown about the x-axis (i) or the y-axis (j) (or the z-axis (k), which we didn't need).

After some algebra we obtain

$$\tan(\theta/2) = \frac{\sin(\frac{\psi-\phi}{2})}{\cos(\frac{\psi+\phi}{2})}.$$

Note that the braid limit is the angle $\pi/4$ (i.e. $\theta = \psi = \phi = \pi/2$).

It is remarkable that this relationship between the Temperley-Lieb algebra and rotations should give rise to a new algebra, again related to the braid group. This new algebra has a local relation which is higher order than quadratic, determined by the lowest order multiplicative relation at the transfer matrix level - in this case the quartic relation

$$t_i^4 = 1.$$

Multiplicative relations are necessarily carried across to the new solution, but additive ones are not.

We note that this coincidence is apparently uniquely associated with the Ising model, since the condition $w_{ab}^4(\pi/2) = 1$ is uniquely associated with this model.

We note further that there is a relationship between the eigenvalues of the Transfer Matrix and those of the corresponding rotation matrix. If we take a single rotation in the plane of one of our pairs of axes the eigenvalues of the transfer matrix are, up to multiplicity, $\exp(\pm i\theta/2)$, while those of the rotation matrix are 1 or $\exp(\pm i\theta)$. If we take several commuting plane rotations by θ_j ($j = 1, ..., n$ say, with $n \leq D/2$, or equivalently $n = D/2$ with some $\theta_j = 0$) then the eigenvalues are

$$\exp\left((i/2)\sum_j \pm\theta_j\right) \tag{4.3}$$

(for all 2^n possible choices of the signs) and

$$\exp(\pm i\theta_1),\ \exp(\pm i\theta_2), ...$$

($2n$ distinct possibilities) respectively.

In order to compute the spectrum of a full transfer matrix it is thus only necessary to compute the spectrum of the corresponding rotation matrix, provided that this may be put in the above form. Since the rotation matrix is exponentially smaller than the transfer matrix this is a potentially vast simplification of the problem.

Arranging the m-site layer transfer matrix in the symmetric form of equation 2.11, that is, in the form

$$T = V_1^{1/2} V_2 V_1^{1/2} \tag{4.4}$$

we can write the factor $V^{1/2}$ in the corresponding rotation matrix in the form $J(\theta) \otimes 1_m$ where, from equation 4.2 and below

$$J(\theta) = \begin{pmatrix} \cosh(\theta) & i\sinh(\theta) \\ -i\sinh(\theta) & \cosh(\theta) \end{pmatrix}. \tag{4.5}$$

Let us introduce a matrix S, which achieves a single cyclic shift of basis states, for example

$$S = \begin{pmatrix} 0 & 1 & 0 & 0 & .. & 0 \\ 0 & 0 & 1 & 0 & .. & 0 \\ 0 & 0 & 0 & 1 & .. & 0 \\ \vdots & & & & & \\ 0 & 0 & 0 & 0 & .. & 1 \\ 1 & 0 & 0 & 0 & .. & 0 \end{pmatrix}.$$

If required, S could also introduce some sign changes at the boundary, which we have assumed periodic. Provided we stay away from the boundary it will not be necessary to specify the precise boundary conditions in order to illustrate the method, as we will see. We can then write the second factor in equation 4.4, up to an overall scalar, as the statistical mechanical ***dual*** squared of equation 4.5 (with a different angle parameter), that is,

$$S(J(2\phi) \otimes 1_m)S^{-1}.$$

The angles θ and ϕ are determined by the usual horizontal and vertical lattice coupling parameters β_1 and β_2. We note, by comparing the definition of the $Q = 2$ state Potts transfer matrix from chapter 2 with the definition of $J(\theta)$ above, that

$$\phi = \beta_2/2 \tag{4.6}$$

and

$$\theta = \tanh^{-1}(e^{-\beta_1}).$$

The resultant matrix for T is clearly invariant under double cyclic shifts of basis states (conjugation by S^2), so its diagonalisation can be greatly facilitated by the (fourier) similarity transformation

$$F.[(J(\theta) \otimes I_m).\left(S(J(2\phi) \otimes I_m)S^{-1}\right).(J(\theta) \otimes I_m)].F^{-1}$$

where, with $z = e^{2i\pi/m}$,

$$F = \begin{pmatrix} 1 & 0 \\ 0 & 1 \end{pmatrix} \otimes \begin{pmatrix} 1 & 1 & 1 & \dots & 1 \\ 1 & z & z^2 & \dots & z^{n-1} \\ 1 & z^2 & z^4 & \dots & z^{2n-2} \\ \vdots & & & & \\ 1 & z^{n-1} & z^{2n-2} & \dots & z^{(n-1)^2} \end{pmatrix}.$$

After a modicum of algebra we find that the resultant 2×2 blocks are each of the form

$$\begin{pmatrix} \begin{array}{c} \cosh(2\phi)\cosh(2\theta)- \\ (z^k + z^{-k})[\sinh(2\phi)\sinh(2\theta)] \end{array} & \begin{array}{c} -i\cosh(2\phi)\sinh(2\theta) \\ +(z^k)[i\sinh(2\phi)\sinh^2(\theta)] \\ +(z^{-k})[i\sinh(2\phi)\cosh^2(\theta)] \end{array} \\ & \\ \begin{array}{c} i\cosh(2\phi)\sinh(2\theta) \\ -(z^{-k})[i\sinh(2\phi)\sinh^2(\theta)] \\ -(z^k)[i\sinh(2\phi)\cosh^2(\theta)] \end{array} & \begin{array}{c} \cosh(2\phi)\cosh(2\theta)- \\ (z^k + z^{-k})[\sinh(2\phi)\sinh(2\theta)] \end{array} \end{pmatrix} \tag{4.7}$$

where the complete set of m blocks is given by $k = 0, 1, 2, ..., m-1$.

Notice that the determinant of each block is 1, as required, so the spectrum *will* have the form

$$l_{\pm k} = e^{\pm x_k}$$

suitable for reading off the spectrum of the transfer matrix (cf. equation 4.3). In fact it must have this form, since an arbitrary rotation can be written as a product of commuting rotations. To see this note that a real matrix has eigenvalues which are either real or in complex conjugate pairs.

By taking the trace of each block we deduce that the exponents x_k are just the solutions of

$$\cosh(x_k) = \cosh(2\phi)\cosh(2\theta) - (z^k + z^{-k})[\sinh(2\phi)\sinh(2\theta)].$$

Each of these is positive in the physical region, so the largest eigenvalue of the *transfer matrix* from equation 4.3, i.e. the one appropriate for obtaining the free energy, will be

$$\lambda_0 = \exp\left((1/2)\sum_{k=1}^{m} x_k\right)$$

giving, in the thermodynamic limit,

$$f \sim \lim_{m\to\infty}(1/m)\ln(\lambda_0) = \lim_{m\to\infty}(1/2m)\left(\sum_{k=1}^{m} x_k\right)$$

The equivalence '$\sim$' here is to remind us that we have ignored a regular additive contribution. This comes from neglecting a scalar factor in the transfer matrix, which can be eliminated in any case by introducing a trivial additive term in the Hamiltonian.

Noting equation 4.6 we find that in the *isotropic* case ($\beta = \beta_1 = \beta_2$) we have

$$\cosh(x_k) = \cosh(\beta)\coth(\beta) - \cos(k\pi/m).$$

Using the integral representation

$$x_k = (1/\pi)\int_0^\pi dy \ln(2(\cosh(\beta)\coth(\beta) - \cos(k\pi/n)) - 2\cos(y))$$

we can thus write the limit free energy as

$$f \sim \frac{1}{2\pi^2}\int_0^\pi dy \int_0^\pi dy' \ln(2(\cosh(\beta)\coth(\beta) - \cos(y')) - 2\cos(y)) \quad (4.8)$$

or equivalently, after some manipulation,

$$f \sim \frac{1}{2\pi}\int_0^\pi dz \ln\left(\cosh(\beta)\coth(\beta)\left(1+\sqrt{1-\left(\frac{2\cos(z)}{\cosh(\beta)\coth(\beta)}\right)^2}\right)\right).$$

To get the last form we used the identity

$$\frac{1}{2}\int_{-\pi}^{\pi} dw \ln(1+t\cos(w)) = \int_0^\pi dw \ln(1+t\cos(w)) = \pi\ln(1+\sqrt{1-t^2}).$$

The dominant contribution to the specific heat close to the critical point $k=1$, where

$$2/k = \cosh(\beta)\coth(\beta),$$

is thus given by

$$S(\beta) \sim \int_0^{\pi/2} \frac{dz}{\sqrt{1-k^2 sin^2(z)}}$$

which has a logarithmic singularity at the critical point, that is $\alpha = 0$.

What has happened here is that the complexities of the Ising representation of the braid group and the Yang-Baxter equation have been miraculously pulled through onto a smaller and more manageable realisation. This has happened in such a way that the spectrum of the full model may be recovered. In general there is no obvious reason why different realisations should have such closely related spectra (unless they are actually isomorphic representations). It is intriguing to wonder if any of the multitude of other tensored space (and hence exponentially growing) representations occuring in statistical mechanics may be similarly mimicked.

4.1.2 By translation in the layer

An alternative solution proceeds as follows (Schultz, Mattis and Lieb 1964). Consider the Clifford algebra generated by operators $C_{\pm i}$ $(i = 1,..m)$ obeying the relations

$$\{C_i, C_j\} = \delta_{i,-j}.$$

This has a representation of dimension 2^m with basis generated by the action of C_{-i} $(i > 0)$ on a vector $|0>$ defined by

$$C_i|0>=0 \qquad i>0$$

where we use the notation

$$C_{-i}|0>=|i>$$

and

$$\left(\prod_{i=1}^{k} C_{-j_i}\right)|0>=|j_1,...,j_k> \qquad j_1<j_2<..<j_k \tag{4.9}$$

so

$$C_i|i>=C_iC_{-i}|0>=(1-C_{-i}C_i)|0>=|0>$$

and so on.

Note that in the definition of the algebra each pair $(\pm i)$ of generators is of equal standing. The numbering is incidental and does not imply any ordering among the generators. On the other hand the *representation* requires us to introduce a total order on positive indexed generators through equation 4.9. We have arbitrarily chosen this to be given by the numbering.

Writing

$$|i>=\begin{pmatrix}0\\1\end{pmatrix}\otimes\begin{pmatrix}0\\1\end{pmatrix}\otimes\begin{pmatrix}0\\1\end{pmatrix}\otimes...\otimes\begin{pmatrix}1\\0\end{pmatrix}\otimes...\otimes\begin{pmatrix}0\\1\end{pmatrix}$$

where the flipped 'spin' is in the ith position, we find

$$C_i=\begin{pmatrix}-1&0\\0&1\end{pmatrix}\otimes\begin{pmatrix}-1&0\\0&1\end{pmatrix}\otimes...\otimes\begin{pmatrix}0&0\\1&0\end{pmatrix}\otimes 1_2\otimes 1_2\otimes...\otimes 1_2.$$

and $C_{-i}=C_i^\dagger$.

We note that the effect of the ordering has been to introduce a sort of linear arrangement (with a boundary) into the generators.

Considering the subalgebra generated by all bilinears in the generators we see that the representation above splits into two blocks. The basis states for the blocks are those with k (from equation 4.9) odd and even respectively.

The independent linear combinations of generators n_j (integer j) defined by

$$n_j=\frac{1}{\sqrt{m}}\sum_{k=1}^{m}\exp(-2\pi ijk/m)C_k$$

for $j > 0$ and $n_{-j} = n_j^\dagger$, obey

$$C_k = \frac{1}{\sqrt{m}} \sum_{r=1}^{m} \exp(2\pi i k r/m) n_r .$$

These alternative generators for the algebra thus have the same anticommutation relations as the objects C_j.

The linear combinations N_j (integer $2j$) defined by

$$N_j = \frac{1}{\sqrt{m}} \sum_{k=1}^{m} \exp(-2\pi i j k/m) C_k$$

for all j (so $N_{-j} \neq N_j^\dagger$), obey

$$C_k = \frac{1}{\sqrt{m}} \sum_{r} \exp(2\pi i k r/m) N_r .$$

provided that the latter sum is over either m integer or m half integer values (in steps of 1) only. There are thus two non-independent sets of N-operators here. The set with half integer indices gives the 'periodicity' condition $C_{m+1} = C_1$, the other gives $C_{m+1} = -C_1$. They obey the anticommutation relations

$$\{N_j, N_k\} = 0$$

$$\{N_j, N_k^\dagger\} = \delta_{j,k} .$$

Putting

$$S_{2i-1} = 2C_{-i}C_i - 1$$

$$S_{2i} = (C_{-i} - C_i)(C_{-i-1} + C_{i+1})$$

(i positive) and

$$U_i = \frac{1}{\sqrt{2}}(1 + S_i)$$

we find that the operators U_i obey the Temperley-Lieb relations with $Q = 2$, together with the additional relations defining the $Q = 2$ unitarisable quotient. Hence, for the purpose of calculating eigenvalues, we can safely use them in place of the usual U_i matrices when constructing the $Q = 2$ Potts transfer matrix. In the representation above they take the form

$$U_{2i-1} = \sqrt{2}(1_2 \otimes .. \otimes \begin{pmatrix} 1 & 0 \\ 0 & 0 \end{pmatrix} \otimes 1_2 \otimes ...)$$

$$U_{2i} = 1/\sqrt{2}(1_2 \otimes .. \otimes \begin{pmatrix} 1 & 0 & 0 & 1 \\ 0 & 1 & 1 & 0 \\ 0 & 1 & 1 & 0 \\ 1 & 0 & 0 & 1 \end{pmatrix} \otimes 1_2 \otimes ...).$$

Note that they do not couple the subspaces of odd and even k. This means that we can extend the definition to a periodic completion U_0 in two ways, corresponding to $C_{m+1} = \pm C_1$.

In either case the transfer matrix becomes, up to the usual unimportant factors,

$$T \sim V_1 V_2$$

where

$$V_2 = \exp\left(\frac{\beta_2}{2}\sum_{i=1}^{m}(C_{-i} - C_i)(C_{-i-1} + C_{i+1})\right)$$

and

$$V_1 = \exp\left(\tanh^{-1}(e^{-\beta_1})\sum_{i=1}^{m}(2C_iC_{-i} - 1)\right).$$

These factors may immediately be rewritten in terms of the operators N_j as

$$V_2 = \prod_p \exp(\beta_2(\cos(p)(N_p^\dagger N_p + N_{-p}^\dagger N_{-p}) + i\sin(p)(N_p N_{-p} - N_{-p}^\dagger N_p^\dagger)))$$

$$V_1 = \prod_p \exp(-2\tanh^{-1}(e^{-\beta_1})(N_p^\dagger N_p + N_{-p}^\dagger N_{-p} - 1))$$

where the products are over the appropriate non-negative values of the indices (depending on the boundary conditions).

The advantage of this form is that each bilinear in the operators N_q commutes with any other with different q, so the various factors in the transfer matrix may be rearranged as follows

$$T \sim \prod_p \Big[\left(\exp(-2\tanh^{-1}(e^{-\beta_1})(N_p^\dagger N_p + N_{-p}^\dagger N_{-p} - 1))\right)$$
$$.\exp(\beta_2(\cos(p)(N_p^\dagger N_p + N_{-p}^\dagger N_{-p}) + i\sin(p)(N_p N_{-p} - N_{-p}^\dagger N_p^\dagger))) \Big].$$

Here each factor in square brackets commutes with every other one, so they may be simultaneously diagonalised.

We may construct a representation for the objects N_q in an analogous way to that for the objects C_m (again the representations for the bilinear subalgebras are pure odd and even). The algebras are isomorphic and we may use the same vacuum $|0>$.

Writing $N^\dagger_{\pm j}|0>=|\pm j>$ we note that each factor in the transfer matrix is diagonal on the $\{|+j>,|-j>\}$ subspace, but not on the $V^{(j)}=\{|0>,|-j+j>\}$ subspace. In the overall even subspace $\otimes^j V^{(j)}$ we have

$$T \sim \otimes^j A(j)$$

where the matrix $A(j)$ is given by

$$A(j)=\exp(\beta_2\cos(j))\begin{pmatrix}\exp(-2\tanh^{-1}(e^{-\beta_1})) & 0\\ 0 & \exp(2\tanh^{-1}(e^{-\beta_1}))\end{pmatrix}$$

$$\cdot\begin{pmatrix}\cosh(\beta_2)+\sinh(\beta_2)\cos(j) & \sinh(\beta_2)\sin(j)\\ \cosh(\beta_2)+\sinh(\beta_2)\cos(j) & \sinh(\beta_2)\sin(j)\end{pmatrix}$$

Each 2×2 direct product factor is thus isomorphic to a direct summand from the rotation realisation of the previous section. Diagonalising each such matrix we hence obtain the same answer as in section 4.1.1, that is, equation 4.8.

The final *temperature dependent* basis change required for the complete diagonalisation above may again be regarded as a linear transformation to a final set of fermion operators ψ_j. The transfer matrix may thus be written in the form

$$T \sim \exp\left(-a\sum_k(\frac{x_k}{a}(\psi^\dagger_k\psi_k-1/2)\right)$$

where the sum is over all allowed momenta (depending on the boundary conditions) and all the temperature dependence is in x_k. We have introduced a physical lattice spacing a, so that $\frac{x_k}{a}$ has units of energy.

The resultant momentum space Hamiltonian density is that of a massive fermion in a box (Kogut 1979, and c.f. Ramond 1989). Recall that the field theoretic Hamiltonian density is

$$E_p\psi^\dagger_p\psi_p$$

where with m the mass E_p is the relativistic energy

$$E_p=\sqrt{p^2+m^2}.$$

On the other hand, by considering lattice momenta near m for m large, $k'=k+m$, with k/m small, our result may be written *at criticality* in the form

$$x_k \sim k'/m.$$

At this point we can then introduce the physical (dimensionful) momentum $p = k'/ma$, take $a \to 0$ and $m \to \infty$ with the ratio fixed, and formally recover a free massless self-charge conjugate Majorana fermion field (c.f. Bjorken and Drell 1964). We will return to this observation shortly.

4.1.3 By translation in the plane

There is some use to be made of the translation symmetry not just in the transfer matrix layer direction, but in the perpendicular direction as well. See Stanley 1971, Landau and Lifshitz 1980, and references therein.

The trick is to note, from equation 1.8, that yet another way to write the isotropic 2 state Potts partition function on an arbitrary M site lattice is in the form

$$Z \sim \exp\left((-1/2)\sum_{n=1}^{\infty} C(n)\left(\frac{\exp(\beta)-1}{\exp(\beta)+1}\right)^n\right)$$

where the function $C(n)$ depends on the lattice. For the square lattice

$$C(n) = \sum_{loops\ l}\left(\prod_{v=1}^{n}\exp(iR(v)/2)\right)$$

where the sum is over all closed loops of length n on the lattice; the products is over the n vertices passed through by the loop; and $R(v)$ is the change in direction of the loop at the vertex v on the loop.

Consider V_{4M} as the space of directed bonds of the lattice (i.e. two per bond). For $i, j \in End(V_{4M})$ define $R(i,j)$ as $R(v)$ if (i,j) defines a path through a vertex (i.e. a segment of a loop); and $R(i,j) = i\infty$ otherwise. Then define a matrix W with elements

$$W_{ij} = \exp(iR(i,j)/2).$$

Since

$$C(n) = \left(\frac{1}{n}\right) Tr(W^n) = (1/n)\sum_{i=1}^{4M}(e_i)^n$$

where $\{e_i\}$ are the eigenvalues of W, we have

$$Z \sim \prod_{i=1}^{4M}\sqrt{\left(1 - e_i\,\frac{\exp(\beta)-1}{\exp(\beta)+1}\right)}.$$

Let us make the indices i and j a little more explicit. We will write i_1,i_2 for the cartesian coordinates of the site at the end of the directed bond, and $i_0 = 1,2,3,4$ for the direction from which it is approached by the bond. Assuming that the lattice is on a torus we have two directions of translational symmetry in W, so we may fourier transform with respect to both horizontal and vertical bond coordinates. If the periodic cycle lengths are the same in either direction, L say, we can write

$$W'_{mn} = \sum_{jk} z^{m_1 j_1 + m_2 j_2} W_{jk} z^{n_1 k_1 + n_2 k_2}.$$

where $z = \exp(2i\pi/L)$. The matrix W' is block diagonal, mixing only among sets of four states of given momentum. Within these blocks we have

$$W'_{m_0 n_0} = \begin{pmatrix} z^{m_1} & i^{-1/2} z^{m_2} & 0 & i^{1/2} \\ i^{1/2} z^{m_1} & z^{m_2} & i^{-1/2} z^{-m_1} & 0 \\ 0 & i^{1/2} z^{m_2} & z^{-m_1} & i^{-1/2} z^{-m_2} \\ i^{-1/2} z^{m_1} & 0 & i^{1/2} z^{-m_1} & z^{-m_2} \end{pmatrix}.$$

This may be diagonalised directly, whereupon we recover the same free energy as before (equation 4.8).

Once again I recommend that you take another look at the approach in chapter 3 at this point.

4.2 On conformal field theory

Following our discussion of block spin transformations in chapter 1 we note that systems with second order phase transitions exhibit a kind of scale invariance at their critical points. Since they also have short range interactions it is plausible that they will have invariance under locally varying scale transformations, or conformal transformations. A general effect of this is to restrict the possibilities for critical exponents, at least among those models with unitary field theory limits (Cardy 1987). In two dimensional field theories with local conformal invariance a set of generators of the infinitessimal conformal transformations appear as coefficients in the Laurent expansion of the light cone diagonal components of the energy-momentum stress tensor. In two dimensions these generators must give a representation of generators for (two commuting copies of) the Virasoro algebra (Belavin, Polyakov and Zamolodchikov 1984).

At time of going into press there were at a similar stage a number of books on precisely this subject, so we will not go into details. Essentially we proceed as follows:

The Virasoro algebra $\mathcal{V}$ may be defined abstractly by generators L_n (n integer) and central element c, and commutation relations

$$[L_n, L_m] = (n-m)L_{n+m} + \frac{c}{12}m(m+1)(m-1)\delta_{n+m,0}.$$

Since c is a central element it may be regarded as a scalar in any irreducible representation. To see the connection with conformal transformations in two dimensions consider the quotient $c = 0$, where we have the differential representation on the space of analytic functions of $z = x + iy$

$$L_n = -z^{n+1}\frac{d}{dz} \qquad (n \textit{ integer}).$$

The action of these generators is to induce infinitessimal conformal transformations

$$f(z) \to f(z + \epsilon(z))$$

for any infinitessimal analytic function

$$\epsilon(z) = -\sum_{n=-\infty}^{\infty} z^{n+1}\epsilon_n.$$

Specifically

$$\left(1 + \sum_{n=-\infty}^{\infty} \epsilon_n L_n\right) f(z) = f(z) + \epsilon(z)\frac{df(z)}{dz} \sim f(z + \epsilon(z)).$$

In quantum conformal field theory we can allow transformations generated by the Virasoro algebra with finite but *scalar* c, as an order $\hbar$ effect.

Since statistical mechanical models with second order phase transitions are built using the Temperley-Lieb algebra with rational r we expect that quotients of $\mathcal{V}$ appear as subalgebras of quotients of the notional limits $T_\infty(e^{i\pi/r})$.

For example, recall that in the unitarisable Ising model transfer matrix the representation of the Temperley-Lieb generators obeys the additional relations

$$1 - \sqrt{2}(U_i + U_{i+1}) + U_iU_{i+1} + U_{i+1}U_i = 0.$$

We have used this, for example in section 4.1.2, to define generators of a Clifford algebra by

$$S_i = 1 - U_i\sqrt{2}.$$

Matrices obeying the anticommutation relations

$$\{b_n, b_m\} = \delta_{n+m,0} \tag{4.10}$$

can then be built from the $\{U_i\}$ in a large number of possible ways. For example for $n > 0$

$$b_n = 1/2\left(\prod_{j=1}^{4j-1<2n+2} S_{4j-1} + \prod_{j=0}^{4j+1<2n+2} S_{4j+1}\right) S_{2n+2};$$

$b_{-n} = b_n^T$, and

$$b_0 = S_2/\sqrt{2}.$$

It follows that the objects

$$M_n = (1/2)\left(\sum_{k>n} + \sum_{k>0}\right)(k - n/2)b_{n-k}b_k$$

with the sums running over half integers, obey the relations

$$[M_n, b_m] = -(m + n/2)b_{n+m}$$

and hence give a representation of the Virasoro algebra $L_n \mapsto M_n$, with $c = 1/2$ (see, for example, Tsuchiya and Kanie 1986, Goddard, Nahm and Olive 1985).

The 'vacuum' vector $|0>$ defined by

$$b_n|0>= 0 \qquad (n > 0)$$

obeys

$$M_n|0>= 0 \qquad (n \geq -1)$$

and the 'one particle' state

$$|1/2>= b_{-1/2}|0>$$

obeys

$$M_0|1/2>= 1/2|1/2>$$

and

$$M_n|0>= 0 \qquad (n \geq 1).$$

The objects

$$N_n = 1/2\left(\sum_{k\geq n} + \sum_{k>0}\right)(k - n/2)b_{n-k}b_k + (1/16)\delta_{n,0}$$

obey the same relations, with only one highest weight representation corresponding to

$$L_0|0>= 1/16|0> .$$

In fact this exhausts the possibilities for unitary representations at $c = 1/2$, and the possibilities for critical exponents are restricted accordingly (Friedan, Qiu and Shenker 1984).

We can understand this construction from a field theory point of view by recalling from section 4.1.2 that the two dimensional Ising model at its critical point (and thermodynamic limit) can be written as a free massless two component Majorana fermion field $\psi = \begin{pmatrix} \psi^{(+)} \\ \psi^{(-)} \end{pmatrix}$ (the Majorana condition means $\overline{\psi} = \psi^\dagger\gamma_0 = \psi^T\gamma_0$).

The physical space Lagrangian density (Ramond 1989, Bjorken and Drell 1964) is then

$$L = \overline{\psi} i\gamma_\mu\partial_\mu\psi$$

which may be written in light cone coordinates as

$$L = i\psi^{(+)}\partial_+\psi^{(+)} + i\psi^{(-)}\partial_-\psi^{(-)}$$

exhibiting the conformal invariance of the Action:

$$S = \int dx^+dx^-(i\psi^{(+)}\partial_+\psi^{(+)} + i\psi^{(-)}\partial_-\psi^{(-)})$$

The canonically quantised fields may be written

$$\psi^{(\pm)}(t,\theta) = \frac{1}{\sqrt{2\pi}}\sum_n e^{-in(t\pm\theta)}b_n^{(\pm)}$$

where the sums are over integers or half integers depending on the choice of periodic or antiperiodic boundary conditions (respectively called Ramond and Neveu-Schwartz boundary conditions in field theory terminology). The anticommuting constructs from the Temperley-Lieb generators (equation 4.10) are essentially the anticommuting operators used to second quantise the Dirac field.

In this framework the Virasoro generators are the fourier components of the light cone momentum density. That is, considering the Neveu-Schwartz case, and defining a linear *normal ordering* bracket by

$$:b_i b_{-i}: \ = b_i b_{-i} \quad if \quad i < 0$$

$$:b_i b_{-i}: \ = -b_{-i} b_i \quad if \quad i > 0$$

we form

$$L_n = \frac{1}{4}\int_0^{2\pi} e^{in\theta} : \psi^{(-)}(\theta)\left(-i\frac{\partial}{\partial\theta}\psi^{(-)}(\theta)\right) - \left(-i\frac{\partial}{\partial\theta}\psi^{(-)}(\theta)\right)\psi^{(-)}(\theta) : \; d\theta.$$

We immediately recover the same expression as for M_n above.

It may be deduced using the charged Coulomb gas approach (see for example Cardy 1987, Nienhuis 1987) that systems in the same universality class as the Potts models with $r-3 \in \mathbf{N}$ give rise to conformal field theories with

$$c = 1 - 6/r(r-1).$$

More abstractly, a unitarity requirement on the Virasoro algebra restricts the possible choices for c to the set above and $c > 1$ (Friedan, Qiu and Shenker 1984). The number of inequivalent irreducibles is also restricted. This is highly reminiscent of the situation in Temperley-Lieb algebras (chapters 6 and 7). Unfortunately there is no transfer matrix formulation for the Coulomb gas approach, and no direct constructions from the Temperley-Lieb algebras (other than the one given above) are yet known.

This is a very interesting problem (see chapter 13, Luther and Peschel 1975, Thacker and Itoyama 1988). The approach which partly motivates chapter 13 is to interpret the cabling transformations of braids as certain limits of block spin transformations. The fixed points of the cabling morphisms (if any) correspond to possible fixed points in the associated models. We can then look for an action of the Virasoro algebra in this framework....

Chapter 5

Algebra: general principles

Introduction

One objective of this book is to examine an area of the 'no mans land' between Maths and Physics. Both camps run sortees in this area, and would like to claim it for themselves. However, neither can travel under their own flag in any safety! Taking Baxter's philosophy on this matter (see his book, p.1) to extreme, our internationalist creed is to plunder techniques from either camp, however difficult, to achieve our objectives.

On the other hand, if you are a physicist, it is possible that aquiring familiarity, in depth, with the algebraists ambit would take more time than you care to spend. So, rather than merely direct your background reading in this huge subject area, we will mention *here* a few of the general principles most useful for our purposes. More *technical* ideas which we need will then be introduced as and when required.

For those finding sufficient time on their hands I suggest Hamermesh (1962), Fraleigh (1978), MacLane and Birkoff (1979), Anderson and Fuller (1974) and Cohn (1989) for further reading.

5.1 Algebras

Consider a vector space A (over the field of complex numbers), together with a map $A \times A \rightarrow A$ called multiplication such that for all $a, b, c \in A$

$$(a + b)c = ac + bc$$

$$c(a+b) = ca + cb$$

$$(ab)c = a(bc).$$

This is an associative algebra (over the field of complex numbers), usually also called A. The map ensures that each element of A may simultaneously be regarded as a linear operator on the vector space.

If an algebra has an identity element under multiplication this is called the *unit* or 1, and the algebra is *unital*. If it has no unit then one may be added in a natural way.

An example of an algebra is the matrices.

An algebra homomorphism (or just a *morphism*) from algebra A to algebra B is a map

$$f : A \to B$$

such that

$$fa\, fb = fab.$$

An n dimensional matrix representation of A is a homomorphism from A to $n \times n$ matrices. If the morphism is injective then the representation is said to be faithful. That is, a representation is faithful if every distinct element of the algebra is represented by a distinct matrix. If the morphism is surjective the representation is said to be irreducible. For example, any one dimensional representation is irreducible.

An algebra A is *isomorphic* to an algebra B (written $A \cong B$) if there are morphisms $f : A \to B$ and $g : B \to A$ such that the composite maps fg and gf are both identity maps. In practice we will usually simply regard isomorphic A and B as the *same* algebra.

5.1.1 On specifying algebras

For our purposes an algebra A (over the field of complex numbers) is usually specified by giving a subset of A whose elements are called generators, together with a set of relations, which are equations relating linear combinations (over the field of complex numbers) of products of generators. The elements of the algebra, $a \in A$, are all linear combinations (with coefficients in the complex numbers) of arbitrary products of generators. A straight product of generators is called a *word*.

A spanning set $\{v_i\}$ of elements of the algebra is a set in terms of which *any* element of the algebra may be written as a linear combination, that is,

$a \in A$ implies that there exist complex numbers C_i such that

$$a = \sum_i C_i v_i.$$

If the numbers C_i are unique for all $a \in A$ then $\{v_i\}$ is a basis for A. The defining relations will serve to reduce the number of elements required in a basis set by providing linear relations between words.

The dimension of an algebra is the same as the dimension of the vector space, i.e. the order of any basis $\{v_i\}$.

Example 1.

For q a non-zero complex parameter consider the family of unital algebras $T(q)$ defined by generators $\{1, U\}$, together with relations

$$11 = 1$$

$$1U = U1 = U$$

$$UU = dU$$

where

$$d = q + q^{-1}.$$

A viable basis for each such algebra is the generators.

In our example, $T(q_1)$ is isomorphic to $T(q_2)$ if $d_1, d_2 \neq 0$ (i.e. $q_1, q_2 \neq \pm i$), since we can define a function

$$f : T(q_1) \rightarrow T(q_2)$$

such that

$$f1 = 1$$

$$fU = \frac{d_1}{d_2} U$$

with the properties

$$f1\, f1 = f1$$

and

$$fU\, fU = \left(\frac{d_1}{d_2}\right)^2 UU = \left(\frac{d_1}{d_2}\right)^2 d_2 U = d_1 \left(\frac{d_1}{d_2}\right) U = d_1\, fU$$

and vice versa.

An algebra may alternatively be specified by giving a set of $n \times n$ matrices, or linear transformations of a vector space with coefficients in the complex numbers. These matrices generate a subspace of the $n \times n$ matrix algebra which is the smallest space which contains them and is closed under matrix multiplication. The connection is that these matrices may be thought of as giving a faithful linear representation of a set of abstract generators. For example,

$$1 = \begin{pmatrix} 1 & 0 \\ 0 & 1 \end{pmatrix}$$

$$U = \begin{pmatrix} q & 1 \\ 1 & q^{-1} \end{pmatrix}$$

specify the same algebra as in example 1.

5.1.2 Subalgebras and quotients

An algebra B is a subalgebra of an algebra A (denoted $B \subset A$) if the elements of B are a subset of the elements of A, such that the insertion $B \rightarrow A$ is a homomorphism. For example, we can obtain a subalgebra of an algebra defined by generators and relations by simply removing some generators (and any associated relations).

A left sided ideal B of A is a subalgebra such that

$$ab \in B \quad \forall\, a \in A, b \in B.$$

The left sided ideal of A generated by an element $a \in A$ is the algebra with vector space spanned by the set $\{ba \text{ for all } b \in A\}$. For example, the left sided ideal generated by $U \in T(q)$ is just spanned by U itself.

The double sided ideal generated by $a \in A$ is the subalgebra with vector space spanned by $\{bac \text{ for all } b, c \in A\}$.

The double sided ideal generated by B a subset of A is the subalgebra with vector space spanned by $\{abc \text{ for all } a, c \in A,\ b \in B\}$.

Suppose we have a morphism s from algebra B to A. Then the *image* of s is the subalgebra of A

$$Im(s) = \{s(b)\ \ \forall b \in B\}$$

and the *kernel* of s is the double sided ideal of B

$$\ker(s) = \{b|\ b \in B\ ,\ s(b) = s(0)\}.$$

For example, the kernel of an injection is zero.

Suppose we have a further morphism p from A to algebra G, then the sequence

$$B \xrightarrow{s} A \xrightarrow{p} G$$

is *exact at* A when $Im(s) = \ker(p)$. For example, if

$$A \xrightarrow{p} G \rightarrow 0$$

is exact then p is a surjection.

A quotient algebra G of A is an algebra with a surjective homomorphism from A to G.

Suppose an algebra A is specified by generators and relations, then any algebra which may be generated by the same set of generators, but extending the set of relations by some additional (non-trivial) relations, is a (proper) quotient algebra of A.

If we form the quotient of A by subalgebra B (denoted A/B) this means we form the algebra obtained from A by introducing the additional relations that every element of B be set to zero.

For any quotient algebra G of A there is at least one $B \subset A$ such that G is isomorphic to A/B. To see this note that all the additional relations for A/B may be written in the form $a = 0$, where $a \in A$. Such a relation implies that all the elements of the double sided ideal generated by a must be set to zero.

The surjective homomorphism (or epimorphism) p from A to the quotient A/B has kernel the double sided ideal generated by B. If we call this double sided ideal D, then

$$A/B \cong A/D.$$

Note that defining the coset $D + a$ for $a \in A$ as the subset of A given by

$$D + a = \{d + a \;\forall\; d \in D\}$$

then we may equivalently take

$$p : a \mapsto D + a.$$

Altogether we have that

$$0 \rightarrow D \xrightarrow{s} A \xrightarrow{p} A/D \rightarrow 0$$

is exact at D, A and A/D.

5.1.3 Linear representation theory

If we have a basis of elements $\{v_i\}$ of the algebra A, multiplication of any of these by a generator on the left (or right) gives a new element of the algebra which must be a linear combination of elements of the basis set again. Restricting to multiplication on the left, if $a \in A$ then there exists a unique matrix $C(a)$ with complex elements $C(a)_{ij}$ such that

$$av_i = \sum_j C(a)_{ij} v_j .$$

The matrix $C(a)$ *represents* an element a by giving the transformation of each element of the basis set into the appropriate linear combination. The morphism here is actually

$$a \mapsto C^T(a)$$

since

$$bav_i = b\sum_j C_{ij}(a)v_j = \sum_j C_{ij}(a)\sum_k C_{jk}(b)v_k = \sum_k \left[\sum_j C_{ij}(a)C_{jk}(b)\right] v_k$$

so we then get

$$ba \mapsto [C(a)C(b)]^T = C(b)^T C(a)^T .$$

This faithful representation of A is called the regular representation.

In general there will be subspaces of A which are also closed under left (right) action of A, such as the left (right) ideals. Each gives rise to representations of A in the same way as above.

More generally still, any vector space V with a bilinear map $A \times V \to V$ such that for all $a, b \in A$ and $v, w \in V$

$$(a + b)v = av + bv$$

$$a(v + w) = av + aw$$

$$(ab)v = a(bv)$$

$$(\ 1v = v\ )$$

is called a left A module, and gives rise to an isomorphism class of representations of A (one for each basis of V) in the same way.

Here we usually regard isomorphic left A modules as being the *same* module. We do not regard isomorphic but distinct representations as being the same, since in transfer matrix construction such representations would give rise to distinct Boltzmann weights. On the other hand, isomorphic representations would give rise to very strongly equivalent statistical mechanical models, since the spectrum of the transfer matrix would be identical.

Right A modules may be similarly defined. A left *and right* A module M which also obeys the condition

$$(am)b = a(mb)$$

for all $m \in M$ and $a, b \in A$ is called an A bimodule. Note that A is itself an A bimodule, by the associativity condition.

These other representations of the algebra may not be faithful. They will, however, be faithful representations of some quotient algebra.

An invariant subspace of a left A module is any subspace which is closed under the left action of A.

If a module contains no invariant subspaces under left action of A then it is called a simple (left) A module. The consequent representations will be irreducible.

Equivalently, as we have said, a d-dimensional representation is irreducible if every d-dimensional matrix is the representative of some algebra element.

A direct summand of a left A module V is any subspace which is closed and whose complement in A is also closed.

Any subspace of a finite dimensional algebra A which is a direct summand of A under left action of the generators is called a projective (left) A module. Bases for such a module will give rise to representations which are direct summands of the regular representation.

We can always produce a representation of an algebra A by picking an element $a \in A$ and forming the associated left ideal. A basis for the ideal may be obtained by running through $b \in \{v_i\}$ in some order and including ba in each instance in the basis if it cannot be written as a linear combination of existing elements. This basis will then give rise to a representation, in the same way as the basis set $\{v_i\}$, except that it will not, in general, contain a spanning set for the whole algebra A. The ideal *will* contain a spanning set if the chosen element $a \in A$ was the unit.

Direct sums of distinguished, but possibly isomorphic, vector spaces formed in this way will clearly give rise to further representations of A.

A representation is decomposable if the basis vector space may be written as the direct sum of 2 subspaces, both of which are invariant. Otherwise it is indecomposable. For an example, consider the case $q = i$ in the representation given for example 1.

Indecomposable projective (left) A modules of dimension d will be denoted P_d, while simple modules will be denoted S_d.

5.2 Physics and the radical

An element of an algebra $a \in A$ is said to be nilpotent if repeated multiplication of a by itself eventually yields zero. An algebra is nilpotent if every element is nilpotent. The radical J of an algebra A is the maximal nilpotent double sided ideal of A.

A finite dimensional algebra is said to be semi-simple if it has no proper nilpotent double sided ideal. In this case every indecomposable representation is irreducible, and each simple module is a projective module.

In many of the physically interesting cases, however, the relevant algebras are not semi-simple and an arbitrary representation cannot necessarily be written as a direct sum of irreducibles. In general this complication may be overcome by quotienting an algebra A by its radical J (see, for example, Anderson and Fuller 1974). Because this radical is nilpotent it contributes only non-propagating contributions to the transfer matrix, which do not affect the bulk properties (that is, they will disappear when T is raised to a high power, or equivalently they do not affect the characteristic polynomial or spectrum of T). The quotienting leaves a semi-simple 'top' algebra A/J, which is the maximal semi-simple quotient of A.

In the $Q = 0, 1, 2, 3$ state Potts cases the algebra is further quotiented to leave an algebra with only unitarisable representations (i.e. representations in which the generators $\{U_i\}$ are self-adjoint, or Hermitian matrices). This combined quotienting is relatively easy (see chapter 6), but unfortunately it has the effect of eliminating some propagating contributions, i.e. long distance properties, in more general models based on the Temperley-Lieb algebra. That is, it eliminates operators which would not normally disappear when T is raised to a high power. For example, with $Q = 1$ it leaves only a one dimensional quotient algebra whereas, as we shall see, there are an infinite number of distinct irreducible representations in the thermodynamic limit. These are physically relevant, for example, in the percolation problem (see Baxter 1982). We will discuss the minimal quotienting procedure which leaves all of them intact.

Perhaps the key reasons, however, for concerning ourselves with the more esoteric aspects of the structure of algebras are:

i) the fact that the Ising model was *solved* because a key simplification (equation 3.8) arose out of the apparent complication of the extra structure of the Temperley-Lieb algebra at $Q = 2$;

ii) because it seems to be the easiest way to get at the part of the structure which *is* physical.

There is also the structural similarity with Virasoro algebras (the algebras of conformal transformations in quantum field theory), which was a large part of the original interest in these studies, and which is still unexplained (see later).

5.3 Induction and restriction

Suppose we have an algebra homomorphism

$$\psi : B \rightarrow A$$

(for example, through $B \subset A$ or $A = B/J$), then any representation of A, $R : A$ say, gives rise to a representation of B by using only the representatives of the images under ψ (for example, if $B \subset A$ then keep only the representatives of the appropriate subset). We call this process restriction of $R : A$ to B, and denote such a representation by $R : A \downarrow B$.

At the level of modules, let us denote an arbitrary left A module by ${}_AM$. If $B \subset A$ then ${}_AM$ is already a perfectly good B module by the insertion $B \rightarrow A$. The restriction of ${}_AM$ to a B module is written ${}_B\downarrow_A M$.

More generally, if

$$\psi : B \rightarrow A$$

then ${}_AM$ may be regarded as a B module by defining the multiplication $B \times {}_AM \rightarrow {}_AM$ by

$$bm = \psi(b)m$$

for all $b \in B$ and $m \in {}_AM$.

Now A itself is an A bimodule (we can write ${}_AA_A$), therefore it is also a left A right B module, a (A, B)-module, by restriction (let us write ${}_AA_B$). Let us define the tensor product over B of ${}_AA_B$ with an arbitrary left B module ${}_BN$ as the ordinary tensor product quotiented by a certain set of equivalence relations:

$${}_AA_B \otimes_B {}_BN = {}_AA_B \otimes {}_BN/\sim$$

where the equivalence relations are

$$a\psi(b) \otimes n \sim a \otimes bn$$

for all $n \in N$, $a \in A$ and $b \in B$ (if $B \subset A$ then $\psi(b) = b$). We then define the module *induced* on A by ${}_BN$ as

$${}_A\uparrow_B N = {}_AA_B \otimes_B {}_BN.$$

For example, if J is the radical of B and

$$\psi : B \rightarrow B/J$$

then

$$_{B/J}B/J_B \otimes_B {}_BN = N/JN.$$

For B an algebra we define a morphism of B modules to be any map

$$\psi : {}_BN \rightarrow {}_BM$$

taking

$$x \mapsto \psi(x)$$

such that for all $b \in B$ and $x \in {}_BN$

$$b\psi(x) = \psi(bx).$$

A *Category of (left) A modules* is a set of (left) A modules together with the morphisms between modules in the set.

Let us restrict attention to left modules M of algebras A which have the property that the map

$$A \times M \rightarrow M$$

is surjective (we write $AM = M$). We will call the category of such modules the surjective category. Note that *all* the modules for algebras with a unit have this property.

Furthermore

Proposition 1 *For A any algebra and e an idempotent in A then the surjective category of left AeA modules includes all quotients of AeA as a left module.*

Proof:

Let us write $B = AeA$, then

$$B \times B \rightarrow B \tag{5.1}$$

is surjective, since for all $a, b \in A$

$$aeb = aee.eeb.$$

Now suppose I an invariant subspace of B (as a left module), then

$$B(B/I) = (BB)/I = B/I$$

i.e. B/I is in the surjective category.

This concludes the proof of proposition 1.

More explicitly, consider

Proposition 2 *Let B be an algebra, and M a left B module, then*

$$B \times M \rightarrow M$$

is surjective if and only if there exists at least one injective set map

$$c : M \rightarrow B \times M$$

$$c : x \mapsto b \times m$$

such that $x = bm$.

The proof follows immediately from the definition of a surjection.

Corollary 2.1

$$B \otimes_B M \cong M.$$

Proof:
The isomorphism takes

$$b \otimes m \leftrightarrow bm.$$

Then for I an invariant subspace of B the surjective property $BB = B$ implies that every element of B, and hence every $a \in B/I$ may be written

$$a = bh$$

for some $h, b \in B$. On the other hand, there exists a suitable $h \in B/I$ itself, since I is an invariant subspace. Therefore from proposition 2 the map

$$B \times B/I \rightarrow B/I$$

is surjective.
This concludes the explicit proof of proposition 1.

If $Hom_A({}_AV, {}_AW)$ is the vector space of A homomorphisms from ${}_AV \rightarrow {}_AW$, then Frobenius reciprocity is the isomorphism

$$Hom_A({}_A\uparrow_B N, {}_AM) \cong Hom_B({}_BN, {}_B\downarrow_A M).$$

This is a standard result for unital algebras (see, for example, Anderson and Fuller 1974). That it holds for surjective categories has been pointed out by Westbury.

A proof is obtained as follows. First we construct a map α from the right hand side to the left hand side; then we will construct another map β which reverses the arrow; and finally we will show that the composites $\alpha\beta$ and $\beta\alpha$ are both identity maps.

For every morphism of B modules

$$\psi : {}_BN \to {}_B\!\downarrow_A M$$

define a map

$$\alpha(\psi) : {}_AA_B \otimes_B {}_BN \to {}_AM$$

by

$$\alpha(\psi) : a \otimes x \mapsto a\psi(x)$$

where the quotient is to be understood. Note that

$$\alpha(\psi)(ab \otimes x) = \alpha(\psi)(a \otimes bx)$$

since

$$\alpha(\psi) : ab \otimes x \mapsto ab\psi(x) = a\psi(bx).$$

This $\alpha(\psi)$ is a morphism of A modules since for all $a' \in A$

$$a'\alpha(\psi)(a \otimes x) = a'a\psi(x) = \alpha(\psi)(a'a \otimes x).$$

For every morphism of A modules

$$\theta : {}_AA_B \otimes_B {}_BN \to {}_AM$$

taking

$$a \otimes x \mapsto \theta(a \otimes x)$$

such that for all $a' \in A$

$$a'\theta(a \otimes x) = \theta(a'a \otimes x)$$

define a map

$$\beta(\theta) : {}_BN \to {}_B\!\downarrow_A M$$

by the following construction.

Recall that

$$B \times {}_BN \to {}_BN$$

surjective implies that for all $x \in {}_BN$ there exists some $y \in B \subset A$ and $z \in {}_BN$ such that $yz = x$ (y and z are not necessarily unique, but for example if B has a unit we may take $y = 1$, $z = x$). In any case, to each

x we can think of associating a specific pair y, z such that $yz = x$, then we take

$$\beta(\theta) : x \mapsto \theta(y \otimes z).$$

This is a morphism of B modules since for all $b \in B$

$$b\beta(\theta)(x) = b\theta(y \otimes z) = \theta(by \otimes z) = \theta(b \otimes x)$$

while with $y'z' = bx$

$$\beta(\theta)(bx) = \theta(y' \otimes z') = \theta(b \otimes x)$$

by definition of the tensor product.

We find

$$\alpha(\beta(\theta))(a \otimes x) = a\beta(\theta)(x) = a\theta(y \otimes z) = \theta(a \otimes x)$$

so β is 1-to-1 and α is onto, and

$$\beta(\alpha(\psi))(x) = \alpha(\psi)(y \otimes z) = y.\psi(z) = \psi(yz) = \psi(x).$$

QED.

Consider the following example:

For algebra $B = \mathbf{C}$ take ${}_BN$ to be the unique simple module, and for $A = T(i)$ (i.e. $d = 0$) take ${}_AM$ the simple module spanned by U. We have, denoting modules by their generators,

$$Hom_A(\{1, U\}, U) \cong Hom_B(1, 1).$$

There is clearly a basis of exactly one morphism on the right. On the left

$$f : \{1, U\} \to U$$

defined by

$$f : a \mapsto Ua$$

is the only possibility. Is this a morphism? Well,

$$1.f1 = 1.U = U = f1.1$$

$$U.f1 = U.U = 0 = fU.1$$

$$U.fU = U.0 = 0 = fU.U$$

and

$$1.fU = 1.0 = 0 = f1.U$$

so yes!

Because lattice models can usually be formulated on lattices of various widths, the corresponding transfer matrices may often be associated with towers of algebras $A_1 \subset A_2 ... \subset A_\infty$. Let J_k be the radical of A_k, then a *Bratelli diagram* describes the irreducible content of

$$R : A_n/J_n \downarrow A_{n-1}/J_{n-1}$$

for all n and all simple (or all projective, or all generically simple) R.

5.4 On the structure of algebras

The regular representation contains copies of all the irreducible representations of a finite dimensional unital algebra (semi-simple or otherwise) as simple modules. By definition the regular representation may be written as a direct sum of indecomposable projective left modules (which give indecomposable representations). There is a bijection between indecomposable projectives P and irreducibles S which sends P to P/JP. Furthermore, the multiplicity of P in the regular representation is the dimension of the associated irreducible.

To see this consider the example of an application of Frobenius reciprocity given above. More generally we see that if ${}_AM$ is a simple module it will appear as a quotient of ${}_AA_A$ (which is the A module induced from the simple module for $B = 1$) exactly $dim({}_AM)$ times:

$$Hom_A(A, {}_AM) \cong Hom_{\mathbf{C}}(\mathbf{C}, {}_{\mathbf{C}} \downarrow_A M).$$

The claimed result then follows from the bijection $S \leftrightarrow P/JP$ (which is not so easy to prove.....).

In the next chapter we thus start by writing down the regular representation for the Temperley-Lieb algebra. We will give various bases for this representation, which will allow us to discuss the structure of the algebra, and its top, and hence label possible long distance properties of statistical mechanical models.

Every algebra of dimension d is a subalgebra of the algebra of $d \times d$ matrices $M_d(\mathbf{C})$, which has dimension d^2.

The structure of an algebra as a subalgebra of $M_{\mathrm{dim}(A)}(\mathbf{C})$ may be exhibited directly with elements of the algebra. For each independent elementary matrix (matrix with one non-zero entry) there is a corresponding elementary operator.

An element of an algebra $e \in A$ is idempotent if $ee = e$. The double sided ideal generated by an idempotent is projective. To see this consider the decomposition

$$A = Ae \oplus A(1 - e).$$

An idempotent $e \in A$ is primitive if there exist no non-trivial pair of idempotents e_1 and e_2 in A such that $e_1 + e_2 = e$ and $e_1e_2 = e_2e_1 = 0$. For example, the primitive idempotents in the algebra generated by $\{1, U\}$ for $d \neq 0$ are

$$1 - U/d$$

and

$$U/d.$$

When $d = 0$ neither of these is well defined and the only primitive idempotent is 1 itself.

Equivalently, in the case of semi-simple algebras, an idempotent e is primitive if, for every element in the algebra W, we have $eWe = \chi(W)e$, where $\chi(W)$ is a complex scalar depending only on W. In this case the left ideal generated from a primitive idempotent is indecomposable and hence irreducible. To see this suppose that the left ideal generated from e is decomposable. Then by the definition there exists for some $b \in A$ an element be such that $ebe \neq \chi(b)e$ for any complex number $\chi(b)$.

5.4.1 Morita equivalence

A weaker property of an algebra than equivalence to a known algebra, but which is nonetheless a significant first step in determining the structure, is Morita equivalence.

What follows is standard for unital algebras (e.g. in Anderson and Fuller 1974). The generalisation to surjective categories was suggested by Westbury.

Surjective categories of modules for algebras A and B are said to be Morita equivalent if there exist left projective bimodules ${}_AV_B$ and ${}_BW_A$ such that

$${}_AV_B \otimes_B {}_BW_A = A$$

and

$${}_BW_A \otimes_A {}_AV_B = B.$$

The significance of Morita equivalence may be understood in terms of morphisms of modules:

We can use the ${}_AV_B$ and ${}_BW_A$ bimodules to define set maps between the sets of left A and left B modules as follows. Let us consider the left A modules ${}_AM$ and ${}_AN$, and the left B modules ${}_BM'$ and ${}_BN'$, . Then

$$F : M \mapsto {}_BW_A \otimes_A {}_AM.$$

The map is defined similarly on N.

The map defined using ${}_AV_B$ is

$$G : M' \mapsto {}_AV_B \otimes_B {}_BM'.$$

We will give examples when we come to talk about specific algebras (see chapter 9).

We can extend F to define a map of algebra homomorphisms complimentary to the set map. Suppose there is an algebra homomorphism $\psi : M \to N$. Let us consider $x \otimes y \in F : M$, i.e. $x \in {}_BW_A$ and $y \in {}_AM$, such that

$$\psi : y \mapsto \psi(y),$$

then

$$F(\psi) : x \otimes y \mapsto x \otimes \psi(y).$$

A B module homomorphism map extending G may be similarly defined. The direction of homomorphisms is preserved (with respect to the set map) by construction in either case, as is the identity map. Furthermore, suppose we have two composable algebra homomorphisms ψ and ψ' such that

$$y \overset{\psi}{\mapsto} \psi(y) \overset{\psi'}{\mapsto} \psi'(\psi(y)) = (\psi'\psi)(y),$$

then the composite of morphisms maps under F to $F(\psi')F(\psi) = F((\psi'\psi))$, again by construction.

Such a joint map of modules and morphisms is called a functor. Composing the functors F and G we find that both FG and GF act as the identity on the appropriate modular categories. For example,

$$\begin{aligned} {}_AM \overset{F}{\mapsto} {}_BW_A \otimes_A {}_AM \overset{G}{\mapsto} & {}_AV_B \otimes_B ({}_BW_A \otimes_A {}_AM) \\ &= ({}_AV_B \otimes_B {}_BW_A) \otimes_A {}_AM \quad \cong \quad {}_AM \end{aligned}$$

where, since the product of bimodules is the regular representation bimodule, the last isomorphism uses the property of surjective categories in corollary 2.1.

The images of M and N under F are distinct if and only if M and N are distinct, and similarly for M', N' under G. To see this note that the necessity part is by construction, while sufficiency follows from the 'inverse' relationship of F and G.

Any homomorphism among the A modules thus implies a matching homomorphism among the B modules, and *vice versa*. This is the key consequence of Morita equivalence.

5.5 Centraliser algebras

Let A be an algebra, and M a left A module, then the algebra of endomorphisms of M which commute with the action of A is written

$$B = End_A(M)$$

and is called the centraliser algebra of A with respect to M. If M is a faithful left A module, then

$$A = End_B(M).$$

Let $\{V^{(i)}\}$ be a set of inequivalent indecomposable left A modules, such that M has an A module decomposition

$$M = \oplus_{i=1}^{n} m_i V^{(i)}$$

then B has n irreducible representations, with dimensions m_i. Similarly if $\{W^{(j)}\}$ are indecomposable B modules, and

$$M = \oplus_j n_j W^{(j)}$$

then A has irreducible representations of dimension n_i.

An example of a centraliser algebra will appear in the next section.

5.6 Bialgebras

Any group G has a natural group homomorphism

$$\Delta : G \rightarrow G \otimes G$$

called comultiplication, defined for $a \in G$ by

$$\Delta(a) = a \otimes a.$$

This means that if V and W are representations of G then the representation $V \otimes W$ of $G \otimes G$ may be pulled back along Δ to give a representation of G.

Obviously there is no such natural comultiplication for algebras. An algebra A with a unit is a bialgebra if there is a morphism

$$\Delta : A \rightarrow A \otimes A$$

called comultiplication, which, for all $a \in A$ and $b, c \in A$ such that

$$\Delta(a) = b \otimes c$$

satisifies the coassociativity condition

$$\Delta(b) \otimes c = b \otimes \Delta(c) \quad \in A \otimes A \otimes A.$$

Any bialgebra therefore has tensor product representations. The difficulty is in finding a comultiplication Δ.

The universal enveloping algebras of Lie algebras have comultiplications which are recovered from their close associations with Lie groups. Jimbo 1985 and Drinfeld 1985 found that they could define certain one complex parameter deformations of the relations for these algebras, for which comultiplications could in turn be defined by deformations of the standard ones.

For example, following Lusztig 1989, for q a complex parameter let $U_q Sl(2)$ (q not a root of unity) be the algebra with generators $E, F, K^{\pm 1}$ and relations

$$KE = q^2 EK$$

$$KF = q^{-2} FK$$

$$[E, F] = \frac{K - K^{-1}}{q - q^{-1}}.$$

Defining

$$[N]! = \prod_{s=1}^{N} \frac{q^s - q^{-s}}{q - q^{-1}}$$

then if q is an lth root of unity we include the additional generators

$$E^{(l)} = E^l / [l]!$$

and

$$F^{(l)} = F^l / [l]!.$$

These algebras have comultiplication given by

$$\Delta(E) = E \otimes 1 + K \otimes E$$

$$\Delta(F) = F \otimes K^{-1} + 1 \otimes F$$

and

$$\Delta(K) = K \otimes K.$$

Iterating $k - 1$ times we have

$$E \mapsto \sum_{i=1}^{k} (\otimes^{i-1} K) \quad \otimes E \quad \otimes (\otimes^{k-i} 1).$$

Similarly

$$F \mapsto \sum_{i=1}^{k} (\otimes^{i-1} 1) \ \otimes F \ \otimes (\otimes^{k-i} K^{-1})$$

and

$$K \mapsto \otimes^k K.$$

In particular, $U_q Sl(2)$ has the following representation:

$$E = \begin{pmatrix} 0 & 1 \\ 0 & 0 \end{pmatrix}$$

$$F = \begin{pmatrix} 0 & 0 \\ 1 & 0 \end{pmatrix}$$

$$K = \begin{pmatrix} q & 0 \\ 0 & q^{-1} \end{pmatrix}.$$

This implies, through the iterated comultiplication, an action of $U_q Sl(2)$ on $\otimes^k V_2$. We can then ask, what is the sequence of algebras, indexed by k, defined by $End_{U_q Sl(2)}(\otimes^k V_2)$?

For $0 < i < k$ let us define

$$U_i = \otimes^{i-1} 1_2 \ \otimes \begin{pmatrix} 0 & 0 & 0 & 0 \\ 0 & q & -1 & 0 \\ 0 & -1 & q^{-1} & 0 \\ 0 & 0 & 0 & 0 \end{pmatrix} \otimes \ \otimes^{k-i-1} 1_2.$$

The matrix U_i commutes with every summand of the tensor representation of E, except for the the ith and $(i+1)^{th}$. By direct computation we find that it commutes with these two taken together. We see immediately that $U_i \in End_{U_q Sl(2)}(\otimes^k V_2)$. We will find in chapter 13 that these matrices generate the whole centraliser algebra.

5.7 Algebraic overview of following chapters

Let G be an arbitrary unoriented graph with an A_n subgraph, let G' be the graph obtained by decorating every bond of G with a node, and let nG be the set of nodes of G'. Then GB is the algebra with generators $\{g_i^{\pm 1} | i \in nG\}$ and relations

$$g_i g_j g_i g_j \ldots = g_j g_i g_j g_i \ldots$$

where the number of factors on each side is equal to the number of bonds between i and j, plus 2. By decorating every bond of G we have ensured that the maximum number of factors is 1+2=3 (but see e.g. Hoefsmit 1974).

We have the following table of algebra morphisms:

$$
\begin{array}{ccccc}
GB & \stackrel{m_1}{\longleftarrow} & A_nB = B_{2n} & \stackrel{m_7}{\longrightarrow} & NV_{2n-1}(x) \\
\downarrow m_2 & & \downarrow m_2 & & \\
GH(q) & \stackrel{m_1}{\longleftarrow} & A_nH(q) = H_{2n-1}(q) & & \\
\downarrow m_3 & & \downarrow m_3 & & \\
GNH(q) & \stackrel{m_1}{\longleftarrow} & NH_{2n-1}(q) & \stackrel{m_8}{\longleftarrow} & End_{U_qSl(N)}(\otimes^m V_{N+p}) \\
\downarrow m_4 & & \downarrow m_4 & & \updownarrow (N=2) \\
GT(Q) & \stackrel{m_1}{\longleftarrow} & T_{2n-1}(q) & \stackrel{m_8}{\longleftarrow} & End_{U_qSl(2)}(\otimes^m V_{2+p}) \\
\downarrow m_5 & & \downarrow m_5 & & \\
GT^u(Q) & \stackrel{m_1}{\longleftarrow} & T^u_{2n}(q) & & \\
\downarrow m_6 & & & & \\
G_{2n}(Q, p_f) & & & & \\
\downarrow m_6' & & & & \\
G_{2n}(Q) & & & &
\end{array}
$$

Here we will only give a précise and chapter reference for each morphism:

m_1 is injection of the subalgebra associated with the A_n subgraph (chapter 8 - only the Temperley-Lieb case is dealt with explicitly);

m_2 is the (local) Hecke quotient

$$g_i^2 = q^2 + (1 - q^2)g_i$$

(chapter 9 - only the A_n case is dealt with explicitly);

$m_3(N)$ are further Hecke quotients (chapter 9);

m_4 is the Temperley-Lieb quotient (isomorphism for $N = 2$)

$$\{g_i, g_{i+1}\} = g_i + g_{i+1} + g_i g_{i+1} g_i - 1$$

(chapters 6 and 8 - note that the parameterisations by q and Q are distinct, but readily interchangeable);

m_5 is a quotient, first by the Jacobsen radical and then to unitarisable representations - trivial for some q not roots of unity - which may be done all in one go using the quotient relation

$$E_{r(q)} = 0$$

(chapters 6, 7 and 8);

m_6 is injection of the unitarisable Temperley-Lieb quotient as a subalgebra of the generalised Clifford algebra, corresponding to the occurence of the Potts model as a special case of Z_Q symmetric models (chapter 10);

m_7 are other quotients of the braid group, that is with higher than second order local relations (chapter 12, Baxter 1982, Birman and Wenzl 1988);

$m_8(e)$ is (generically) injection of a cabling subalgebra associated with the cable idempotent e, with p and m depending on e and n (chapter 13, Deguchi, Wadati and Akutsu 1988, Murakami 1989, Date *et al* 1988).

5.7.1 On the braid group

As we can see, the group algebra of the braid group is one of the main source algebras for the algebraic approach to the formulation of statistical mechanical models. At this stage it is worth reviewing some useful features, which carry over into the heirarchy of quotient algebras.

The braid group on n strings B_n may be defined by the generators $\{g_i, g_i^{-1}; i = 1, 2, .., n-1\}$ and relations

$$g_i g_i^{-1} = 1$$

$$g_i g_{i+1} g_i = g_{i+1} g_i g_{i+1}$$

$$g_i g_{i+j} = g_{i+j} g_i \qquad (j \neq \pm 1)$$

(Birman 1974). We also use B_n to denote the associated group algebra.

Of equal standing is the geometrical picture of elements of B_n as *configurations* of n strings traversing an oriented box from top to bottom with no doubling back. Here two configurations are equivalent if one can be moved into the other without cutting any string. The equivalence class of the identity *includes* (but is not restricted to) all n string boxes with no overlaying strings. Group composition of elements $a, b \in B_n$ means connecting the bottoms of the strings in a with the tops of the strings in b (by a representative of the identity). The orienting of the boxes is necessary because some braids look different from the back.

This scheme may be represented conveniently on the plane, provided we preserve the over/under information as in the following representative of the generator g_i:

$g_i \mapsto$ 

The compositions involved in the braid identities then become

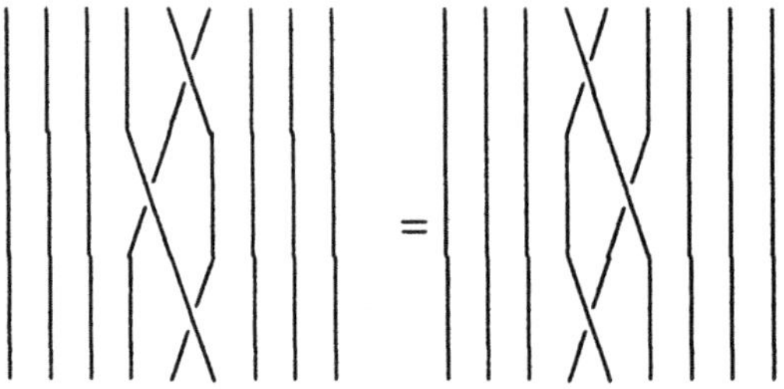

and so on.

5.7.2 Algebra morphisms

The way in which B_n algebra morphisms pull through to the quotient algebras we are interested in makes some of them very useful in statistical mechanics. All the ones we will discuss have an important geometrical significance, which we will return to once we have got the definitions out of the way.

Let $G \in B_n$ be defined by

$$G = \prod_{i=1}^{n-1} g_i$$

i.e.

$G \mapsto$ 

and

$$g_0 = G g_{n-1} G^{-1}$$

Then with the index on generators to be read *mod n*, the *translation* automorphism

$$B_n \overset{(m)}{\rightarrow} B_n$$

is defined by

$$(m) : g_i \mapsto g_{i+m} \quad .$$

There is a similar but distinct 'translation' obtained by replacing g_i with g_i^{-1} in G. Many other automorphisms with the flavour of a translation may be constructed.

An alternative set of generators for B_n is $g^{\pm 1} = g_1^{\pm 1}$ and $G^{\pm 1}$. Proof:

$$G^i \, g \, G^{-i} = g_{i+1}.$$

Because of this, any quotient relation automatically spawns a set of translations of itself, which streamlines presentations. It is also convenient to use g as a notational representative of g_i.

Let $M \in B_n$ be defined by (a braid equivalent to) puting a 180° twist in the identity element, i.e. a braid equivalent to the twist

$M \mapsto$ 

then the reflection automorphism

$$B_n \overset{(-)}{\rightarrow} B_n$$

is defined by

$$(-) : g_i \mapsto g_{n+1-i}.$$

Note that

$$(-) : G \mapsto \tilde{G}$$

(chapter 2) so

$$(-) : g_0 \mapsto M g_0 M^{-1}.$$

It is easy to see from the representation

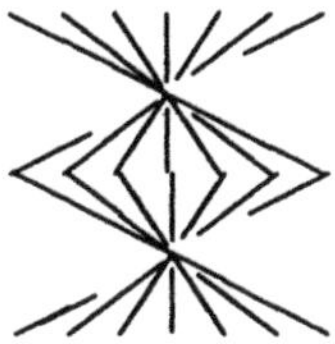

that $M^2 = G^n$ is central.

Chapter 6

Temperley-Lieb algebras: generic cases

On the next 3 chapters

In the next three chapters a compendium of useful results for the Temperley-Lieb algebras is collected. Several representations coming from statistical mechanical models are described, and used to determine the structure of the algebras.

The Temperley-Lieb algebras $T_k(q)$ introduced in chapter 2 are indexed by the number k of Temperley-Lieb generators U_i, and parameterised by a scalar q. In the present chapter we collect those results which hold for all q, and those which hold for all but exceptional values of q. In particular we write down the regular representation of the Temperley-Lieb algebra $T_k(q)$ in the basis of inequivalent reduced words on the k Temperley-Lieb generators.

As we have shown, when $k = 2n - 1$ representations of these generators may be used to construct the transfer matrices for statistical mechanical models on an n site wide lattice. We show here that the generically irreducible representation of $T_{2n-1}(q)$ responsible for the unique free energy in such models restricts to the regular representation of $T_n(q)$. We give equivalent forms for the regular representation, derived from lattice models, which manifest its indecomposable structure when k and q are such that $T_k(q)$ is a semi-simple algebra.

In the next chapter we will generalise to obtain the structure of $T_k(q)$ when not semi-simple. In subsequent chapters we will generalise to algebras associated with higher dimensional statistical mechanical models, and to algebras realising models with more general interactions.

6.1 Review

The n-site layer transfer matrices for a wide range of statistical mechanical models, including the Q-state Potts models (chapter 1), the percolation problem, the ice-type/6 vertex models (Baxter 1982 and chapter 12) and the critical Andrews Baxter Forrester models (ABF 1984), may be written in the form

$$T = \left(\prod_{i=1,..,n} (1 + xU_{2i-1}) \right) \left(\prod_{i=1,..,n-1} (x + U_{2i}) \right) . \tag{6.1}$$

Here the matrices $\{U_i \ : \ i = 1, ..., 2n-1\}$ give a representation for the generators of a Temperley-Lieb algebra $T_k(q)$, where $k = 2n - 1$ and

$$q = \frac{\sqrt{Q} \pm \sqrt{Q-4}}{2}$$

i.e.

$$q + q^{-1} = \sqrt{Q}$$

(Temperley and Lieb 1971). This algebra is abstractly defined by the k generators and relations

$$U_i U_i = \sqrt{Q} U_i \tag{6.2}$$

$$U_i U_{i\pm 1} U_i = U_i \tag{6.3}$$

$$U_i U_{i+j} = U_{i+j} U_i \qquad (j \neq 1) \tag{6.4}$$

which generalise automatically to even k. In most statistical mechanical contexts the variable q can be an arbitrary scalar parameter (the Potts model case is an exception here, since it only makes sense for positive integer Q). This parameter is fixed for a given model. On the other hand the variable x in equation 6.1 is a function of the temperature (see chapter 2). The precise representation of the algebra involved in the transfer matrix depends again on the model, *and* on the boundary conditions.

For various reasons interest in Temperley-Lieb algebras is extremely widespread. In addition to their relevance for statistical mechanics and conformal field theory they have an intimate connection with knots and braids (see below). These are now the focus of physicists attention in their own right, following recent developments in quantum field theory and string theory. For this reason the oportunity will be taken here to provide the tools for a detailed analysis of any Temperley-Lieb algebra.

The regular representation has as a basis any basis of elements of the algebra. Specifically we may use the set of all words in the generators $\{U_i\}$

which are distinct under the relations. We will show how to take linear combinations of these operators in order to exhibit the structure of the algebra (ie. to construct primitive idempotents and so on).

6.1.1 Technical notes

Up to overall factors of $\sqrt{Q}$ the relations 6.2 to 6.4 take words in the generators to other words. The *length* of a word is the number of generators in the word.

We will call a word in the generators a *reduced* word if its length cannot be shortened by applying the relations (ignoring factors of $\sqrt{Q}$). The only ambiguity in the expression of a reduced word then comes from the possible juxtapositions of commuting generators. With this in mind it is useful to introduce the notion of *stacking*. A word including a sequence of mutually commuting generators may sometimes be written with these generators stacked:

$$\begin{matrix} & U_1 & \\ U_2 & & U_2 \\ & U_3 & \end{matrix} = U_2U_3U_1U_2 = U_2U_1U_3U_2$$

As indicated, this means that the relations may be used to write the generators in the stacked part in any order.

The dimension of the algebra, or equivalently the number of words in the generators (including the unit) which are linearly independent under the relations, is clearly independent of q.

The alternative set of generators

$$t_i^{\pm 1} = 1 - q^{\pm 1}U_i \tag{6.5}$$

obey the braid relations (noted in Temperley 1986),

$$t_it_{i\pm 1}t_i = t_{i\pm 1}t_it_{i\pm 1} \tag{6.6}$$

plus

$$t_i^2 = (1 - q^2)t_i + q^2 \tag{6.7}$$

and

$$\{t_i, t_{i+1}\} = t_i + t_{i+1} + t_it_{i+1}t_i - 1. \tag{6.8}$$

When $q = 1$ the relations 6.6 and 6.7 define the group algebra for the symmetric group on $k+1$ objects, S_{k+1}, and 6.8 the quotient corresponding to two row Young tableau.

The algebra $T_k(q)$ has an involution

$$R : U_i \mapsto MU_iM^{-1} = U_{k-i+1}$$

corresponding to the reflection symmetry on the lattice. We desribe the operator M in chapters 3 and 13.

Let us define $U_{k+1} = U_0$ as a certain special element of $T_k(q)$ which obeys the relations as if the indices are defined mod $2n$, but which is not a new generator. Then the algebra has further automorphisms

$$t : U_i \mapsto GU_iG^{-1} = U_{i+1} \tag{6.9}$$

(corresponding to duality/translation on the lattice) where a convenient, but not unique, realisation for G is given by

$$G = \prod_{i=1}^{k} t_i. \tag{6.10}$$

In particular, $U_0 = GU_kG^{-1}$, and $U_1 = GU_0G^{-1}$.

Note that any $T_k(q)$ may, therefore, be alternatively generated by a *pair* of generators $\{U_1, G\}$. We will return to these points later.

We will use the notation W^T for the element of the algebra obtained by writing the generators in element W in reverse order. Note that equation 6.8 is invariant under the reversal $W \rightarrow W^T$, as are the defining relations.

Note that $T_k(q) \subset T_m(q)$ if and only if $k < m$. The subalgebra can be realised in various ways, the most obvious being identification of the first k generators. We will occasionally discuss constructions involving some set of generators in an abstract sense, without specifying k. Such constructions apply to any algebra containing sufficient generators.

6.2 Preliminary remarks

The transfer matrix in equation 6.1 provides for the construction of a lattice model partition vector on a plane rectangular or, equivalently, a cut cylindrical lattice. We can close the cylinder by incorporating the factor $(x + U_0)$ into equation 6.1. However, in order to consider 2 dimensional lattices on surfaces of the more general type described in section 1.2.2 we need to be able to work with lattice 'pants' (a discussion of *higher dimensional* lattices will be given in chapter 8). The lattice in the pants is now to be seen as a graph with many 'internal' nodes and 3 *sets* of 'external' nodes called *boundaries*.

Perhaps the simplest (and the historically correct) place to begin our discussion is the Whitney polynomial representation of the Potts model, described in section 1.5. In this case the most important peices of information for extending the partition vector lattice from a given pants are the connectivities of the nodes, a) within each of the three sets of external

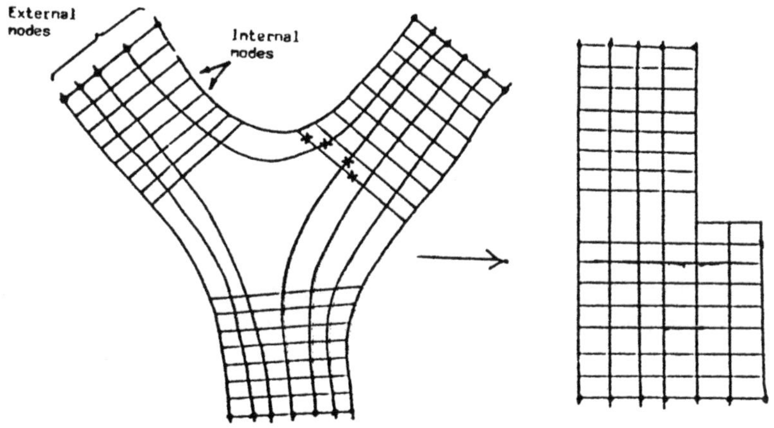

Figure 6.1: Schematic tailoring of lattice pants.

nodes (possibly via internal nodes); and b) between the sets. We will see shortly that this list of connectivities form a basis for representations of distinct Temperley-Lieb algebras (Martin 1986b) simultaneously at each of the three boundaries - a tri-module. Composing boundaries (i.e. connecting nodes in pairs) between 2 pants builds basis states for more complicated modules. By ignoring (or connecting all of the nodes in) one of the boundaries the pants can be used to construct assymetric bimodules - relating pairs of TL algebras with different numbers of generators (see figure 6.1). This is one of the striking ways in which physics gives us a handle on these towers of algebras. Since they describe similar systems (differing only in size) it forces on them a well structured heirarchy.

We will return to the details of the Whitney representation shortly. One idea which is useful to abstract from it first is that of a diagrammatic representation of the algebra for the plane (or cut cylinder).

6.2.1 Two faithful diagrammatic representations

1) Whitney diagrams

Consider the infinite set T_k of square lattices of width (i.e. horizontal length) $[k/2]$ bonds, and arbitrary finite height (vertical length), where the last vertical line of bonds are *marked* if k is even. Associate to each such lattice $L \in T_k$ the set L' of unmarked bonds of L. Then the set $P(L')$ is the set of 'bond coverings', i.e. the power set of L' (the set of all possible subsets of the set of unmarked bonds of L). A set $P(L)$ is obtained from $P(L')$ by adding to each element the set of marked bonds of L. The set T'_k is the union of the sets $P(L)$ over T_k

$$T'_k = \bigcup_{L \in T_k} P(L).$$

Regarding L as a graph, we can form a subgraph of L from $l \in P(L)$, called a Whitney diagram, by drawing only the bonds in this set. We will also call this subgraph l. In this case we say the lattice L *supports* the subgraph l.

Two nodes or sites of the lattice L are *connected* in the subgraph obtained from any $l \in P(L)$ if there is a path from one to the other on bonds of the subgraph. An isolated connected cluster is a connected cluster of nodes not connected to any point on the top or bottom layer of L.

We define a multiplication

$$T'_k \times T'_k \to T'_k$$

by

$$l_1 l_2 \mapsto l_3 \tag{6.11}$$

where l_3 is obtained by adjoining the top layer of the lattice supporting l_2 to the bottom of that supporting l_1 (a bond is present in l_3 in the merged layer at the line of contact if one is present in either l_1 or l_2).

Now form the set R'_k of cosets of T'_k, where a coset consists of all bond coverings which give the same list of connections between nodes i) within the top layer of the lattice, ii) between nodes within the bottom layer of the lattice, and iii) between the top and bottom layers.

Define the order of a subgraph l of $L \in T_k$ as the number of horizontal bonds present plus the number of vertical bonds absent in l when compared to the k full lattice layers present in L. We will (not necessarily uniquely) label each coset by some element of the coset with the property that no element has lower order, and, given this, that no element has fewer lattice layers.

We define an algebra $J_k(q)$ from the set of cosets by defining a multiplication of cosets $C_1, C_2 \in R'_k$ as follows. Take a labeling element of each coset, $c_1, c_2 \in T'_k$, say, and form a new bond covered lattice from these by identifying the bottom layer of c_1 with the top layer of c_2, where any bonds in either of these layers are to be included. This layer is then an internal layer, whose connections can be forgotten (except in so far as they affect the new type (iii) connections), and we have an element c_3 of some new coset C_3. Multiplication is then defined by

$$C_1 C_2 \mapsto Q^n C_3$$

where n is the number of isolated connected clusters in c_3.

For $0 < i < [k/2]$ let us introduce generators $(i\, i+1) \in J_k(q)$. Such a generator is defined by the coset including the subgraph of the lattice of height 0 with only the bond between nodes i and $i+1$ present:

$(i\, i+1) =$ 

For $0 < i \leq [(k-1)/2]$ define generators $(i.) \in J_k(q)$ by the coset including the subgraph of the lattice of height 1 with no horizontal bonds present and only the vertical bond in position i absent:

$(i.) =$ 

All cosets may be generated by these objects. To see this note that the whole of T'_k is generated by them using the multiplication 6.11 together with *occasional* projections to the coset representative, since every row can be built up with any bond present or absent.

A diagrammatic representation of words in the Temperley-Lieb algebra $T_k(q)$ is obtained by the maping

$$\alpha : T_k(q) \rightarrow J_k(q) \tag{6.12}$$

defined by

$$\alpha : U_{2i}/\sqrt{Q} \mapsto (i\, i+1)$$

and

$$\alpha : \sqrt{Q}\, U_{2i-1} \mapsto (i.).$$

It is easy to check that this is an algebra morphism, for example

$$(i.)(i\, i+1)(i.) = (i.)$$

amounts to the statement that the subgraph

has the same list of connections within and between the top and bottom layers as (i.). It follows that α is surjective. We will see in section 6.3 that α is injective and hence an isomorphism. The labelling element of each coset corresponds to a reduced word.

Note that no multiplication C_1C_2 increases the number of *distinct* connected clusters connecting between the top and bottom layers. This means that the subset of cosets with no connections top to bottom (or just the one forced by k even) form a basis for a $J_k(q)$ bimodule. Furthermore, the subset with $\leq p$ distinct connections top to bottom also form a basis for a $J_k(q)$ bimodule.

These powerful properties are extensively generalised to algebras associated with higher dimensional statistical mechanical models in chapter 8.

2) Boundary diagrams

The following construction is similar, but does not distinguish so strongly between odd and even generators. This advantage is offset by the greater versatility of the Whitney picture when it comes to treating more general algebras. The following construction is used, however, in the study of cabling subalgebras (chapter 13).

The procedure here is to represent the Whitney diagrams by the *boundaries* of connected regions. We can then discard the original lattice and work in the abstract, as we will now describe.

Consider the set D_{k+1} of diagrams consisting of 2 parallel horizontal rows of $k+1$ nodes with $k+1$ non-intersecting strings lying in the plane between rows and connecting the nodes in pairs.

The coset $c(d)$ of a diagram d is the subset of all diagrams in the set obtained by continuous deformations of d which hold nodes fixed.

Let us form an algebra $A_k(q)$ from the set of cosets by defining multiplication of cosets. The multiplication of $c(d_1)$ and $c(d_2)$ first requires the forming of a new element d' of D_{k+1} by joining the strings at the bottom nodes of d_1 with those at the top nodes of d_2, and then discarding the internalised nodes and any closed loops of string. We then form the coset $c(d')$. The multiplication in the vector space V with basis the cosets is given by

$$c(d_1)c(d_2) \mapsto Q^{n/2}c(d') \tag{6.13}$$

where n is the number of closed loops discarded.

There is an isomorphism between $A_k(q)$ and $J_k(q)$. To see this number the nodes in each row from 1 to $k+1$ and consider Whitney diagram

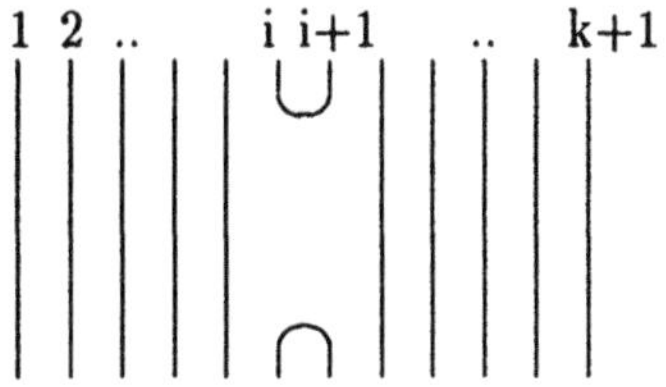

Figure 6.2: Boundary diagram for U_i in $T_k(q)$.

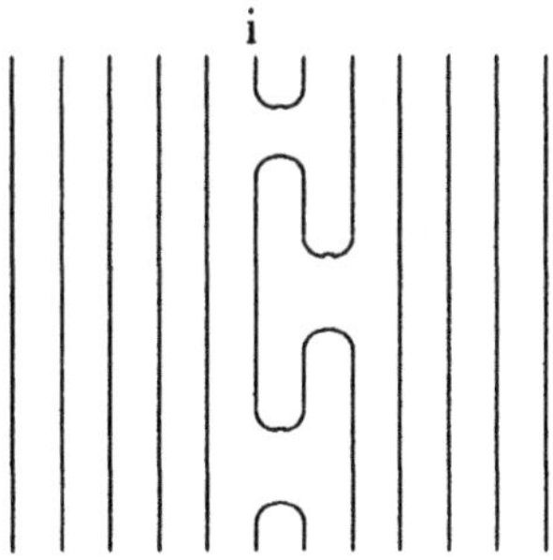

Figure 6.3: Boundary diagram exhibiting the relation $U_i U_{i+1} U_i = U_i$.

lattice sites residing in the rows *between* each pair of adjacent nodes with numbering $(odd, even)$. Then every coset of boundary diagrams implies a distinct list of connections between such objects, and hence a distinct coset of Whitney diagrams, and visa versa.

A boundary diagrammatic representation of words in $T_k(q)$ is obtained as follows. The generator U_i in $T_k(q)$ maps to the coset of the diagram shown in figure 6.2. Generators are then composed by connecting the bottom of the diagram for the first factor with the top of the diagram for the second, so

$$U_i U_{i+1} U_i = U_i$$

is given by figure 6.3.

In such diagrams we simply pull the line starting in the $i+2$ position straight to exhibit the identity. To give a quick procedure for comparing coset representative diagrams we note that every diagram can be deformed so as to consist of lines with at most one stationary point.

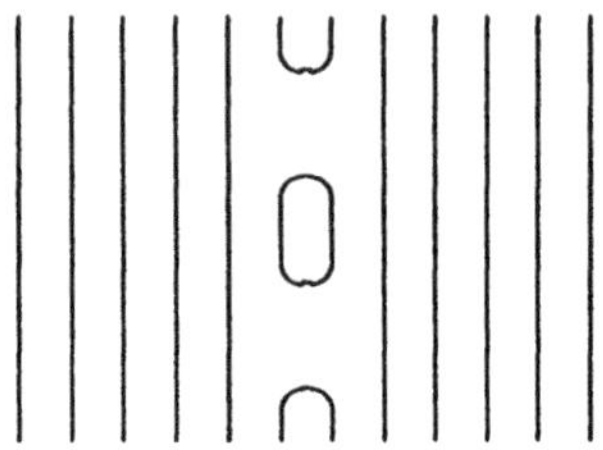

Figure 6.4: Diagram for U_iU_i before removing the closed loop.

Similarly $U_iU_i = \sqrt{Q}U_i$ is represented in figure 6.4. We can see from equation 6.13 that the map is an algebra morphism, and we have from the map 6.12 that it is surjective. We will see in section 6.3 that it is an isomorphism, by a counting argument. Since we interpret closed loops as contributing a factor $\sqrt{Q}$ these diagrams can be used to match any word to a reduced word, with an appropriate scalar factor.

6.3 Generic structure of $T_k(q)$

Preview

The main result of the next two sections is to be found on page 155. For n a positive integer, r a complex number *excluding* the rationals,

$$q = e^{i\pi/r}$$

and

$$Q = 4\cos^2(\pi/r) \tag{6.14}$$

this is a theorem giving the structure of the algebra $T_n(q)$ defined by the generators $\{U_i : i = 1, 2, .., n\}$ and relations

$$\begin{aligned} U_iU_i &= \sqrt{Q}U_i && (6.15) \\ U_iU_{i\pm1}U_i &= U_i && (6.16) \\ U_iU_j &= U_jU_i \qquad i-j \neq \pm 1 && (6.17) \end{aligned}$$

The proof is straightforward, but requires careful organisation. To this end we will adopt a more than usually formal style in its presentation.

We will also need to do some preparatory work:

6.3.1 Combinatorial identities

Definition 1 *For* $n = 0, 1, 2, 3, \ldots$ *define the Catalan numbers* C_n *by:*

$$C_n = \binom{2n}{n} - \binom{2n}{n-1} = \frac{(2n)!}{n!(n+1)!}.$$

For example, the first few are:

$n:$	0	1	2	3	4	5	6
$C_n:$	1	1	2	5	14	42	132

Definition 2 *For* $n, m \in Z_+$, $m+n$ *odd,* $m < n+2$ *and* $p = [(n-m+1)/2]$ *define integers* $C_{n,m}$ *by*

$$C_{n,m} = \binom{n-1}{p} - \binom{n-1}{p-2}.$$

For example, the first few are:

$C_{n,m}$	$m=1$	2	3	4	5	6	7	8	9	10
$n=1$		1								
2	1		1							
3		2		1						
4	2		3		1					
5		5		4		1				
6	5		9		5		1			
7		14		14		6		1		
8	14		28		20		7		1	
9		42		48		27		8		1

The following useful identities arise from the definitions:

$$C_{n,m} = C_{n-1,m-1} + C_{n-1,m+1} \tag{6.18}$$

$$C_r = C_{2r,1} = C_{2r-1,2}. \tag{6.19}$$

Definition 3 *For* n, l, m, r *integers,* $n, l+1, r-1 > 0$, *and* $0 \leq m < r$ *define integers* $C_{n,lr+m,r}$ *as follows.*

If $lr + m > n + 1$

$$C_{n,lr+m,r} = 0$$

otherwise

$$C_{n,lr,r} = C_{n,lr}$$

and for $m \neq 0$:

$$C_{n,lr-m,r} = C_{n,lr-m} - C_{n,lr+m,r}$$

or equivalently

$$C_{n,(l-2)r+m,r} = C_{n,(l-2)r+m} - C_{n,lr-m,r}$$

Alternatively for $m \neq 0$:

For a given n and r, define L by

$$L = \frac{n+1-(n+1)_{mod\ r}}{r}$$

Then for j, i integers define $f_m(j)$ by

$$f_m(2i+1) = (L-2i)r + m$$

and

$$f_m(2i+2) = (L-2i)r - m.$$

We then have

$$C_{n,f_m(2i+1),r} = \sum_{j=0}^{i} C_{n,f_m(2i-2j+1)} - \sum_{j=0}^{i-1} C_{n,f_m(2i-2j)}$$

and

$$C_{n,f_m(2i+2),r} = \sum_{j=0}^{i} C_{n,f_m(2i-2j+2)} - \sum_{j=0}^{i} C_{n,f_m(2i-2j+1)}$$

Definition 4 *Define* $C'_{n,f_m(j),r}$ *by*

$$C'_{n,f_m(j),r} = C_{n,f_m(j+1),r}$$

so that

$$C'_{n,f_m(j),r} + C_{n,f_m(j),r} = C_{n,f_m(j)}$$

6.3.2 Sequence notation

Definition 5 *Define a sequence* $\{s\}$ *to be any ordered set of positive integers* s_i *indexed by* $i = 0, 1, 2, 3, \ldots$

$$\{s\} = s_0 s_1 s_2 \,..\, s_i \,..$$

with the properties

$$s_0 = 1$$

and

$$s_i - s_{i+1} = \pm 1. \tag{6.20}$$

The length *of* $\{s\}$ *is the number of integers in the set.*

Note that $s_1 = 2$ is forced.

A given sequence $\{s\}$ is determined by the ordered list of its entries. An example of such a sequence, of length 20, is

$$\{s\} = 12121212323456567654. \qquad (6.21)$$

Definition 6 *A subsequence of $\{s\}$ of length n is defined to be the ordered set of positive integers obtained by taking any n consecutive elements of $\{s\}$.*

Note that a subsequence obeys equation 6.20 but the first element will not be $s_0 = 1$ in general.

Definition 7 *If $\{s\}$ has a subsequence of the form*

$$g \; \overbrace{g \pm 1}^{s_i} \; g$$

then s_i is said to be a maximum (resp. minimum) of $\{s\}$.

Definition 8 *If s_i is a minimum of $\{s\}$ then the sequence $\{s^i\}$ is defined by*

$$\begin{aligned} s^i_j &= s_j \qquad (j \neq i) \\ s^i_i &= s_i + 2. \end{aligned}$$

For $j, k \in Z_+$ suppose $\{s\}$ has length $l > k + j$ and

$$s_k - s_{k+j} = \pm j.$$

Then the subsequence from s_k to s_{k+j} is uniquely determined by equation 6.20 (for $k < i < k + j$ no s_i is a maximum or minimum). It follows that $\{s\}$ is also uniquely determined by the ordered list of its maxima and minima, encapsulated by its end points.

Definition 9 *For a, b positive integers let ab appearing within a sequence denote the subsequence of length $|b - a + 1|$ defined by first element a and last element b.*

For example the sequence above may be written

$$\{s\} = 1212121326574.$$

Definition 10 *For a, b, j positive integers let $(ab)^j$ denote the subsequence defined in the following way:*

$$(ab)^1 = ab$$

and otherwise

$$(ab)^j = ab\,(ab)^{j-1}$$

For example, the sequence above may be written

$$\{s\} = (12)^3 1326574.$$

Definition 11 *For $n \in Z_+$ define $S(n)$ as the set of all sequences $\{s\}$ having length $n+1$, i.e.*

$$\{s\} = s_0 s_1 s_2 \,..\, s_i \,..\, s_n.$$

Definition 12 *For $n, m \in Z_+$ define $S(n,m)$ as the subset of $S(n)$ consisting of sequences with final entry $s_n = m$.*

The range of values of m for which $S(n,m)$ is not empty is

$$m = n+1, n-1, n-3, ...$$

Proposition 3 *If $S(n,m)$ is not empty then the order of the set is*

$$|S(n,m)| = C_{n,m}.$$

Proof:
By condition 6.20 we have

$$|S(n,m)| = |S(n-1,m-1)| + |S(n-1,m+1)|,$$

c.f. equation 6.18, and $S(1,2) = 1$.

Definition 13 *Define $P(n,m)$ as the set of pairs of sequences $(\{s\},\{t\})$ for all $\{s\},\{t\} \in S(n,m)$.*

Definition 14 *Define $S_{\{t\}}(n,m)$ as the subset of $P(n,m)$ obtained by taking all elements of $P(n,m)$ with the same second sequence, $\{t\}$.*

Definition 15 *Define $P(n)$ as the disjoint union of sets $P(n,m)$ over all values of m. That is the set of pairs of sequences of $n+1$ positive integers*

$$(\{s\},\{t\}) = (s_0 s_1 s_2 .. s_n, t_0 t_1 t_2 .. t_n)$$

with

$$s_0 = t_0 = 1$$

$$s_n = t_n$$

$$s_i - s_{i+1} = \pm 1$$

and

$$t_i - t_{i+1} = \pm 1.$$

Proposition 4 *The order of the set $P(n)$ is*

$$|P(n)| = C_n.$$

Proof:
Note that

$$|P(n)| = \sum_m |S(n,m)|^2.$$

By a reflection symmetry we see that the subset of $S(2n,1)$ with the property that $S_n = m$ has order $|S(n,m)|^2$. Therefore by equation 6.19 we have

$$\sum_m |S(n,m)|^2 = |S(2n,1)| = C_n.$$

Definition 16 *Define a partial order on sequences $\{s\}$ in $S(n,m)$ by*

$$\{s\} \leq \{t\}$$

if and only if

$$s_i \leq t_i$$

for all i.

Equivalently

$$\{s\} \leq \{t\}$$

if and only if $\{t\}$ can be obtained from $\{s\}$ by a sequence of moves of the form

$$\{s\} \rightarrow \{s^j\} \rightarrow \ldots \rightarrow \{v\} \rightarrow \{v^i\} \rightarrow \ldots \rightarrow \{t\}$$

Note that the definition (but not the equivalent form) may be extended to define a partial order on sequences $\{s\}$ in $S(n)$.

Proposition 5 *The poset is a lattice.*

Outline proof:
The meet of $\{s\}$ and $\{t\}$

$$\{v\} = \{s\} \wedge \{t\}$$

is given by

$$v_i = \textit{lesser of } s_i, t_i$$

for all i.

The universal bounds are as follows. Their is a unique 'lowest' of the sequences $\{s\}$ in $S(n,m)$, for which $\{s\} \leq \{t\}$ for all $\{t\}$ in $S(n,m)$. We will denote this sequence by $\{e_m\}$, so that with $p = \frac{n-m+1}{2}$

$$\{e_m\} = (12)^p \; 1 \; m,$$

or simply by $\{e\}$ in cases where no ambiguity arises. With $d = \frac{n+m+1}{2}$ the unique highest sequence is given by

$$\{f\} = 1 \; d \; m \, .$$

Definition 17 *For* $(\{s\},\{t\}),(\{v\},\{w\}) \in P(n,m)$ *we define a partial order by*

$$(\{s\},\{t\}) < (\{v\},\{w\})$$

if and only if either

$$\{t\} < \{w\}$$

or

$$\{s\} < \{v\}$$

and

$$\{t\} = \{w\}.$$

It follows from Proposition 5 that the poset is a lattice with universal bounds $(\{e\},\{e\})$ and $(\{f\},\{f\})$.

It will be useful to be able to emphasise the entries in, and location of, the particular 3 entry subsequence of a sequence $\{s\}$ centered on the element s_i. If

$$s_{i-1} \; s_i \; s_{i+1} = a \; b \; c \; ,$$

say, then we write this subsequence as

$$(abc)_i \quad .$$

In equation 6.21 we have, for instance,

$$\{s\} = \ldots \; (123)_7 \; \ldots \quad .$$

Then again, suppose $\{s\}$ is a sequence with a subsequence of the form

$$\ldots \; (g \; g-1 \; g)_i \; \ldots$$

then $\{s^i\}$ is a sequence differing from $\{s\}$ only in having the subsequence

$$\ldots \; (g \; g+1 \; g)_i \; \ldots$$

6.3.3 A primitive central idempotent in $T_n(q)$

Definition 18 *For n a natural number define a function $P_n(x)$ by*

$$P_n(2\cos(\pi/r)) = \frac{\sin(n\pi/r)}{\sin(\pi/r)}$$

Note that

$$P_0(x) = 0$$

$$P_1(x) = 1$$

and

$$P_{n+1}(x) = xP_n(x) - P_{n-1}(x)$$

For example, the first few are

n	$P_n(x)$
0	0
1	1
2	x
3	$x^2 - 1$
4	$x(x^2 - 1)$
5	$x^4 - 3x^2 + 1$

Hereafter it is to be understood that P_n is a function of $x = 2\cos(\pi/r) = \sqrt{Q}$ as defined in equation 6.14.

Definition 19 *Define a function k_n by*

$$k_n = P_{n-1}/P_n$$

i.e.

$$k_n(2\cos(\pi/r)) = \frac{\sin((n-1)\pi/r)}{\sin(n\pi/r)}$$

Note that

$$1/k_{n+1} = \sqrt{Q} - k_n$$

A sequence of idempotents in $T_n(q)$

Definition 20 *For $m = 1, 2, 3, .., n+2$ define $E_m \in T_n(q)$ by*

$$E_1 = E_2 = 1$$

and then

$$E_m \in T_{m-2}(q) \subset T_n(q)$$

and

$$E_m E_m = E_m$$

and for $i = 1, 2, .., m-2$

$$E_m U_i = U_i E_m = 0.$$

Let us consider the existence and uniqueness of such an element.

Definition 21 *Define $I[m-2] \in T_{m-2}(q)$ by*

$$I[0] = 1$$

and

$$I[m-2] = I[m-3](1 - k_{m-1}U_{m-2})I[m-3]$$

The existence of $I[m-2]$ for a given value of r is guaranteed unless some k_n required in its construction has a pole at that point.

Proposition 6 *If $I[m-2]$ exists then*

$$E_m = I[m-2]$$

Proof:

From the definition E_m is a primitive central idempotent in $T_{m-2}(q)$ containing the unit. Any two such objects must be identical. On the other hand assume that for some non-negative integer b we have

$$E_{b+2} = I[b]$$

and

$$(I[b]U_{b+1})^2 = k_{b+2}^{-1}\, I[b]U_{b+1}$$

(both are clearly true for $b = 0$). Then

$$\begin{aligned} I[b+1]I[b+1] &= (I[b] - k_{b+2}I[b]U_{b+1}I[b])^2 \\ &= I[b] - 2k_{b+2}I[b]U_{b+1}I[b] + k_{b+2}^2(I[b]U_{b+1})^2 I[b] \\ &= I[b+1] \end{aligned} \quad (6.22)$$

and

$$U_i I[b+1] = U_i I[b] I[b+1] = 0$$

for $(1 \leq i \leq b)$ and

$$U_{b+1} I[b+1] = U_{b+1} I[b] - k_{b+2} (U_{b+1} I[b])^2 = 0. \tag{6.23}$$

Furthermore, we have

$$\begin{aligned}(I[b+1]U_{b+2})^2 &= I[b+1]U_{b+2}(I[b] - k_{b+2} I[b] U_{b+1} I[b]) U_{b+2} \\ &= I[b+1]U_{b+2}(\sqrt{Q} - k_{b+2}) \\ &= k_{b+3}^{-1} I[b+1] U_{b+2}. \end{aligned} \tag{6.24}$$

Note that we have used the symmetry, by construction, of $I[b]$ under reversal of operator order

$$I[b] = I[b]^T.$$

This completes the proof of proposition 6.

6.3.4 Translation/reflection notation

Let $m, t \in Z_+$ and $m + t \leq n$ then the subalgebra of $T_n(q)$ generated by $\{U_i \;:\; i = 1, 2, .., m\}$ is $T_m(q)$. Define $T_m^{(t)}(q)$ as the subalgebra generated by $\{U_i \;:\; i = 1+t, 2+t, .., m+t\}$. We have

$$T_m(q) \sim T_m^{(t)}(q).$$

The isomorphism is given by the following bijection between the generators in the 2 subalgebras

$$U_i \xrightarrow{t} U_i^{(t)} = U_{i+t} \qquad (i = 1, 2, .., m) \tag{6.25}$$

Definition 22 *For $Y \in T_m(q)$ define $Y^{(t)} \in T_m^{(t)}(q)$ as the element obtained from Y by applying the map t in equation 6.25 for each U_i in Y.*

There is also an automorphism of $T_n^{(t)}(q)$ realised by

$$U_{i+t} \xrightarrow{(-)} U_{n+1+t-i} \qquad (i = 1, 2, .., n) \tag{6.26}$$

Definition 23 *Suppose $Y \in T_n^{(t)}(q)$ but $Y \notin T_n^{(t+1)}(q)$ and $Y \notin T_{n-1}^{(t)}(q)$. Then define $Y^{(-)} \in T_n^{(t)}(q)$ as the element obtained from Y using equation 6.26 for each U_i in Y.*

Examples

The operator $E_m^{(t)}$ may be obtained from the recursive formula for E_m by translation. The first few are:

$$E_3^{(t-1)} = 1 - Q^{-1/2}U_t \tag{6.27}$$

$$E_4^{(t-1)} = 1 + (U_tU_{t+1} + U_{t+1}U_t - \sqrt{Q}(U_t + U_{t+1}))/(Q-1) \tag{6.28}$$

$$\begin{aligned} E_5^{(0)} = 1 + (& - \sqrt{Q}((Q-1)(U_1+U_3) + QU_2) \\ & + Q(U_1U_2 + U_2U_1 + U_2U_3 + U_3U_2) \\ & - \sqrt{Q}(U_3(U_1U_2 + U_2U_1) + (U_1U_2 + U_2U_1)U_3) \\ & - (Q+1)U_1U_3 \\ & + \left.\frac{(1-\sqrt{Q}U_2)}{\sqrt{Q-1}}U_1U_3\frac{(1-\sqrt{Q}U_2)}{\sqrt{Q-1}}\right) /(Q(Q-2)) \end{aligned} \tag{6.29}$$

and so on. Note the symmetry under $U_i \to U_{b+1-i}$ in E_{b+2}, which is a general consequence of its uniqueness.

These examples illustrate the point that the construction procedure fails for some Q values, $Q = Q_c$, say, where the formal idempotent has terms whose coefficients have a pole. In these cases the appropriate idempotent to provide a basis for the trivial representation is obtained by simply subtracting the potentially divergent part before taking $Q \to Q_c$.

To see this, suppose without loss of generality that

$$E_m = (1/(Q-Q_c)^d)X(Q) + Y(Q)$$

where $d \in Z_+$ and X, Y are elements of the algebra which are finite at $Q = Q_c$. Then $E_mE_m = E_m$ (the formal proof of which was independent of Q) implies that

$$X^2 \propto (Q-Q_c)^d$$

and $U_iE_m = 0$ implies that

$$U_iX \propto (Q-Q_c)^d.$$

Thus (at $Q = Q_c$) X itself is a nilpotent double sided ideal, and hence in the radical, of the $T_{m-2}(q_c)$ algebra. Subtracting then just corresponds to quotienting by (part of) the radical. We will return to this point in the next chapter.

6.4 Bases for $T_n(q)$

Definition 24 *If $\{s\}, \{t\}$ are the sequences in a pair in $P(n+1)$ then (s,t) is an element of $T_n(q)$ obtained iteratively as follows:*

Firstly for m a positive integer, $m+n$ odd, $m < n+2$ and $p = (n-m+1)/2$ define

$$(e_m, e_m) = ((12)^p 1m, (12)^p 1m) = \left(\prod_{i=1}^{p} U_{2i-1}\right) E_m^{(2p)} . \qquad (6.30)$$

A general (s,t) is then obtained by iterative use of the following general identities:

Suppose $s_i = g-1$ a minimum of $\{s\}$, then

$$(s^i\ ,\ t) = \sqrt{k_g k_{g+1}}\,((1 - U_i/k_g))\ (s\ ,\ t) \qquad (6.31)$$

and in general

$$(t,s) = (s,t)^T . \qquad (6.32)$$

It follows from the equivalent statement of Definition 16 that every pair in $P(n+1)$ is covered by this construction. The order in which the identities may be applied in moving from the initially defined operator to (s,t) is not unique. Note from relation 6.17, however, that the definition of (s,t) is unique, i.e. independent of the choice of order.

Note the following useful identity

Proposition 7 *With $\{s\}, \{s^i\}$ defined as in Definition 24 we have*

$$U_i\ (s^i,t) = \sqrt{k_g k_{g+1}}\left(1 - \frac{\sqrt{Q}}{k_g}\right) U_i(s,t) = \frac{1}{k_{g+1}}(s^i,t) - \sqrt{\frac{k_g}{k_{g+1}}}\ (s,t)$$

by direct calculation, whereupon

$$(s,t) = \sqrt{\frac{k_{g+1}}{k_g}}\left(\frac{1}{k_{g+1}} - U_i\right)(s^i,t) = \sqrt{\frac{1}{k_{g+1}k_g}}\,(1 - k_{g+1}U_i)\,(s^i,t) \qquad (6.33)$$

We also have

Proposition 8

$$U_i(...(g\ g\pm 1\ g\pm 2)_i\ ...,\ t) = 0$$

Proof: by induction:

Assume the proposition is true for all occurences of the given subsequences in sequences below a certain sequence $\{w\}$ in the partial order (it is true for all occurences in $\{e\}$ by construction). By the defining relations it is sufficient to prove the proposition in the case

$$\{w\} = \; ..., g, (g+1, g, g-1)_i, ... ,$$

i.e. to prove that here $U_i(w,t) = 0$.

Let $\{x\} = \; ..., g, g-1, g, g-1, ...$ and $\{y\} = \; ..., g, g-1, g-2, g-1, ...$ (where all the unspecified steps are taken to be the same). Since $\{y\}$ is below $\{x\}$ is below $\{w\}$ we have that $U_{i-1}(y,t) = 0$ by assumption. Then

$$\begin{aligned} U_i(w,t) &= U_i\sqrt{k_g k_{g+1}}(1 - U_{i-1}/k_g)\sqrt{k_{g-1}k_g}(1 - U_i/k_{g-1})(y,t) \\ &= U_i k_g \sqrt{k_{g-1}k_{g+1}}(1 - U_i/k_{g-1} + U_{i-1}U_i/(k_g k_{g-1}))(y,t) \\ &= k_g\sqrt{k_{g-1}k_{g+1}}(1 - Q^{1/2}/k_{g-1} + 1/(k_g k_{g-1}))U_i(y,t) \\ &= 0. \end{aligned} \quad (6.34)$$

This completes the proof of proposition 8.

Remark 1 *The proof of this identity requires of the E_m appearing in Definition 24 only that it has the property $U_iE_m = 0$ for $i = 1, 2, .., m-2$.*

Corollary 8.1 *If $\{t\} \in S(n,m)$ then the left ideal generated by (e,t) has a basis given by $\{(v,t)$ for all $\{v\} \in S(n,m)\}$.*

Proof: The action of U_i on (v,t) with $\{v\}$ any sequence is covered by propositions 7 and 8.

Walk diagrams

It is quite useful to have a diagrammatic version of these basis states. We achieve this by depicting the elements of each sequence as heights, as in figure 6.5. The figure exemplifies the diagram for the lowest sequence

$$\{e\} = (12)^4 15.$$

Each element of $T_k(q)$ in definition 24 then corresponds to a pair of walks.

The complete set $S(8,1)$, providing a basis for a representation of $T_7(q)$, is given in figure 6.6. We have indicated by construction how each walk is obtained from its predecessors in the partial order discussed in the previous section. In general such bases come from the configuration spaces for lattice layers in the Andrews Baxter Forrester (1984) model.

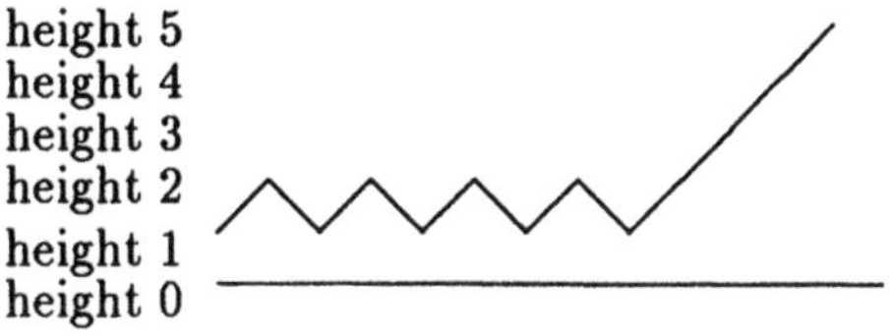

Figure 6.5: The sequence 1212121212345 $\in S(12,5)$.

For any 2 sequences $\{u\}$ and $\{t\}$ define

$$\delta_{ut} = \begin{cases} 1 & u = t \\ 0 & u \neq t \end{cases}$$

where $u = t$ means $u_i = t_i$ for all $i = 1, 2, .., n$.

Note from Definition 24 that the operators (s,t) are well defined for all irrational r. We will call the Temperley-Lieb algebras with this set of r values the generic algebras. Note that *some* such operators are also defined for non-generic, i.e. rational, r.

Theorem 1 (Generic theorem) *(i) If the operators (u,s) and (t,v) are well defined then:*

$$(u,s)(t,v) = \delta_{st}(u,v) \tag{6.35}$$

(ii) When all defined, the set of operators (t,v) for all $(\{t\},\{v\}) \in P(n+1)$ are a basis for, and as elementary operators exhibit the entire (multi-matrix) structure of, the algebra $T_n(q)$.

Proof
part(i): by induction.

Introduce a partial order on the set of pairs of walks $(\{s\},\{t\})$ in equation 6.35 (i.e. not necessarily in $P(n+1)$) by $(\{s\},\{t\}) < (\{u\},\{v\})$ *if* $\{t\} < \{v\}$ on extending the original partial order to any pair of walks by inluding s_{n+1}; and $(\{s\},\{t\}) < (\{u\},\{t\})$ *if* $\{s\} < \{u\}$.

Assume that the identity is proved for all pairs of walks below $(\{s\},\{t\})$, and all $\{u\},\{v\}$. It is trivially true if $\{s\}$ and $\{t\}$ are the lowest walks in their respective partial orders. We stress that the assumption implies that if $\{w\}$ is below $\{t\}$ then the identity holds for pairs $(\{x\},\{w\})$ for *all* $\{x\}$.

Consider $\{w\}$ to be a walk immediately below $\{t\}$, i.e. $\{t\} = \{w^h\}$ for some h, then by definition 24

$$(u,s)(t,v) = (u,s)\sqrt{k_g k_{g+1}}(1 - U_h/k_g)(w,v) \tag{6.36}$$

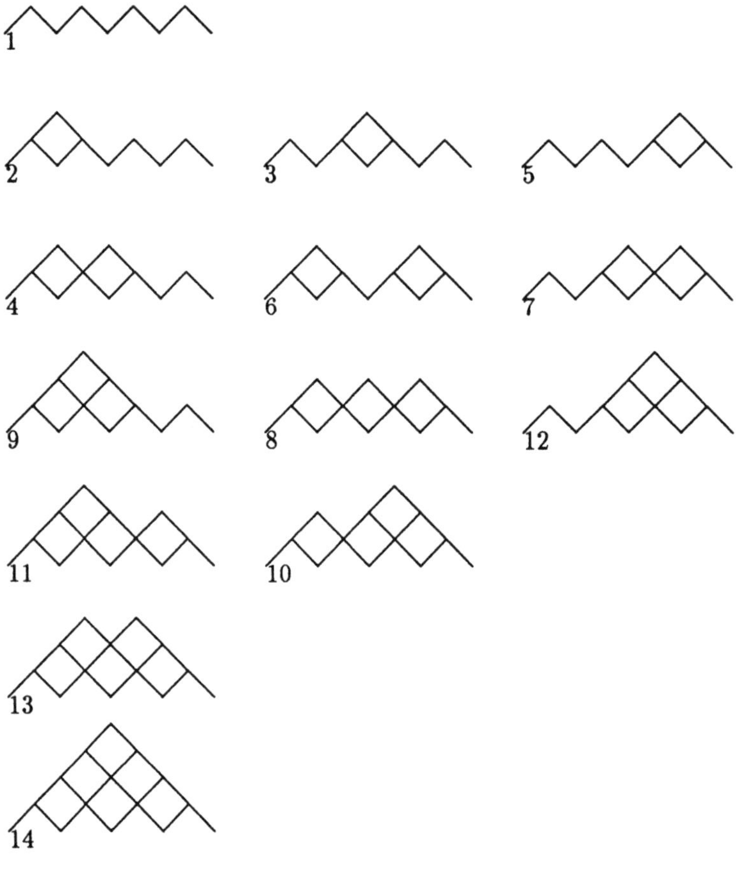

Figure 6.6: Walks in the $S(8,1)$ representation of $T_7(q)$.

for some g. Pictorially this means that t and w differ only in the neighbourhood of $(..., w_{h-1}, w_h, w_{h+1}, ...)$, where they have the walk shapes $\bigwedge$ and $\bigvee$ respectively. For convenience we will call this neighbourhood of the sequence the neighbourhood of h; specifically, the entries numbered $h-1, h, h+1$. There are then three possibilities. The first is that

$$(u,s)U_h = 0 \tag{6.37}$$

in which case s, v and t are all necessarily distinct and the product $(u,s)(w,v)$, and hence $(u,s)(t,v)$, vanishes by assumption.

In the two remaining possibilities $(u,s)U_h$ is expressible as a linear combination of (u,s) and (u,a) (for some sequence $\{a\}$), using

$$(u,a) = (u,s)\sqrt{k_b k_{b+1}}(1 - U_h/k_b) \tag{6.38}$$

for some b, if $\{s\}$ precedes $\{a\}$ in the partial order, and

$$(u,s) = (u,a)\sqrt{k_b k_{b+1}}(1 - U_h/k_b) \tag{6.39}$$

if $\{a\}$ precedes $\{s\}$.

In the former case the shapes of $\{s\}$ and $\{a\}$ in the neighbourhood of h are $\bigvee$ and $\bigwedge$ respectively (in the latter, the roles are reversed). In the former case then, s and t are distinct, a and w are distinct, and $a = t$ if and only if $s = w$, whereupon $g = b$. In general then

$$\begin{aligned}(u,s)(t,v) &= (u,s)\sqrt{k_g k_{g+1}}(1 - U_h/k_g)(w,v)\\ &= \sqrt{k_g k_{g+1}}(u,s)(w,v)\\ &\quad -(\sqrt{k_g k_{g+1}}/k_g)(k_b(u,s) - \sqrt{k_b/k_{b+1}}(u,a)\,)(w,v)\\ &= \sqrt{k_g k_{g+1}}(1 - (k_b/k_g))(u,s)(w,v)\end{aligned} \tag{6.40}$$

which vanishes either identically or by assumption.

In the latter case w and s are distinct, and t and a are distinct, and $s = t$ if and only if $a = w$, whereupon $b = g$. In either situation here we have

$$\begin{aligned}(u,s)(t,v) &= (u,s)\sqrt{k_g k_{g+1}}(1 - U_h/k_g)(w,v)\\ &= \sqrt{k_g k_{g+1}}(u,s)(w,v)\\ &\quad - \left(\frac{\sqrt{k_g k_{g+1}}}{k_g}\right)\left(1 - \frac{\sqrt{Q}}{k_b}\right)\sqrt{k_b k_{b+1}}\\ &\quad \cdot\left(k_b\,(u,a) - \sqrt{\frac{k_b}{k_{b+1}}}\;(u,s)\right)(w,v)\\ &= -\sqrt{\frac{k_{g+1}}{k_g}}\left(k_b - \sqrt{Q}\right)\sqrt{k_b k_{b+1}}(u,a)(w,v).\end{aligned} \tag{6.41}$$

This vanishes by assumption unless $a = w$, in which case $b = g$ (so the coefficient is unity) and

$$(u,a)(a,v) = (u,v)$$

by assumption, since a preceeds s.

Thus, if the identities hold for all pairs of walks below $\{s\},\{t\}$ they also hold for $\{s\},\{t\}$. They therefore hold for all $\{s\}$. If they hold for all $\{s\}$ with $\{t\}$ then we can replace $\{t\}$ with $\{w\}$ immediately above $\{t\}$ and $\{v\}$ with $\{t\}$ in the above cases and the same arguments apply.

This completes the proof of theorem 1(i).

Corollary 1.1 *If the operators (w,s) are well defined for all $\{w\},\{s\} \in S(n+1,m)$ then the left ideal generated by any (u,s) is a simple module.*

Corollary 1.2 *If all the operators (s,t) for a given $T_n(q)$ are well defined then they span a (not necessarily proper) subalgebra with multi-matrix structure.*

Proof: These operators satisfy the definition of elementary operators for an algebra with such a structure.

part (ii):

We need the following definition:

Definition 25 *If $(\{s\},\{t\})$ is in $P(n+1)$ then $(s \circ t)$ is a word in the generators of $T_n(q)$ obtained as follows:*

Firstly

$$((12)^p 1m \circ (12)^p 1m) = \prod_{i=1}^{p} U_{2i-1} \tag{6.42}$$

then iteratively

$$(\ldots (g\; g+1\; g)_i \ldots \circ\; t) = U_i\; (\ldots (g\; g-1\; g)_i \ldots \circ\; t) \tag{6.43}$$

and

$$(t \circ s) = (s \circ t)^T. \tag{6.44}$$

Proposition 9 *The words $(v \circ w)$ $\forall(\{v\},\{w\}) \in P(n+1)$ span the same vector space as the elements (s,t) $\forall(\{s\},\{t\}) \in P(n+1)$ where these latter are defined.*

Proof:

From the definitions 24 and 25.

There is a bijection which takes

$$(s,t) \leftrightarrow (s \circ t).$$

From theorem 1 (i) the elements (s,t) are linearly independent when defined. The reduction of words by applying the defining relations does not depend on q, except in an overall power of $\sqrt{Q}$.

This completes the proof of proposition 9.

Corollary 1: Words corresponding to distinct elements of $P(n+1)$ have distinct reduced words.

Corollary 2: The words $(v \circ w)$ $\forall(\{v\},\{w\}) \in P(n+1, 1 \text{ or } 2)$ span the same space as the elements (s,t) $\forall(\{s\},\{t\}) \in P(n+1, 1 \text{ or } 2)$ where these latter are defined.

Proof: $E_1 = E_2 = 1$.

Proposition 10 *The set of words is a basis for the regular representation of $T_n(q)$.*

Outline Proof:

By a suitably organised application of the defining relations. Consider the double sided ideal generated by

$$((12)^p 1m, (12)^p 1m) = ((12)^p 1m \;\; o \;\; (12)^p 1m) + X$$

where X is an element of the double sided ideal generated by $((12)^p 1\ (m-2)\ \ o\ \ (12)^p 1\ (m-2))$, by the definition of $E_m^{(2p)}$. If at any stage in this procedure we act with U_i on the left, say, on the current word (regarding $E_m^{(2p)}$ as a word, for convenience), we obtain a new word which appears in the iterative definition above *unless* the current subsequence around s_i is of the form

$$(g\ g+1\ g)_i$$

or

$$(g\ g+1\ g+2)_i.$$

However, in the first of these cases we do not get a new reduced word because of relation 1, and in the second case because of relation 2 and or proposition 8. By starting with the minimum value of m (i.e $m = 1$ or 2) and increasing we note, from the definition of E_k, that the double sided ideal generated by the unit is finally included in this scheme.

This completes the proof of proposition 10 and of theorem 1 (ii).

Corollary 10.1 *$T_n(q)$ is isomorphic to the Whitney diagram and boundary diagram algebras.*

To see this note that the number of distinct boundary diagrams equals the number of reduced words, and that we established a surjection in section 6.2.1(2).

6.5 Useful identities

It is worth noting the following identity:

$$U_{c-2}\left(\prod_{j=b}^{1}{}' U_j\right) = \left(\prod_{j=b}^{1}{}' U_j\right) U_c \tag{6.45}$$

for all $3 \leq c \leq b$. Its validity may be readily verified and understood diagrammatically (exercise). Alternatively, we have the following:

Outline Proof

On the right hand side U_c commutes through the product leftward until

$$...U_cU_{c-1}U_{c-2}U_c... = ...U_cU_{c-2}... = ...U_{c-2}U_cU_{c-1}U_{c-2}...$$

then U_{c-2} commutes through leftward to the end.

A similar, T transformed, identity follows immediately.

6.5.1 The word $X(d,c)$

The word $X(d,c)$ is a word in the generators $\{U_i\}$ with several useful properties. Consider $T_{2n-1}(q)$. For convenience we will number the generators

$$U_{1-n}, U_{2-n}, ..., U_{-1}, U_0, U_1, U_2, ..., U_{n-1}$$

in this section, then $X(0,n-1)$ is the word (reading from left to right)

$$\begin{array}{ccccccccccc}
 & & & & & U_{n-1} & & & & & \\
 & & & & \cdots & & \cdots & & & & \\
 & & & U_3 & & \cdots & & U_3 & & & \\
 & & U_2 & & U_2 & & & & U_2 & & \\
 & U_1 & & U_1 & & \cdots & & U_1 & & U_1 & \\
U_0 & & U_0 & & U_0 & & & & U_0 & & U_0 \\
 & U_{-1} & & U_{-1} & & \cdots & & U_{-1} & & U_{-1} & \\
 & & U_{-2} & & U_{-2} & & & & U_{-2} & & \\
 & & & U_{-}3 & & \cdots & & U_{-3} & & & \\
 & & & & \cdots & & \cdots & & & & \\
 & & & & & U_{1-n} & & & & &
\end{array}$$

where we have stacked commuting generators (i.e. groups amongst which the order is unimportant). In the diagrammatic form we have, for example with $n=3$, reading the word from left to right but *drawing* from top to bottom:

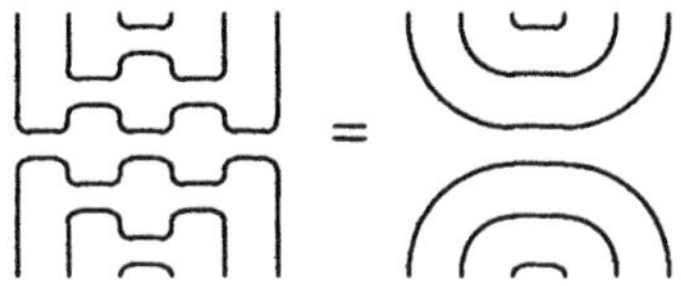

Proposition 11

$$U_a X(0,c) = U_{-a} X(0,c) \tag{6.46}$$

for all $-c \leq a \leq c$.

Outline Proof:
(i) By inspection of the following diagrammatic identity:

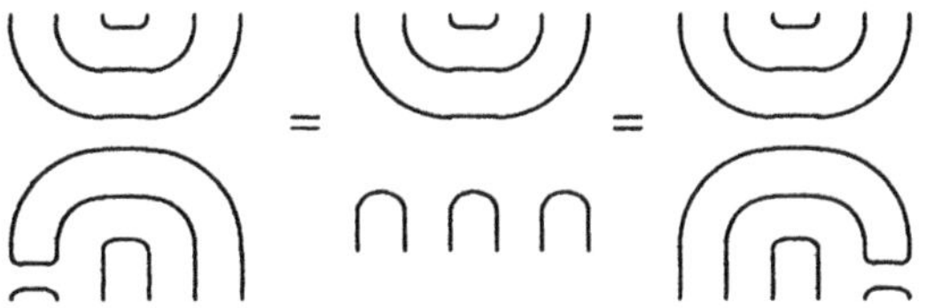

Note from the commuting properties of the generators that another way of arranging (stacking) the generators in $X(0, n-1)$ is:

```
                          n-1
                  n-2                          n-2
               ..                           ..
           2                           2
       1                          1                          1
  0                          0                    ...  0          0
      -1                         -1                         -1
           -2                          -2
               ..                           ..
                  2-n                          2-n
                          1-n
```

where we have simply written j in place of U_j for the sake of neatness. That is,

$$X(0, n-1) = \prod_{i=0}^{n-1} \left(\prod_{j=0}^{i} U_j \prod_{j=-1}^{-i}{}' U_j \right) . \tag{6.47}$$

Noting the T symmetry of $X(0, c)$ we see from equation 6.47 that

$$X(0, c) \propto X(0, c-1)X(0, c) \tag{6.48}$$

whereupon, with the aid of identity 6.45, property 6.46 may be proved by a conventional induction on c:

(ii) Assume property 6.46 true for $X(0, c-1)$ with $c \geq 4$ (it is clearly true for $c = 1, 2, 3$). Then it is only necessary to verify for $a = c$, since all other cases are covered by equation 6.48. We have

$$X(0, c)U_c \tag{6.49}$$

$$= X(0, c-1) \left(\prod_{j=-c}^{-1} U_j \right) \left(\prod_{j=c}^{0}{}' U_j \right) U_c \tag{6.50}$$

$$= X(0, c-1) \left(\prod \right) U_{c-2} \left(\prod \right) \tag{6.51}$$

$$= X(0, c-1)U_{c-2} \left(\prod \right) \left(\prod \right) \tag{6.52}$$

$$= X(0, c-1)U_{2-c} \left(\prod \right) \left(\prod \right) \tag{6.53}$$

$$= X(0, c-1) \left(\prod \right) U_{2-c-2} \left(\prod \right) \tag{6.54}$$

$$= X(0, c-1) \left(\prod \right) \left(\prod \right) U_{-c}. \tag{6.55}$$

Similarly we can show that the word $X(0, n-1)$ is R symmetric, and is the longest reduced word, in $T_{2n-1}(q)$.

The word $X(d,c)$ is $X(0,c)$ translated so that the 'middle' generator is U_d rather than U_0 and hence

$$U_{a+d}X(d,c) = U_{-a+d}X(d,c).$$

6.5.2 Longest words, module by module

Note that the walk-walk/word correspondences of section 6.4 give a systematic way of writing the set of distinct words in $T_k(q)$, although not necessarily in the form of *reduced* words. For example, recalling that diagrams are read from bottom to top and that stacked words are read from left to right we see that

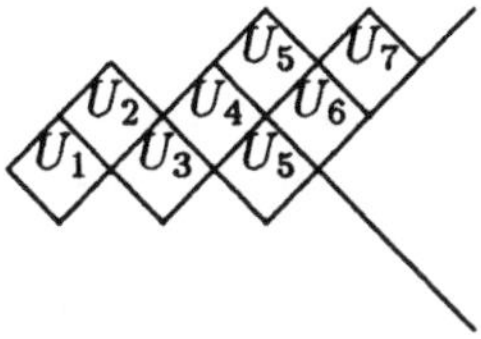

$$\rightarrow \left|\begin{array}{c|cc} U_1 & & \\ & U_2 & \\ U_3 & & \\ & U_4 & \\ U_5 & & U_5. \\ & U_6 & \\ & & U_7 \end{array}\right.$$

is reduced, but

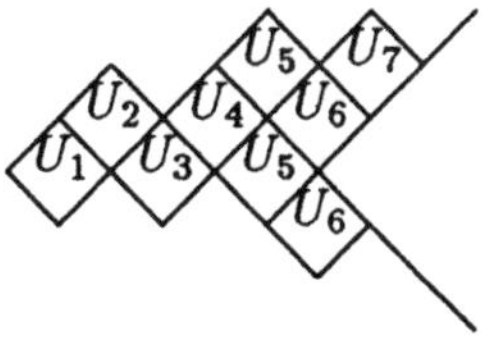

$$\rightarrow \begin{array}{c|c|cc} & U_1 & & \\ & & U_2 & \\ & U_3 & & \\ & & U_4 & \\ & U_5 & & U_5 \\ U_6 & & U_6 & \\ & & & U_7 \end{array} = \begin{array}{ccc} U_1 & & \\ & U_2 & \\ U_3 & & \\ & U_4 & \\ & & U_5 \\ & U_6 & \\ & & U_7 \end{array}$$

on applying the defining relations.

Note specifically, however, that their is an association of a certain subset of words to boundary diagrams which takes

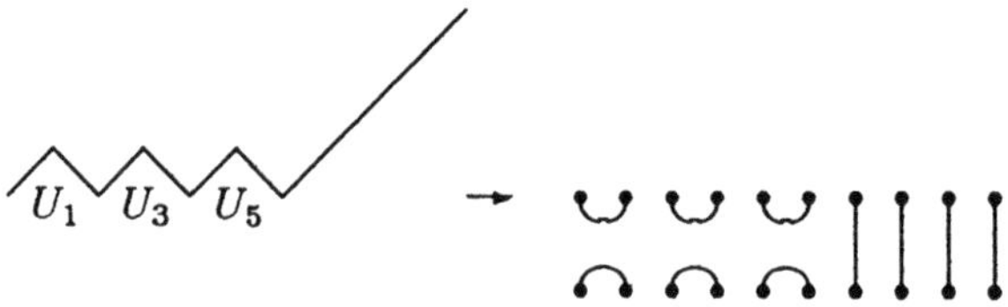

and then involves writing only on the left or only on the right as in the scheme illustrated above. Note that only the walk on the left (resp. right) in the walk/walk-word correspondence is changed in this procedure. The words obtained thus, i.e. those obtained by acting on the initial word above with further generators from the left only (say), employing the appropriate boundary diagram module quotient, and as prescribed by the correspondence, are always in reduced form. Note further that each such set of words has a unique longest word associated with, in our example, the (left) walk diagram

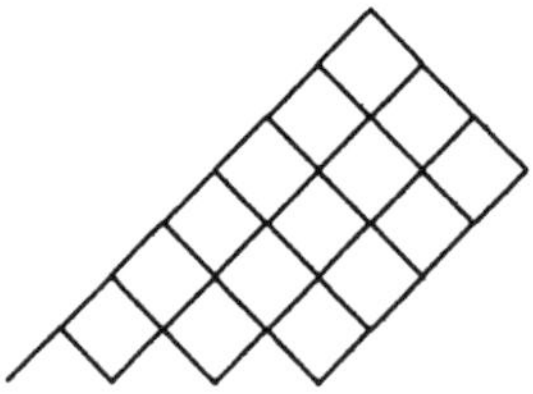

and, for a general initial walk of the form shown above, W_I, with the (left) walk diagram built from W_I which is unique in having no minimum in its upper envelope.

For any other walk diagram built from the same initial walk the upper envelope has at least one minimum. Therefore any other word W in such a set has some U_a such that, with $L(W)$ the length of the reduced form of W,

$$L(U_a W) = L(W) + 1 \tag{6.56}$$

and $U_a W$ is itself a reduced word.

It follows that

Proposition 12 *If*

$$U_a W = U_a W'$$

for all a then

$$W = W'. \tag{6.57}$$

Corollary 12.1 *Let I be any invariant subspace of the left ideal generated by the (idempotent) word corresponding to the initial walk above. Then I contains an element with a non-vanishing contribution of the associated unique longest word.*

Proof:

Suppose without loss of generality that

$$Y = \sum_w C_w W_w$$

is in I, where the sum is over distinct words W_w, with positive coefficients C_w. Then $U_a Y$ is in I and of the same form. Furthermore, either the longest word is in Y or there is a choice of a such that at least one of the words W_w will be longer. By equation 6.57 there is a sequence of choices for a on iterating this process such that the longest word appears in the sum.

Note also that all the words generated when acting on *both* sides as prescribed starting from

$$U_1 U_3 U_5 \ldots U_{2n-1}$$

in $T_{k=2n-1}(q)$ are reduced. It follows then that

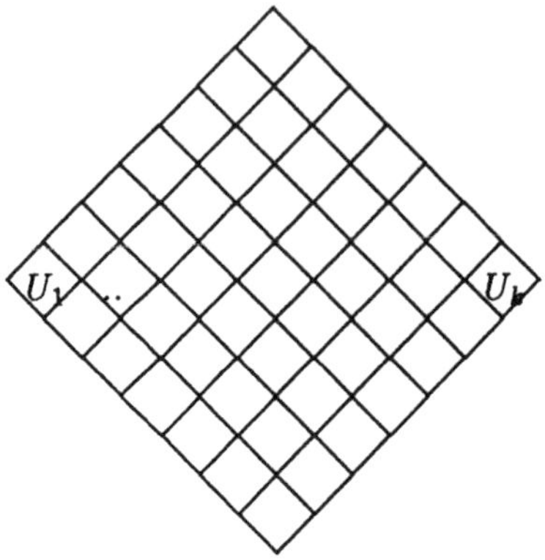

$$= X(n, n-1)$$

is the longest reduced word in the algebra overall.

Since the boundary diagram modules give (with quotient) a basis for what we have established in section 6.4 as being generically irreducible representations it follows that we can get from any word associated with such a module to any other by acting with further generators. In other words we can get from any word whose full diagram has a given number of lines passing from top to bottom to another word with the same number of lines in this way.

Finally note that any word except 1 can be written in the form

$$W = XU_1Y$$

where X, Y are words not involving U_1 (see also chapter 9).

6.5.3 The ABF regular representation

Note that the basis coming from $S(2n, 1)$ gives a generically irreducible representation of dimension C_n ($k = 2n - 1$). The basis is generated from the same primitive idempotent as is the Whitney diagram representation in which the top and bottom layers are disconnected (the idempotent is (e_1, e_1) from section 7.8). The Andrews-Baxter-Forreter (ABF or *walk*) basis for this representation with $k = 2n - 1 = 7$, for example, consists of 14 cases. In figure 6.6 we listed them.

If r is not rational then the representation is irreducible and thus equivalent to the Whitney representation (which is well defined for all q). The restriction to $n-1$ operators gives the regular representation of this subalgebra. This may be seen directly from the Whitney basis. Alternatively, note that in the ABF basis the irreducible content of the representation is totally transparent.

To see this, recall that the basis states are associated with distinct possible walks, representing configurations of ABF variables, $s_i > 0$ in a lattice layer, subject to the constraint that the boundary variables are set to $s_1 = s_{2n+1} = 1$. More generally, as we have seen, if $s_1 = 1$ then the possible values of s_{2n+1} (s_{2n+2}) give bases for the various irreducible representations of the $2n-1$ (resp. $2n$) operator algebra. In the ABF model itself there is the additional constraint that $s_i < r'$, which makes no difference to us here unless r is rational, in which case it restricts to an irreducible invariant subspace (corresponding to unitarisable representations) as we will see in chapter 7. This simplification is important in statistical mechanics, as it leads to an enhancement of the Yang-Baxter equations (see chapter 3 and ABF 1984).

From definition 24 we note that, on restriction to the first $n-1$ operators, basis elements corresponding to configurations which differ in any of the last $n+1$ variables are not coupled. Thus the representation breaks up into blocks labelled by the $(n+1)^{th}$ and subsequent variables. But it is the $(n+1)^{th}$ variable alone of these which distinguishes these representations. By reflection symmetry the multiplicity of each irreducible representation is thus equal to its dimension. This describes the regular representation for a semi-simple algebra. It hence confirms that the set of irreducible representations described above is complete.

The generic Bratelli diagram follows similarly.

Writing $q = \exp(i\pi/r)$, and considering the case $r = r'/r''$ with r', r'' coprime, the representations coming from definition 24 have some divergent matrix elements if $r' < 2n+1$. Correspondingly, in these special cases some of the basis states constructed in the proof of generic irreducibility are not well defined. However, the trace of a generator is equal to (for example) $C_{n-1}\sqrt{Q}$ generically, and so is formally finite for all r. A well defined representation may be recovered by initially regarding r as an indeterminate complex variable, and making certain similarity transformations before taking r to its rational value. However, the irreducibility property may be lost. We give full details in chapter 7.

Chapter 7

Special cases

In this chapter we will prove a theorem (stated on page 173) giving the structure of the Temperley-Lieb algebras for all r.

We will need some more preparations:

7.1 More combinatorics and sequences

Definition 26 *For $n, l, r \in Z_+$ and for $0 < m < r$ define $S(n, lr+m, r)$ as the subset of $S(n, lr+m)$ containing only those sequences with the property that for each i such that*

$$s_i = (l+1)r$$

there exists some $j > i$ such that

$$s_j = lr.$$

In other words $S(n, lr+m, r)$ is the set of sequences which never touch $(l+1)r$ after they last touch lr.

Definition 27 *Define $S'(n, lr+m, r)$ as the subset consisiting of those sequences with the property that for each j' such that*

$$s_{j'} = lr$$

there exists some $i > j'$ such that

$$s_i = (l+1)r.$$

In other words $S'(n, lr+m, r)$ is the set of sequences which never touch lr after they last touch $(l+1)r$.

Proposition 13 *The orders of the sets are*

$$|S(n, lr + m, r)| = C_{n,lr+m,r} \tag{7.1}$$

and

$$|S'(n, lr + m, r)| = C'_{n,lr+m,r}. \tag{7.2}$$

Proof:
We need

Proposition 14 *The complement of* $S(n, lr+m, r)$ *in* $S(n, lr+m)$ *is just* $S'(n, lr + m, r)$.

Proof:
By definition *every* sequence in $S(n, lr + m)$ has the property that, for each i such that

$$s_i = (l + 1)r$$

then there exists $j' < i$ such that

$$S_{j'} = lr.$$

Now consider the largest such i and suppose that the sequence is not in $S(n, lr + m, r)$. In this case there does not exist $j > i$ such that

$$s_j = lr$$

so each sequence not in $S(n, lr + m, r)$ is in $S'(n, lr + m, r)$. By a similar argument if a sequence is in $S(n, lr+m, r)$ then it is not in $S'(n, lr+m, r)$.

Corollary 14.1

$$|S(n, lr + m, r)| + |S'(n, lr + m, r)| = C_{n,lr+m}.$$

Proposition 15

$$|S'(n, lr + m, r)| = |S(n, (l + 2)r - m, r)|. \tag{7.3}$$

Outline Proof:
There is a bijection between elements $\{s\} \in S'(n, lr + m, r)$ and $\{t\} \in S(n, (l + 2)r - m, r)$ which takes $\{s\}$ to $\{t\}$ in the following way:
By definition there exists some integer j such that s_j is the last occurence of lr in the sequence $\{s\}$. Take $t_i = s_i$ for $i < j$ and $t_i = 2lr - s_i$ otherwise.

In the pictorial representation of sequences we can introduce 'critical' lines into the picture background for given r, which are horizontal lines at heights which are integer multiples of r. In terms of pictures, then, reflect the subsequence of each sequence after its last crossing of a critical line ($s_i = lr$) in that line.

This completes the proof of Proposition 15, equation 7.1 and equation 7.2.

7.2 Towards the main theorem

Definition 28 *For $m \in Z_+$ and $i = 1, 2, .., m$ let a_i be an (isomorphism class of) indecomposable projective left modules, and for $i = 1, 2, .., m-1$ let α_i, β_i and γ_i denote morphisms of modules. Then define $Q(m)$ as the ordinary quiver diagram*

$$a_1 \underset{\xrightarrow{\beta_1}}{\overset{\xleftarrow{\alpha_1}}{}} a_2 \underset{\xrightarrow{\beta_2}}{\overset{\xleftarrow{\alpha_2}}{}} a_3 \underset{\xrightarrow{\beta_3}}{\overset{\xleftarrow{\alpha_3}}{}} \cdots \underset{\xrightarrow{\beta_{m-1}}}{\overset{\xleftarrow{\alpha_{m-1}}}{}} a_m$$

with the composition relations for morphisms

$$\alpha_i \beta_i = \gamma_{i+1} \qquad (i = 1, 2, .., m-1)$$

and

$$\gamma_i = \beta_i \alpha_i \qquad (i = 2, 3, .., m-1)$$

and all other compositions of morphisms going to zero. In the diagram it is to be understood that there are arrows γ_i associated with each class a_i (for $i = 2, 3, .., m$) pointing from a_i to itself, obtained as in the composition relations.

For $i = 1, 2, .., m$ let S_i denote the simple module at the top of a_i. It then follows that the Loewy decomposition of each such projective is:

$$\begin{matrix} S_1 \\ S_2 \end{matrix} \qquad\qquad \begin{matrix} & S_i & \\ S_{i-1} & & S_{i+1} \\ & S_i & \end{matrix} \qquad\qquad \begin{matrix} S_m \\ S_{m-1} \\ S_m \end{matrix}$$

$$(i = 2, 3, .., m-1)$$

Definition 29 *Let $m \in Z_+$ and $\mathbf{c}$ a multi-component object with components $c_i \in Z_+$ for $i = 1, 2, .., m$. Then define the algebra $Q_{\mathbf{c}}(m)$ by the following generators and relations:*

Generators:

For $n = 1, 2, .., m$ and $i, j = 1, 2, .., c_n$ and and $l = 1, 2, .., c_{n+1}$ we have

$$(e_n)_{ij}$$

$$(\alpha_n)_{li} \qquad (\beta_n)_{il} \qquad (n \neq m)$$

$$(\gamma_n)_{ij} \qquad (n \neq 1).$$

Relations:

For $n = 1, 2, .., m$ and $i, j, k = 1, 2, .., c_n$

$$(e_n)_{ij}(e_n)_{jk} = (e_n)_{ik}.$$

and $(n \neq 1)$

$$(\gamma_n)_{ij}(e_n)_{jk} = (e_n)_{ij}(\gamma_n)_{jk} = (\gamma_n)_{ik}.$$

For $n = 1, 2, .., m$ and $i, j = 1, 2, .., c_n$ and $k = 1, 2, .., c_{n-1}$ and $l = 1, 2, .., c_{n+1}$

$$(\alpha_n)_{li}(e_n)_{ij} = (\alpha_n)_{lj} \qquad (n \neq m)$$

$$(e_n)_{ij}(\alpha_{n-1})_{jk} = (\alpha_{n-1})_{ik} \qquad (n \neq 1)$$

$$(\beta_{n-1})_{kj}(e_n)_{ji} = (\beta_{n-1})_{ki} \qquad (n \neq 1)$$

$$(e_n)_{ij}(\beta_n)_{jl} = (\beta_n)_{il} \qquad (n \neq m)$$

$$(\alpha_{n-1})_{ik}(\beta_{n-1})_{kj} = (\gamma_n)_{ij} \qquad (n \neq 1)$$

$$(\beta_n)_{jl}(\alpha_n)_{li} = (\gamma_n)_{ji} \qquad (n \neq m)$$

and all other bi-linear products zero.

This algebra has quiver diagram $Q(m)$. To see this note that the required morphisms are realised here by, for example, the α_{n-1} morphisms (indexed by k):

$$(e_n)_{ij} \stackrel{\alpha_{n-1}}{\rightarrow} (e_n)_{ij}(\alpha_{n-1})_{jk} = (\alpha_{n-1})_{ik}.$$

$$(\alpha_n)_{lj} \stackrel{\alpha_{n-1}}{\rightarrow} (\alpha_n)_{lj}(\alpha_{n-1})_{jk} = 0$$

$$(\beta_{n-1})_{k'j} \stackrel{\alpha_{n-1}}{\rightarrow} (\beta_{n-1})_{k'j}(\alpha_{n-1})_{jk} = (\gamma_{n-1})_{k'k}.$$

$$(\gamma_n)_{i'j} \stackrel{\alpha_{n-1}}{\rightarrow} (\gamma_n)_{i'j}(\alpha_{n-1})_{jk} = 0.$$

Note that with $m = 1$ then $Q_{\mathbf{c}}(1) = M_{c_1}(\mathbf{C})$, the algebra of c_1 dimensional matrices over the complex numbers.

7.2.1 Indexing $T_n(q)$ modules

Remark 2 *Since $S_{\{t\}}(n+1,p)$ provides a basis of an irreducible representation of generic $T_n(q)$ via the corresponding set of words (s,t) from definition 24 then the integer p labels the isomorphism classes of simple modules for generic $T_n(q)$.*

Proposition 16 *Taking the set of words $(s \circ t)$ obtained from $S_{\{t\}}(n+1,p)$ by definition 25 and quotienting by words $(w \circ v)$ arising from sequences with lower final elements, i.e. such that $w_n < s_n$, then $(s \circ t)$ with s varying and t fixed provides the basis for an indecomposable representation for all r.*

Proof (for $r \neq 2$):
This is the basis for an irreducible representation for generic r, by proposition 17 and the definition of E_k. From this case, therefore, the idempotent $(e \circ t)$ is primitive (using the quotient) if it is defined. On the other hand, this idempotent is well defined for all $r \neq 2$.
The generalisation to include $r = 2$ will become apparent as we proceed. In this case there is no indecomposable representation for $p = 1$.

Corollary 16.1 *The integer p labels isomorphism classes of indecomposable projective left modules for all r.*

7.3 The main theorem

Theorem 2 (Main Theorem) *Let $r = r_c$ where $r_c = r'/r''$, $r'-2, r'' \in Z_+$ with r' and r'' coprime and $0 < r'' < r'$. The algebra $T_n(e^{i\pi/r})$ has an isomorphism class of indecomposable projective left modules for each isomorphism class, labelled by p, in the generic or semi-simple case. The projective* remains *simple if p is an integer multiple of r', say*

$$p = lr' \qquad (l \in Z_+).$$

If p is congruent to s mod r', i.e.

$$p = lr' + s$$

$(0 < s < r')$, then the projective enters a block with quiver diagram of the form $Q(m+1)$. The adjacent isomorphism classes in the quiver are given, if they exist, by

$$p = lr' - s$$

and

$$p = (l+2)r' - s.$$

The adjacent classes exist if they are defined, i.e. if $0 < p < n+3$. The integer m for each block is determined by this condition. The case $r = 2$ differs only in that the leftmost isomorphism class in each quiver for odd n is absent.

This determines the algebra up to Morita equivalence, and, when taken in conjunction with Theorem 1, completely describes the representation theory of the algebra.

An equivalent statement of the Main Theorem is as follows:

Definition 30 *Define a multi-component object $\mathbf{c}^j(n, r')$ by components*

$$\left(\mathbf{c}^j(n,r')\right)_i = C_{n+1,F_j(i),r'}$$

Definition 31 *Define a subset of the positive integers*

$$A(n,r') = \{i \mid 0 < i \leq n+2;\ i_{mod\ 2} = n_{mod\ 2};\ i_{mod\ r'} = 0\}$$

Definition 32 *Define another subset of the positive integers*

$$B(n,r') = \{i \mid 0 < i < r';\ i \leq n+2\ ;\ i_{mod\ 2} = n_{mod\ 2}\}$$

Recalling that $n + 2 = Lr' + s$, we introduce

$$m(j) = L+1 \qquad j < s$$

$$m(j) = L \qquad j \geq s$$

if L is even, and

$$m(j) = L+1 \qquad j < r' - s$$

$$m(j) = L \qquad j \geq r' - s$$

if L is odd.

Then $T_n(q)$ is isomorphic to

$$\left[\bigoplus_{i \in A(n,r')} M_{C_{n+1,i}}\right] \oplus \left[\bigoplus_{j \in B(n,r')} Q_{\mathbf{c}^j(n,r')}(m(j))\right]$$

To see that this is an equivalent statement to the main theorem note that the quiver diagrams are the same and that the dimensions may be determined recursively by reference to Theorem 1.

Examples

The cases $r = 3$ and $r = 4$ are illustrated below by giving the dimensions of the simple components in the Loewy decomposition of each indecomposable projective. The first row is $n = 0$, the second $n = 1$ and so on. Some dimensions have been dotted to distinguish inequivalent modules with the same dimension.

$p =$

1	2	3	4	5	6	7	8	9
	1							
1.		1						
	1. 1		1 1. 1					
1. 1		3		1 1. 1				
	1. 4		4 1. 4		1			
1. 4		9		4 1. 1 4		1 4 1		
	1. 13		13 1. 1 13		6		1 13 1	
1. 13		28		13 1. 7 13		7 13 7		1

$p=$ 1	2	3	4	5	6	7	8	9
	1							
1.		1						
	2		1					
2		2 1		1 2 1				
	4 1		4		1 4 1			
4 1		4 5		4 5 4		1 4 1		
	8 6		14		6 8 6		1	
8 6		8 20		20 8 20		6 8 1 6		1 6 1

7.4 Proof of main theorem

The idea of this proof is to write down (primitive) idempotents and examine the (indecomposable) projective left modules they generate. Intermediate results which apply specifically to $r = r_c$ appear as lemmas, and general or generic results as propositions.

We will consider the case

$$r_c = r'$$

i.e. $r'' = 1$ explicitly. The same arguments work in general, with all references to r_c requiring an integer simply replaced by r'.

Definition 33 *Define $u_p \in T_n(q)$ for $p = 3, 4, .., n+2$ by*

$$u_p = -k_{p-1}U_{p-2}$$

Definition 34 *For $p = 3, 4, .., n+2$ and $p+1 \geq s \geq 3$ define $W_{p,s} \in T_n(q)$ by $W_{p,p+1} = 1$ and*

$$W_{p,s} = 1 + \sum_{i=0}^{p-s} \left(\prod_{j=0}^{i} u_{p-j} \right)$$

That is

$$W_{p,s} = 1 + u_p + u_p u_{p-1} + \ldots + u_p u_{p-1} u_{p-2} \ldots u_s. \tag{7.4}$$

Definition 35 *With $X(d+1, c)$ defined as in section 6.5.1 $x(d+1, c)$ is obtained from $X(d+1, c)$ by replacing each U_i with u_i (i.e. an overall translation and renormalisation).*

Definition 36 *Let d be a positive integer multiple of r_c, i.e. $d = mr_c$ ($m \in Z_+$), and $0 < s \leq r_c$, such that $d+s \leq n+2$ and $(d+s)_{mod\ 2} = n_{mod\ 2}$, then define $E^s_{d+s} \in T_n(q)$ by*

$$E^s_{d+s} = E_{d+s} + \left(\prod_{k=1}^{s} k_{d-k+1} \right) E_d \; X(d+1, s-1) \; E_d.$$

Lemma 2.1 *When written as a linear combination of reduced words E^s_{d+s} remains well defined at*

$$r = r_c.$$

Proof:

The idea of the proof is to rewrite E_k as a part which remains finite plus a part which diverges as $r \rightarrow r_c$.

We proceed by proving a series of propositions involving E_k and $W_{p,s}$. Impatient readers may skip to equation 7.8 for the denouement.

Proposition 17

$$E_p = E_{p-1} W_{p,q} E_{q-1} = E_{q-1} W^T_{p,q} E_{p-1} \tag{7.5}$$

for all $p \geq q \geq 3$.

Proof (First identity):

Write $p - q = s \geq 0$ and proceed by induction on s. The identity is true for $s = 0$ by definition of E_p (section 6.3.3). Assume true for $p - q = s$ then

$$\begin{aligned} E_{q+s} &= E_{q+s-1} W_{q+s,q} E_{q-1} \\ &= E_{q+s-1} W_{q+s,q} E_{q-2}(1 + u_{q-1}) E_{q-2} \\ &= E_{q+s-1}(1 + u_{q+s} + .. + u_{q+s}..u_{q+1} + u_{q+s}..u_q)(1 + u_{q-1}) E_{q-2} \\ &= E_{q+s-1} W_{q+s,q-1} E_{q-2}. \end{aligned}$$

The second identity follows immediately from the T (word reversal) symmetry of the definition of E_p.

This completes the proof of proposition 17.

Proposition 18

$$E_p = E_{p-j} \prod_{i=1,j} [W_{p-j+i,q} E_{q-1}]. \tag{7.6}$$

for all $p > j \geq 0$.

Proof:

True for $j = 1$ by equation 7.5. Then by induction, since

$$\begin{aligned} & E_{p-j} \prod_{i=1,j} [W_{p-j+i,q} E_{q-1}] \\ = \; & E_{p-(j+1)} W_{p-j,q} E_{q-1} \prod_{i=1,j} [W_{p-j+i,q} E_{q-1}] \\ = \; & E_{p-(j+1)} \prod_{i=1,(j+1)} [W_{p-(j+1)+i,q} E_{q-1}]. \end{aligned}$$

This completes the proof of proposition 18.

Proposition 19 *Putting* $t = p - s$

$$E_p = E_t \prod_{j=1,n} [W_{t+j,t+1} W^T_{t,t-j+1}] \prod_{j=n+1,s} [W_{t+j,t+1} E_t] \tag{7.7}$$

for all n *such that* $0 \leq n < s$.

Proof:
by induction on n. It is true for $n = 0$ by proposition 18. Then

$$E_t \prod_{j=1,n} [W_{t+j,t+1} W^T_{t,t-j+1}] \prod_{j=n+1,s} [W_{t+j,t+1} E_t]$$

$$= E_t \prod_{j=1,n} [W_{t+j,t+1} W^T_{t,t-j+1}] \; W_{t+n+1,t+1} \; E_t \prod_{j=n+2,s} [W_{t+j,t+1} E_t] \, E_t$$

$$= E_t \prod_{j=1,n} [W_{t+j,t+1} W^T_{t,t-j+1}] \; W_{t+n+1,t+1}$$

$$.E_{t-n-1} \, W^T_{t,t-n} \, E_{t-1} \prod_{j=n+2,s} [W_{t+j,t+1} E_t] \; E_t$$

$$= E_t \prod_{j=1,n+1} [W_{t+j,t+1} W^T_{t,t-j+1}] \prod_{j=n+2,s} [W_{t+j,t+1} E_t] \, E_t$$

where we have used equation 7.5.

This completes the proof of proposition 19.

Sublemma 2.1.1 *Putting $d = mr_c$ then $W_{d+j,d+1}$ remains finite for all $1 < j \leq r$, and $W^T_{d,d-j+1}$ remains finite for all j, as $r \to r_c$.*

Proof:
Note that with $d = mr_c$ then u_{d+1} is not defined (it is divergent) when $r \to r_c$, but $u_{d+2} = 0$ and all other such operators are finite. Note further that

$$u_{d+2} u_{d+1} = -U_d U_{d-1}$$

in the limit. Sublemma 2.1.1 then follows from the definition of $W_{p,s}$.

Let us consider the case of proposition 19 in which $t = d$ and $0 < s \leq r$. Using $E_d u_d = 0$ we can write this case as follows:

$$E_{d+s}$$

$$= E_d(1 + u_{d+1})(1 + u_d) \prod_{j=2,n} [W_{d+j,d+1} W^T_{d,d-j+1}]$$

$$\prod_{j=n+1}^{s} [W_{d+j,d+1} E_d] \; E_d$$

$$= E_d \prod_{j=2,n} [W_{d+j,d+1} W^T_{d,d-j+1}] \prod_{j=n+1}^{s} [W_{d+j,d+1} E_d] \; E_d$$

$$+ \; E_d(u_{d+1})(1+u_d) \prod_{j=2,n} [W_{d+j,d+1} W^T_{d,d-j+1}] \prod_{j=n+1}^{s} [W_{d+j,d+1} E_d] \; E_d \tag{7.8}$$

(for $1 \leq n < s$).

Sublemma 2.1.2 *Assuming that E_d is well defined then the* only *divergence in the right hand side of equation 7.8 as $r \to r_c$ is in the factor u_{d+1} in the second term.*

Proof: From sublemma 2.1.1 all the other elements of the expression are finite.

In fact we can simplify this term considerably, as we will see shortly. First we need

Sublemma 2.1.3 *$k_{d+1}x(d+1, c-1)$ remains well defined as $r \to r_c$ for all $c > 0$;*

$x(d+1, r_c - 1)$ remains finite as $r \to r_c$.

These results follow immediately from the definition of $x(d, c)$.

Let us recall the notation that in a *primed* product $\prod'_j$ the variable is incremented negatively so that, for example,

$$\prod_{j=2}^{1} u_j = 1,$$

but

$$\prod_{j=2}^{1}{}' \; u_j = u_2 u_1.$$

Then we have:

Sublemma 2.1.4 *For all positive integer $c \leq r_c$*

$$R(c) = x(d+1, c-1) W_{d+c+1,d+3} + x(d+1, c-1) W^T_{d,d-c+2} \prod_{b=d+c+1}^{d+1}{}' \; u_b. \tag{7.9}$$

remains well defined as $r \to r_c$.

Proof:

Note that each of the two terms on the right is itself a sum of c terms. The ith term of $W_{d+c+1,d+3}$ is

$$u_{d+c+1}\ u_{d+c}\ \cdots\ u_{d+c+1-(i-2)}$$

(eg. the 1st term is 1). The $(c+1-i)^{th}$ term of $W^T_{d,d-c+2}\prod_{b=d+c+1}^{d+1}{}' u_b$ is

$$\prod_{j=d-(c+1-i)+2}^{d} u_j \prod_{b=d+c+1}^{d+1}{}' u_b.$$

Using the properties of $X(d,c)$ we have, generically,

$$\begin{aligned} & x(d+1,c-1)\left(\prod_{j=d-(c+1-i)+2}^{d} u_j\right)\left(\prod_{b=d+c+1}^{d+1}{}' u_b\right) \\ = \quad & x(d+1,c-1)\prod_{k=i}^{c-1}\frac{k_{d-(c-k)}}{k_{d+(c-k)}}\prod_{j=d+2}^{d+(c-i)+1} u_j\left(\prod_{b=d+c+1}^{d+1}{}' u_b\right) \qquad (7.10)\end{aligned}$$

Proof:

For $a=-1,0,1,..,c-i-2$ we have

$$\begin{aligned} & x(d+1,c-1)\prod_{k=i}^{i+a}\frac{k_{d-(c-k)}}{k_{d+(c-k)}}\left(\prod_{j=d-(c+1-i)+a+3}^{d} u_j\right)\left(\prod_{b=d+c-i+1-a}^{d+(c-i)+1} u_b\right) \\ = \quad & x(d+1,c-1)\prod_{k=i}^{i+a}\frac{k_{d-(c-k)}}{k_{d+(c-k)}}\left(u_{d-(c+1-i)+a+3}\prod_{j=d-(c+1-i)+a+4}^{d} u_j\right) \\ & \left(\prod_{b=d+c-i+1-a}^{d+(c-i)+1} u_b\right) \\ = \quad & x(d+1,c-1)\left(\prod_{k=i}^{i+a}\frac{k_{d-(c-k)}}{k_{d+(c-k)}}\right)\frac{k_{d-(c-i)+a+1}}{k_{d+(c-i)-a-1}}u_{d+(c-i)-a} \\ & \prod_{j=d-(c+1-i)+a+4}^{d} u_j\left(\prod_{b=d+c-i+1-a}^{d+(c-i)+1} u_b\right) \\ = \quad & x(d+1,c-1)\left(\prod_{k=i}^{i+a}\frac{k_{d-(c-k)}}{k_{d+(c-k)}}\right)\frac{k_{d-(c-i)+a+1}}{k_{d+(c-i)-a-1}}\left(\prod_{j=d-(c+1-i)+a+4}^{d} u_j\right)\end{aligned}$$

$$u_{d+(c-i)-a}\left(\prod_{b=d+c-i+1-a}^{d+(c-i)+1} u_b\right)$$

$$= x(d+1,c-1)\left(\prod_{k=i}^{i+a}\frac{k_{d-(c-k)}}{k_{d+(c-k)}}\right)\frac{k_{d-(c-i)+a+1}}{k_{d+(c-i)-a-1}}\prod_{j=d-(c+1-i)+a+4}^{d} u_j$$

$$\left(\prod_{b=d+c-i+1-(a+1)}^{d+(c-i)+1} u_b\right)$$

$$= x(d+1,c-1)\prod_{k=i}^{i+a+1}\frac{k_{d-(c-k)}}{k_{d+(c-k)}}\left(\prod_{j=d-(c+1-i)+a+4}^{d} u_j\right)$$

$$\left(\prod_{b=d+c-i+1-(a+1)}^{d+(c-i)+1} u_b\right)$$

Then

$$x(d+1,c-1)\prod_{k=i}^{c-1}\frac{k_{d-(c-k)}}{k_{d+(c-k)}}\prod_{j=d+2}^{d+(c-i)+1} u_j\left(\prod_{b=d+c+1}^{d+1}{}' u_b\right)$$

$$= x(d+1,c-1)\prod_{k=i}^{c-1}\frac{k_{d-(c-k)}}{k_{d+(c-k)}}$$

$$\cdot\left(\prod_{k=i}^{c-1}[k_{d+c-k}k_{d+c-k+1}]\right)\left(u_{d+c+1}\cdots u_{d+c+1-(i-2)}\ u_{d+2}\ u_{d+1}\right) \tag{7.11}$$

Proof:

From the defining relations we have

$$\prod_{j=d+2}^{d+(c-i)+1} u_j\left(\prod_{b=d+c+1}^{d+1}{}' u_b\right)$$

$$= \left(\prod_{b=d+c+1}^{d+c+1-(i-2)'} u_b\right)\prod_{j=d+2}^{d+(c-i)+1} u_j\left(\prod_{b=d+c-i+2}^{d+1}{}' u_b\right)$$

and for $a = 0, 1, .., c-i-1$ we have

$$\left(\prod_{b=d+c+1}^{d+c+1-(i-2)'} u_b\right)\prod_{j=d+2}^{d+(c-i)+1-a} u_j\left(\prod_{b=d+c-i+2-a}^{d+1}{}' u_b\right)$$

$$
\cdot \prod_{k=i}^{i+a-1} [k_{d+c-k} k_{d+c-k+1}]
$$

$$
= \left(\prod_{b=d+c+1}^{d+c+1-(i-2)'} u_b \right) \left(\prod_{j=d+2}^{d+(c-i)-a} u_j \right) u_{d+(c-i)+1-a}
$$

$$
\cdot \left(\prod_{b=d+c-i+2-a}^{d+1}{}' u_b \right) \prod_{k=i}^{i+a-1} [k_{d+c-k} k_{d+c-k+1}]
$$

$$
= \left(\prod_{b=d+c+1}^{d+c+1-(i-2)'} u_b \right) \left(\prod_{j=d+2}^{d+(c-i)-a} u_j \right)
$$

$$
u_{d+(c-i)+1-a} \; u_{d+c-i+2-a} \; u_{d+c-i+1-a}
$$

$$
\left(\prod_{b=d+c-i+2-a}^{d+1}{}' u_b \right) \cdot \prod_{k=i}^{i+a-1} [k_{d+c-k} k_{d+c-k+1}]
$$

$$
= \left(\prod_{b=d+c+1}^{d+c+1-(i-2)'} u_b \right) \left(\prod_{j=d+2}^{d+(c-i)-a} u_j \right)
$$

$$
u_{d+(c-i)+1-a} \; k_{d+c-i+1-a} \; k_{d+c-i-a}
$$

$$
\left(\prod_{b=d+c-i+2-a}^{d+1}{}' u_b \right) \cdot \prod_{k=i}^{i+a-1} [k_{d+c-k} k_{d+c-k+1}]
$$

$$
= \left(\prod_{b=d+c+1}^{d+c+1-(i-2)'} u_b \right) \prod_{j=d+2}^{d+(c-i)-a} u_j \left(\prod_{b=d+c-i+1-a}^{d+1}{}' u_b \right)
$$

$$
\cdot \prod_{k=i}^{i+a} [k_{d+c-k} k_{d+c-k+1}] \tag{7.12}
$$

This completes the proof of equation 7.11.

Then following on from equation 7.11

$$
x(d+1, c-1) \prod_{k=i}^{c-1} \frac{k_{d-(c-k)}}{k_{d+(c-k)}}
$$

$$
\cdot \prod_{k=i}^{c-1} [k_{d+c-k} k_{d+c-k+1}]
$$

$$
\left(u_{d+c+1} \dots u_{d+c+1-(i-2)} \; u_{d+2} \; u_{d+1} \right)
$$

$$= k_{d+1}k_d \prod_{k=i}^{c-1}[k_{d-(c-k)}k_{d+(c-k)+1}]x(d+1, c-1)u_{d+c+1}\dots u_{d+c+1-(i-2)}$$

The last expression is proportional to the ith term of

$$x(d+1, c-1)W_{d+c+1,d+3}$$

above.

Note that when $r \to r_c$ the prefactor in the last expression becomes -1 (using $k_{-b} = 1/k_{b+1}$ for $b \neq 0$).

This completes the proof of sublemma 2.1.4.

Using $u_d E_d = 0$ we have, generically, that

$$\begin{aligned}
& W^T_{d,d-c+1} W_{d+c+1,d+1} E_d \\
= \ & W^T_{d,d-c+1} \left(W_{d+c+1,d+3} + \left(\prod_{b=d+c+1}^{d+2}{}' u_b \right) + \left(\prod_{b=d+c+1}^{d+1}{}' u_b \right) \right) E_d \\
= \ & \left(W_{d+c+1,d+3} + \left(\prod_{b=d+c+1}^{d+2}{}' u_b \right) + \left(W^T_{d,d-c+1} \prod_{b=d+c+1}^{d+1}{}' u_b \right) \right) E_d \\
= \ & \left(W_{d+c+1,d+3} + \left(\prod_{b=d+c+1}^{d+2}{}' u_b \right) \right. \\
& \left. + \left(W^T_{d,d-c+2} + \prod_{b=d-c+1}^{d} u_b \right) \prod_{b=d+c+1}^{d+1}{}' u_b \right) E_d. \qquad (7.13)
\end{aligned}$$

Proposition 20 *We have the following set of identities for all* $0 < c \leq r_c$*:*

$$\begin{aligned}
& E_d x(d+1, c-1) E_d \left(\prod_{j=c+1}^{s} [W_{d+j,d+1} E_d] \right) E_d \\
= \ & E_d x(d+1, c) E_d \left(\prod_{j=c+2}^{s} [W_{d+j,d+1} E_d] \right) E_d \\
& + \ \textit{terms finite as } r \to r_c. \qquad (7.14)
\end{aligned}$$

Proof:
By proposition 17 we can rewrite the left hand side as

$$E_d x(d+1,c-1)\left[W^T_{d,d-c+1}W_{d+c+1,d+1}E_d\right]\left(\prod_{j=c+2}^{s}\left[W_{d+j,d+1}E_d\right]\right)E_d$$

$$= E_d x(d+1,c-1)\left[\left(W_{d+c+1,d+3}+\left(\prod_{b=d+c+1}^{d+2}{}' u_b\right)\right.\right.$$

$$\left.\left.+\left(W^T_{d,d-c+2}+\prod_{b=d-c+1}^{d}u_b\right)\prod_{b=d+c+1}^{d+1}{}' u_b\right)E_d\right]\left(\prod_{j=c+2}^{s}\left[W_{d+j,d+1}E_d\right]\right)E_d$$

(by equation 7.13)

$$= E_d\left(R(c)+x(d+1,c-1)\left(\left(\prod_{b=d+c+1}^{d+2}{}' u_b\right)\right.\right.$$

$$\left.\left.+\left(\prod_{b=d-c+1}^{d}u_b\right)\prod_{b=d+c+1}^{d+1}{}' u_b\right)\right)E_d\left(\prod_{j=c+2}^{s}\left[W_{d+j,d+1}E_d\right]\right)E_d$$

$$= E_d\left(R(c)+x(d+1,c-1)\left(\prod_{b=d+c+1}^{d+2}{}' u_b\right)+x(d+1,c)\right)$$

$$E_d\left(\prod_{j=c+2}^{s}\left[W_{d+j,d+1}E_d\right]\right)E_d$$

$$= E_d\left(R(c)+k_{d+1}\,x(d+1,c-1)\left(\prod_{b=d+c+1}^{d+1}{}' u_b\right)U_d+x(d+1,c)\right)$$

$$E_d\left(\prod_{j=c+2}^{s}\left[W_{d+j,d+1}E_d\right]\right)E_d \tag{7.15}$$

This completes the proof of proposition 20.

Recall that we are concerned with the term containing u_{d+1} in equation 7.8, which is the only potentially divergent contribution to E_{d+s}. Using proposition 20 repeatedly we can now rewrite this term as follows:

$$E_d(u_{d+1})(1+u_d)\prod_{j=2}^{n}\left[W_{d+j,d+1}W^T_{d,d-j+1}\right]\prod_{j=n+1}^{s}\left[W_{d+j,d+1}E_d\right]\;E_d$$

$$\begin{aligned} &= E_d\, x(d+1,0)\, W^T_{d,d} \prod_{j=2}^{s}[W_{d+j,d+1}E_d]\; E_d \\ &= E_d\, x(d+1,s-1)\; E_d \quad + \textit{ terms finite as } r \to r_c\,. \end{aligned} \tag{7.16}$$

Thus applying equation 7.16 to equation 7.8 we find, *subject to the assumption that E_d is finite,* that the divergent part of E_{d+s} as $r \to r_c$ is equal to the divergent part of

$$E_d x(d+1,s-1)E_d.$$

Sublemma 2.1.5 *E_d is finite at $r = r_c$.*

Proof:

For $m \in Z_+$ assume E_{mr_c} is finite. This is true from the definition for $m = 1$. Then, with $d = mr_c$, E_{d+s} is *not* divergent when $s = r$, since u_{d-r+2} then appears in $x(d+1,s-1)$ (sublemma 2.1.3), cancelling the divergence.

This completes the proof of Sublemma 2.1.5.

Finally we have

Sublemma 2.1.6 *At $r = r_c$*

$$\left(\prod_{k=1}^{s} k_{d-k+1}\right) X(d+1,s-1) + x(d+1,s-1)$$

is well defined.

Outline Proof:

The two terms differ by a factor which may be simplified at $r = r_c$ using

$$k_d k_{d+1} = -1$$

and (for $a \neq 0$)

$$k_{d-a}k_{d+a+1} = 1.$$

This completes the proof of Lemma 2.1.

The remainder of the proof of Theorem 2 hinges on the fact that many of the elements of $T_n(q)$ obtained iteratively through Definition 24 remain well defined at $r = r_c$, even when intermediate elements, i.e. those lower in the partial order, are not well defined. Consider sequences $\{s\}, \{t\}$ and $\{w\}$ with $\{s\}$ more than one step lower than $\{w\}$ in the partial order. We will look at the properties of the composite of factors, L say, obtained by repeated use of definition 24 as required in building (w,t) from (s,t), i.e. in

$$(w,t) = L \quad (s,t).$$

Definition 37 *For $i, j, x \in Z_+$ define $L_{i+1,j+1}(x)$ by*

$$L_{i+1,j+1}(x) = \left(\prod_{k=1}^{x} \sqrt{k_{j+k}k_{j+k+1}}\right)\left(1 + \sum_{a=1}^{x} \prod_{m=a}^{1}{}' \left(\frac{-U_{i+m}}{k_{j+m}}\right)\right). \quad (7.17)$$

and define $R_{i+1,j+1}(x)$ by

$$R_{i+1,j+1}(x) = L^T_{i+1,j+1}(x).$$

For example

$$L_{2,2}(3) = \left(\prod_{k=1}^{3} \sqrt{k_{1+k}k_{2+k}}\right)\left(1 - \frac{U_2}{k_2} + \frac{U_3U_2}{k_3k_2} - \frac{U_4U_3U_2}{k_4k_3k_2}\right).$$

Proposition 21 *Let $k, k', s, s' \in Z_+$, $s, s' \leq r$ and $kr + s > k'r + s'$. Without loss of generality consider $j = kr + s$ and $j + x = k'r + s'$, then $L_{i+1,j+1}(x)$ is finite at $r = r_c$ provided*

$$s, s' < r - 1$$

Proof:

First note that the value of the index i does not affect the scalar factors appearing in $L_{i+1,j+1}(x)$. For the given range of j, x values all the factors are well defined or occur in well defined pairs. To see this note that we can exhibit the pole structure of $L_{i+1,j+1}(x)$ directly by writing it in terms of the polynomials P_b. Recalling that $k_b = P_{b-1}/P_b$, and that $P_b = 0$ if $b = mr$ and is finite otherwise, we have:

$$L_{i+1,j+1}(x) = \sqrt{\frac{P_{j+1}}{P_jP_{j+x+1}P_{j+x}}}(P_j + \sum_{a=1}^{x}(-1)^a P_{j+a}U_{i+a}U_{i+a-1}...U_{i+1}). \quad (7.18)$$

We have only to avoid the zeros of the factors in the square root.

This completes the proof of the proposition.

Proposition 22 *Suppose $\{s\} \in S(n+1)$ has a subsequence of length $x+2$, commencing from s_i ($0 \leq i \leq n-x$), of the form*

$$j+1 \quad j \quad j+x$$

Suppose also that $\{t\} \in S(n+1)$ is given by

$$t_j = s_j \qquad\qquad for\ \ i \geq j \geq i+x+1$$

$$t_j = s_j + 2 \qquad\qquad for\ \ i < j < i+x+1.$$

Then

$$(e,t) = (e,s)\ R_{i+1,j+1}(x).$$

Note that the subsequence of $\{t\}$ commencing from t_i is

$$j+1 \quad j+x+1 \quad j+x\ .$$

In other words the effect of $R_{i+1,j+1}(x)$ here is to modify x consecutive entries in the Right hand sequence:

$$(e\ ,\ \ldots.\ \underbrace{j+1}_{t_i}\ \underbrace{j+x+1}_{t_{i+x}}\ \underbrace{j+x}_{t_{i+x+1}}\ \ldots.)$$

$$= (e\ ,\ \ldots.\ \underbrace{j+1}_{s_i}\ \ \underbrace{j}_{s_{i+1}}\ \ \underbrace{j+x}_{s_{i+x+1}}\ \ldots.)R_{i+1,j+1}(x)$$

Proof of Proposition 22:

Note that the sequence $\{s\}$ in Proposition 22 above has subsequences of the same form, starting from s_i, of length $l+2 < x+2$ entries. Assume the proposition true for $x = l$ (it is true for $x = 1$ by definition) and denote by $\{t_l\}$ the sequence obtained at this level, then, since this has the relevant subsequence

$$\ldots j+l \quad j+l-1 \quad j+l$$

we have, from Definition 24

$$(e,t_{l+1}) = (e,t_l)\sqrt{k_{j+l+1}k_{j+l+2}}\left(1 - \frac{U_{i+l+1}}{k_{j+l+1}}\right)$$

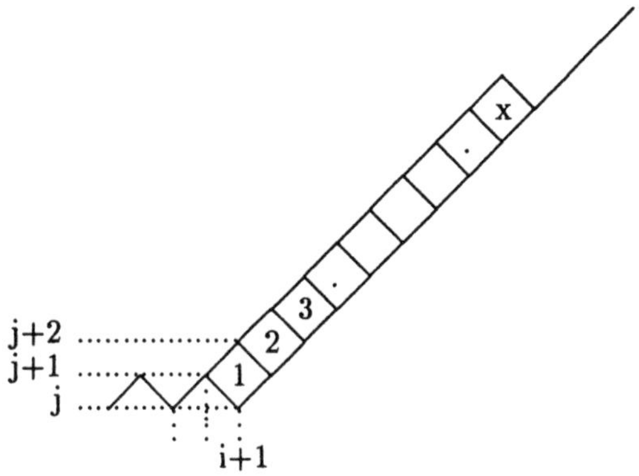

Figure 7.1: Adding diamonds in a strip.

which, by assumption,

$$= (e,s)\ R_{i+1,j+1}(l)\left(1 - \frac{U_{i+l+1}}{k_{j+l+1}}\right)$$

$$= (e,s)\ R_{i+1,j+1}(l+1)$$

where we have used Proposition 8, i.e. $(e,s)U_{i+l+1} = 0$, and Relation 6.17.
This completes the proof of Proposition 22.

In discussing the well defined elements (s,t) it may be useful to think again of the sequences as walks on a square lattice. With the lattice oriented at 45° the entries in the sequence become heights on the lattice with respect to some horizontal base line at height 0, as in chapter 6.

Walks modified by strips

In this picture Proposition 22 says that the product of factors corresponding to adding 'diamonds' to a walk as in the example shown in figure 7.1 simplifies greatly when acting on operators of the general form

$$(w,e) =$$

$$(\ldots, s_i = j+1, s_{i+1} = j, s_{i+2} = j+1, s_{i+3} = j+2, \ldots, s_{i+x+1} = j+x, \ldots\ , e),$$

as exemplified by the *original* walk (the lower envelope) shown in the figure.

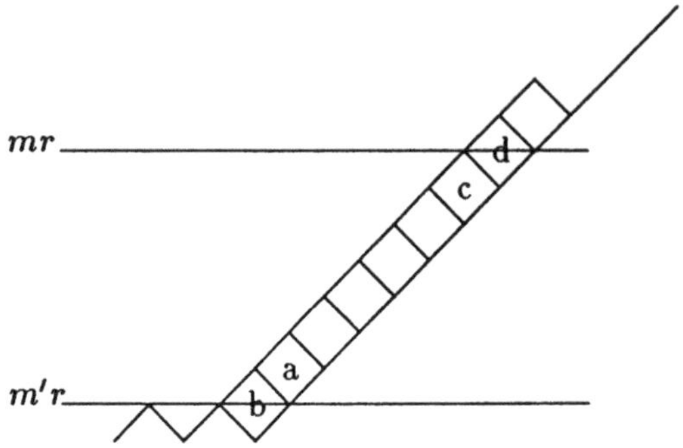

Figure 7.2: Dangerous strip endings.

Using the orthogonality property, the elementary operator for a walk modified from (w, e) by the addition of a strip of x diamonds, with the *first* diamond centered at height $j+1$ and involving the generator U_{i+1} (as shown), becomes:

$$L_{i+1,j+1}(x) \quad (w, e)$$

From Proposition 21 the new operator is finite at $r = r_c$ provided that the first added diamond is not at a height

$$(mr+1, mr, mr+1)$$

i.e. $j = mr$ (which would give a divergence); or at a height

$$(mr, mr-1, mr)$$

i.e. $j = mr - 1$ (which would give zero) unless $x = nr$ or $x = nr + 1$ ($n \in Z$); and the *last* added diamond is not at a height $(m'r, m'r-1, m'r)$ i.e. $j + x = m'r$ or at $(m'r-1, m'r-2, m'r-1)$ i.e. $j + x = m'r - 1$ (divergences) unless $x = n'r$ or $x = n'r + 1$ ($n' \in Z$) respectively. We have indicated the dangerous starting and finishing positions (marked a,b,c and d respectively) in an example with $r = 7$ in figure 7.2.

Definition 38 *For* $d > s \in Z_+$ *and* $d + s - 2 \leq n$ *define* $X_2(d+1, s-1) \in$

$T_n(q)$ *by*

$$X_2(d+1, s-1) = \prod_{k=s}^{1}{}' (U_{2k-1}U_{2k}...U_{d+k-2}).$$

Note

$$X_2^T(d+1, s-1)\, X_2(d+1, s-1) = Q^{s/2}\, X(d+1, s-1)$$

Proof: By iteration using Relations 6.15 and 6.16.

For s, d integers define $F(s,d)$ by

$$F(even, d) = 1$$

$$F(s, odd) = 1 \qquad (s)_{mod\ 4} = 1$$

$$F(s, even) = 1 \qquad (s)_{mod\ 4} = 3$$

$$F(s,d) = 0 \qquad otherwise.$$

Then

Proposition 23 *For* $0 \leq s < d$

$$(e\ ,\ 1\ \ d\ \ d-s) = (-1)^{F(s,d)}\, Q^{-s/2} \left(\prod_{j=1}^{s} \sqrt{\frac{k_{d-j+1}}{k_2}} \right) X_2(d+1, s-1)\, E_d$$

Proof:
By reorganising the definition of $(e, 1\ d\ d-s)$.
Note that

$$\begin{aligned}(e, 1\ \ d\ \ d-s) &= (e, 1\ \ d\ \ d-s)E_d \\ &= E_{d-s}^{(2s)} \left(\prod_{i=1}^{s} \frac{U_{2i-1}}{Q^{1/2}} \right) \left(\prod_{k=s}^{1}{}' R_{2k,2}(d-k-1) \right) E_d\end{aligned} \tag{7.19}$$

Proof:
For $0 \leq j < s$ we have, from proposition 22,

$$(\ e\ , (12)^j\ \ 1\ \ d-j\ \ d-s) = (\ e\ , (12)^{j+1}\ \ 1\ \ d-(j+1)\ \ d-s)\ R_{2j+2,2}(d-2-j).$$

Then

$$(e_{d-s}, e_{d-s}) = E^{(2s)}_{d-s} \prod_{i=1}^{s} \frac{U_{2i-1}}{Q^{1/2}}$$

by definition.

This completes the proof of equation 7.19.

We then write

$$\begin{aligned}
&\left(E^{(2s)}_{d-s} \prod_{i=1}^{s} \frac{U_{2i-1}}{Q^{1/2}}\right) \left(\prod_{k=s}^{1}{}' R_{2k,2}(d-k-1)\right) E_d \\
&= E^{(2s)}_{d-s} Q^{-s/2} \prod_{k=s}^{1}{}' \left(U_{2k-1}\ R_{2k,2}(d-k-1)\right) E_d \qquad (7.20)\\
&= E^{(2s)}_{d-s} Q^{-s/2} \prod_{k=s}^{1}{}' \left(U_{2k-1} \left(\prod_{l=1}^{d-k-1}{}' \sqrt{k_{1+l}k_{2+l}}\right)\right. \\
&\qquad \left.\left(1 - \frac{U_{2k}}{k_2} + .. + (-1)^{d-k-1} \frac{U_{2k}U_{2k+1}..U_{2k+(d-k-1)-1}}{k_2 k_3 .. k_{d-k}}\right)\right) E_d \\
&= E^{(2s)}_{d-s} Q^{-s/2} \prod_{k=s}^{1}{}' \left(U_{2k-1}\ U_{2k}\ ..\ U_{d+k-2}\right) E_d \qquad (7.21)
\end{aligned}$$

The proof of the last equality proceeds as follows. For $a = 1, 2, .., s$ we have

$$\begin{aligned}
&E^{(2s)}_{d-s} Q^{-s/2} \prod_{k=s}^{a}{}' \left(U_{2k-1}\ R_{2k,2}(d-k-1)\right) \\
&\qquad \left(\prod_{k=a-1}^{1}{}' \left(U_{2k-1}\ U_{2k}\ ..\ U_{d+k-2}\right)\right) E_d \\
&= E^{(2s)}_{d-s} Q^{-s/2} \left(\prod_{k=s}^{a+1}{}' \left(U_{2k-1}\ R_{2k,2}(d-k-1)\right)\right) \left(U_{2a-1}\ R_{2a,2}(d-a-1)\right) \\
&\qquad \left(\prod_{k=a-1}^{1}{}' \left(U_{2k-1}\ U_{2k}\ ..\ U_{d+k-2}\right)\right) E_d \\
&= E^{(2s)}_{d-s} Q^{-s/2} \prod_{k=s}^{a+1}{}' \left(U_{2k-1}\ R_{2k,2}(d-k-1)\right)
\end{aligned}$$

$$\left(U_{2a-1}\left(R_{2a,2}(d-a-2)+(-1)^{d-a-1}\sqrt{\frac{k_{d-a+1}}{k_2}}U_{2a}U_{2a+1}..U_{d+a-2}\right)\right)$$

$$\cdot\prod_{k=a-1}^{1}{}' \left(U_{2k-1}\, U_{2k}\, ..\, U_{d+k-2}\right)\, E_d$$

$$= E^{(2s)}_{d-s} Q^{-s/2} \prod_{k=s}^{a+1}{}' \left(U_{2k-1}\, R_{2k,2}(d-k-1)\right)$$

$$.(-1)^{d-a-1}\sqrt{\frac{k_{d-a+1}}{k_2}} \prod_{k=a}^{1}{}' \left(U_{2k-1}\, U_{2k}\, ..\, U_{d+k-2}\right)\, E_d$$

where we have used $U_i E_d = 0$ for $0 < i < d-1$ and the defining relations. Altogether we have, from equation 7.21,

$$(-1)^{F(s,d)}\, E^{(2s)}_{d-s} Q^{-s/2} \left(\prod_{k=1}^{s} \sqrt{\frac{k_{d-k+1}}{k_2}}\right) X_2(d+1, s-1)\, E_d$$

where the first factor may be replaced by 1 using

$$E^{(2s)}_{d-s}\, X_2(d+1, s-1) = X_2(d+1, s-1)\, E_{d-s}$$

which follows from the defining relations.

This completes the proof of proposition 23.

Lemma 2.2 *For* $r = r_c$

$$\sqrt{k_{d+1}}\, (e, 1\;\; d\;\; d-s)$$

is finite for $0 < s < r$*; and*

$$(e, 1\;\; d\;\; d-s)$$

is finite for $s = r$.

Proof:

(i) The case $s \neq r$.

Note than when $r = r_c$ the element $(e, 1\ \ d\ \ d-s)$ becomes undefined (divergent) due to a factor of $\sqrt{k_d}$. Specifically,

$$(e, 1\ \ d\ \ d-s) = \sqrt{k_d k_{d-1}}\,(1 - U_i/k_{d-1})\,(e, 1\ \ d-1\ \ d-2\ \ d-1\ \ d-s)$$

where the *last* factor is well defined by proposition 21.

(ii) The case $s = r$.

In this case the factor $\sqrt{k_d}$ is compensated by a factor of $\sqrt{k_{d-r+1}}$ (in the sense that the product $k_d k_{d-r+1}$ is finite at $r = r_c$) provided that $(e, 1\ \ d-r\ \ d-2r\ \ d-r)$ is finite. This is finite (by an induction) if $(e, 1\ \ 2r\ \ r)$ is finite. To see that this is finite note that we can formally use definition 24, together with proposition 7, to obtain $(e, 1\ \ r\ \ 0\ \ r)$, which is finite at $r = r_c$. This object is not in our usual basis set because we do not allow any sequences touching zero, and indeed it is not linearly independent of the usual basis set. However, it is a well defined operator, which is all we need here.

Corollary 2.2.1 *For $r = r_c$ and integer j such that $0 < jr < d$ then $(e, 1\, d\, d-jr)$ is finite.*

Proof: by induction on j.

Corollary 2.2.2 *For p a natural number the operator*

$$\sqrt{k_{d+1}}\,(\underbrace{(12)^p\ 1\ d-s}_{e}, (12)^{p-s}\ 1\ d\ d-s),$$

when written as a linear combination of reduced words, remains finite as

$$r \to r_c.$$

Proof:

$$(e, (12)^p 1\ \ d\ \ d-s) = \left(\prod_{i=1}^{p} U_{2i-1}\right)\ \ (e, 1\ d\ \ d-s)^{(2p)}$$

by definition.

We also have

Proposition 24

$$(1\ d\ d-s, 1\ d\ d-s) = \left(\prod_{k=1}^{s} k_{d-k+1}\right)\ \ E_d\, X(d+1, s-1)\, E_d.$$

Proof:
Consider

$$(1\ \ d\ \ d-s,e)(e,1\ \ d\ \ d-s) = (1\ \ d\ \ d-s,1\ \ d\ \ d-s)$$

$$= \frac{Q^{-s/2}}{k_2^s}\left(\prod_{k=1}^{s} k_{d-k+1}\right) E_d X(d+1,s-1)E_d$$

where we have used

$$X(d+1,s-1)_2^T\ X(d+1,s-1)_2 = Q^{s/2}X(d+1,s-1).$$

Corollary 24.1 *E^s_{d+s} is idempotent.*

For $2p+d+s = n+1$ consider the left ideal I generated by the idempotent

$$\left(\prod_{i=1}^{p} U_{2i-1}\right) E^{s(2p)}_{d+s}$$

that is by

$$((12)^p\ \ 1\ \ d+s,(12)^p\ \ 1\ \ d+s) + ((12)^p\ \ 1\ \ d\ \ d-s,(12)^p\ \ 1\ \ d\ \ d-s)\ \ .$$

Remark 3 *It is possible to prove at this stage that when $r = r_c$ this idempotent is primitive. In any case the result will come out indirectly in what follows.*

We will now show that I is an indecomposable projective left module of $T_n(q)$.

Lemma 2.3 *Let $x = 2p+d-1$, and let J be the left ideal invariant subspace of I generated by*

$$U_x\left(\prod_{i=1}^{p} U_{2i-1}\right) E^{s(2p)}_{d+s}.$$

Then, generically and in the case $r = r_c$, J is isomorphic to the left ideal generated by

$$\sqrt{k_{d+1}}\ (\underbrace{(12)^{p+s}\ 1\ d-s,}_{e}\underbrace{(12)^p\ 1\ d\ d-s}_{g}).$$

Proof:
With $p = 0$

$$U_x E^s_{d-s} = U_x\ E_d\ x(d+1, s-1)\ E_d = U_x\ (1\ \ d\ \ d-s, 1\ \ d\ \ d-s)$$

so that in general

$$U_x \left(\prod_{i=1}^{p} U_{2i-1} \right) E^{s(2p)}_{d+s}$$

$$= \frac{-1}{k_d}\,(g, g) + \sqrt{\frac{k_{d-1}}{k_d}}\,((12)^p\ \ 1\ \ d-1\ \ d-2\ \ d-1\ \ d-s,\ \ g)$$

by Proposition 7. Note that both terms are finite at $r = r_c$. Since both terms are also manifestly in the left ideal generated by

$$\sqrt{k_{d+1}}\,(\underbrace{(12)^{p+s}\ 1\ d-s}_{e}, (12)^p\ 1\ d\ d-s).$$

it remains to show that this element is in the left ideal generated by the sum.

Let

$$v = \frac{-1}{k_d}(g, g)$$

and

$$w = \sqrt{\frac{k_{d-1}}{k_d}}((12)^p\ \ 1\ \ d-1\ \ d-2\ \ d-1\ \ d-s,\ \ g)$$

then

$$(1 + \frac{1}{k_{d-1}}\ w^T)(v + w) = w$$

by Theorem 1.

Now note from lemma 21 (on well defined 'strips') that there exists a well defined operator R such that

$$w = R\ \sqrt{k_{d+1}}(e, g)\ ;$$

and from the invertability property in equation 6.33 that there exists a well defined operator R' such that

$$R'R = 1$$

Then

$$R'w = \sqrt{k_{d+1}}(e, g).$$

This completes the proof of Lemma 2.3.

Suppose j, t are natural numbers, $t < r$, and $2(jr + t) + d = n + 1$, and consider the sequence

$$\{e_d\} = (12)^t\,(12)^{jr}\ 1\ d$$

Lemma 2.4 *The left ideal of $T_n(q)$ generated by (e_d, e_d) is irreducible at $r = r_c$.*

Proof:

1. The case $t = 0$

The statement is true for $\{e_d\}$ of the form

$$\{e_d\} = (12)^{jr}\ \ 1\ \ d$$

(i.e. $t = 0$) since in this case the highest sequence obtained from $\{e_d\}$ is

$$\{f_d\} = 1\ \ d + jr\ \ d$$

and hence, by Lemma 2.2, the element (f, e) is finite at $r = r_c$. We then have the following

Sublemma 2.4.1 *If (e, e) and (f, e) are well defined then the left ideal generated by (e, e) is irreducible.*

Proof:

Firstly, (f, e) is in every invariant subspace of the left ideal generated by (e, e). To see this, note that it is unique among the basis states generated from (e, e) in containing a finite contribution of the longest possible word, then use the corollary to proposition 12. On the other hand

$$(e, f)(f, e) = (e, e)$$

so (e, e) is in every invariant subspace!

This completes the proof of Sublemma 2.4.1 and of the special case of the Lemma 2.4.

2. The case $t > 0$

Let t be a positive integer less than r. Then for $\{e_d\}$ of the more general form

$$\{e_d\} = (12)^t\ \ (12)^{jr}\ \ 1\ \ d$$

we proceed as follows. Consider the sequence

$$\{g\} = (12)^t \ 1 \ \ d + jr \ \ d$$

The operator (g, e_d) is finite, from above. It is therefore sufficient to consider the case of

Sublemma 2.4.2 *For all positive integers m, $d' = mr_c$ and*

$$\{e'_d\} = (12)^t \ 1 \ \ d'$$

the left ideal generated by $(e_{d'}, e_{d'})$ is irreducible.

This is sufficient since if the element (f, e) associated with the highest sequence starting from (e_{d+jr}, e_{d+jr}) in this case is well defined then so is the highest starting from (g, e_d) above.
Proof of Sublemma 2.4.2:
We will need:

Definition 39

$$W_t = (1 \ \ d + t \ \ d, e)$$

and
Definition 40

$$N_t = (1 \ \ d \ \ d - t \ \ d, e)$$

and

Definition 41 *Define $V_t \in T_n(q)$ as the product of factors required by definition 24 in the construction of W_t from N_t:*

$$W_t = V_t N_t$$

Note that the longest possible word in V_t is that obtained by replacing the consruction in definition 24 with that in definition 25. We then have

Sublemma 2.4.3 *For X_t some linear combination of words, including the longest possible word, with coefficients finite at $r = r_c$, then V_t may be written in the form*

$$V_t = \underbrace{[(\overbrace{k_d k_{d+1}}^{=-1+\sqrt{Q}k_{d+1}})^{2t-1} \prod_{a=1}^{t-1} (\overbrace{k_{d-a} k_{d+a+1}}^{=1})^{2t-2a-1}]^{1/2}}_{[]_t} \ (1 + X_t / k_d)$$

Proof:

By induction on t. The statement is true for $t = 1$ by definition 24. Assume true at some level t, then consider building W_{t+1} in the following way:

$$W_{t+1} = L^{(-)}_{d+2t,d}(t+1)\ \ V_t\ \ L_{d+t,d-t}(t)\ \ N_{t+1}$$

$$= \left(1 - \frac{U_{d+2t}}{k_d} + \underbrace{\sum_{a=2}^{t+1} \prod_{m=1}^{a} \frac{-U_{d+2t+m-a}}{k_{d-1+m}}}_{Z_t} \right) \Box_{t+1}\ (1 + X_t/k_d)$$

$$\cdot \left(1 + \sum_{a=1}^{t-1} \prod_{j=a}^{1\ \prime} (-U_{d+t-1+j}/k_{d-t-1+j}) + \prod_{j=t}^{1\ \prime} (-U_{d+t-1+j}/k_{d-t-1+j}) \right) N_{t+1}$$

Note that the largest possible word is present, and all the terms are either finite or vanishing like $1/k_d$ as $r \to r_c$. We therefore need to keep track only of the terms:

$$\underbrace{\left(\sum \prod (-U/k)\right)}_{Z_t} \prod_{j=t}^{1\ \prime} (-U_{d+t-1+j}/k_{d-t-1+j})$$

$$+ \sum_{a=1}^{t-1} \prod_{j=a}^{1\ \prime} (-U_{d+t-1+j}/k_{d-t-1+j}).$$

Expanding and using the Temperley-Lieb relations these combine in pairs between the two sums to give terms with the required factor of $1/k_d$.

This completes the proof of Sublemma 2.4.3.

Sublemma 2.4.4 *There exist scalar functions of r, a_i, b_i $(i = 1, 2)$ such that*

$$G_i = a_i W_s + b_i N_s \qquad (i = 1, 2) \tag{7.22}$$

are well defined at $r = r_c$, with at least one having finite coefficient of the longest word, and such that

$$G_i^T G_j = \delta_{ij}(e, e)$$

Proof:

Note that

$$N_s = \sqrt{k_d k_{d-1}}\,(1 - U_{d-1}/k_{d-1})(\underbrace{1\ d-1\ d-2\ d-1\ d-s\ d}_{h},\ e)$$

where (h,e) is finite at $r = r_c$ by repeated application of Proposition 21. Then

$$a_i W_s + b_i N_s = (a_i V_s + b_i)\sqrt{kk}\,(1 - U/k)\,(h,e)$$

$$= \{(\square a_i + b_i) + a_i \square X_s/k_d\}\sqrt{kk}\,(1 - U/k)\,(h,e)$$

From this we require

$$\square a_i + b_i = K_i/\sqrt{k_d}$$

where K_i is well defined at $r = r_c$. Putting $K_1 = 0$ then

$$b_1 = -\square a_1$$

and equation 2.4.4 with $i = j = 1$ then requires

$$a_1^2 = 1/(1 + \square\square)$$

Now $i \neq j$ in equation 2.4.4 requires

$$a_1 a_2 + b_1 b_2 = 0$$

which implies

$$a_2 = \square b_2$$

but then

$$b_2^2 + \square\square b_2^2 = 1$$

using $i = j = 2$, so

$$b_2^2 = 1/(1 + \square\square).$$

Finally, we still require

$$\square\square a_2^2 + 2\square a_2 b_2 + b_2^2 = K_2/k_d$$

Now this gives

$$\frac{\square^4}{1+\square\square} + 2\frac{\square\square}{1+\square\square} + \frac{1}{1+\square\square} = K_2/k_d$$

so

$$K_2 = k_d(1 + \square\square)$$

which is finite at $r = r_c$ as required.

This completes the proof of Sublemma 2.4.4.

Let us define

$$(1\ d + t\ d,\ e)_r = G_1 = a_1 W_t + b_1 N_t$$

and

$$(1\ d\ d - t\ d,\ e)_r = G_2 = a_2 W_t + b_2 N_t.$$

Now suppose there exists an invariant subspace of the ideal generated by (e, e). Consider any vector V in this subspace. By proposition 12 the invariant subspace generated by V necessarily includes a vector containing a finite contribution of the longest possible word, and hence of the form

$$Z = \sum_i C_i\, Y_i + C\ (1\ d + t\ d, e)_r + C'\ (1\ d\ d - t\ d, e)_r$$

where the elements Y_i are words other than the longest possible word, and C_i, C, C' are constants such that $C + C' \neq 0$. Therefore, at least one of G_1 and G_2 is in the subspace . Therefore (e, e) is in it (using Sublemma 2.4.4) and therefore the whole space is in it.

This completes the proof of Sublemma 2.4.2 and of Lemma 2.4.

Remarks on basis vectors

With reference to sublemma 2.4.4, and in particular equation 7.22 we introduced

$$(1\ \ d + t\ \ d, e)_r = G_1$$

and

$$(1\ \ d\ \ d - t\ \ d, e)_r = G_2$$

More generally, if a sequence $\{s\}$ has a subsequence of the form

$$s_i = d \quad d + r > s_{k>j>i} > d \quad s_k = d$$

then define a 'reflection' $\{s'\}$ differing only in

$$s'_i = d \quad s'_{k>j>i} = 2d - s_j \quad s'_k = d.$$

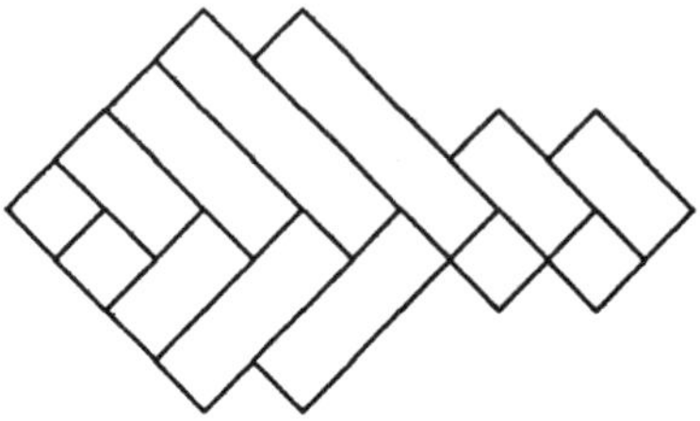

Figure 7.3: Building a 'reflection' by adding strips.

We generalise sublemma 2.4.3 by defining $V_{\{s\}} \in T_n(q)$ as any operator such that

$$(s,e) = V_{\{s\}}\ (s',e)$$

whereupon:

Lemma 2.5 *For $X_{\{s\}}$ some linear combination of words, including the longest possible word, with coefficients finite at $r = r_c$, and $[]_{\{s\}}$ is some scalar function of r which takes the value 1 at $r = r_c$, then $V_{\{s\}}$ may be written in the form*

$$V_{\{s\}} = [k_d k_{d+1}]^{1/2}\ []_{\{s\}}\ \ (1 + X_{\{s\}}/k_d)$$

Proof: similar to sublemma 2.4.3. Consider the order of construction of (s,e) from (s',e) exemplified, in the case

$$\{s\} = ...d\ \ d+1\ \ d+2\ \ d+3\ \ d+4\ \ d+3\ \ d+4$$

$$d+3\ \ d+2\ \ d+1\ \ d+2\ \ d+1\ \ d+2\ \ d+1\ \ d...,$$

by the diagram in figure 7.3. Now assume the proposition true for some pair s, s' corresponding to part of this diagram on the left (it is clearly true for $\{s\} = ...d\ \ d+1\ \ d...$). Work by induction on the number of pairs of strips $L, L^{(-)}$ (of whatever length) added to the right.

Corollary 24.2 *Let $\{s\}$ have no other reflections, then there exist scalar functions of r, a_i and b_i, such that*

$$(s,e)_r = a_1\ (s,e) + b_1\ (s',e)$$

and

$$(s',e)_r = a_2\ (s,e) + b_2\ (s',e)$$

are well defined at $r = r_c$, with at least one having finite coefficient of the longest possible word, and such that with $\{w\}, \{w'\}$ each either $\{s\}$ or $\{s'\}$ then

$$(w,e)^T\ (w',e) = \delta_{w,w'}\ (e,e).$$

Proof: As for sublemma 2.4.4.

Corollary 24.3 *By linearity if $\{s\}$ has many reflections (i.e. many subsequences of the type indicated) then repeated use of these transformations gives well defined operators.*

The new notations for G_1 and G_2 above signify that the well defined vectors have been obtained by linear transformations among pairs of operators (s,t) whose sequences are related by reflection. An arbitrary sequence can have many such reflectable subsequences and hence many reflections. We then have

Definition 42 *Define $(s,t)_r$ and its various reflections (in the sense of the preceding discussion) as the well defined elementary operator obtained from (s,t) and its various reflections by applying concommitant linear transformations such as in equation 7.22.*

The existence of well defined $(s,t)_r$ is a consequence of Lemma 2.4, independently of the explicit construction in equation 7.22.

Definition 43 *I_{d+s} is the left ideal and D_{d+s} the double sided ideal generated by*

$$(e_{d+s}\ o\ e_{d+s}).$$

Lemma 2.6 *The quotient I/J is isomorphic to the quotient I_{d+s}/D_{d+s-2}.*

Proof:
Up to the respective quotients these ideals have, from their definitions, the same generating elements.

Corollary 2.6.1 *I/J is indecomposable.*

Proof: Recall that $(e_{d+s} \; o \; e_{d+s})$ is a primitive idempotent up to quotients by D_{d+s-2}

Let us introduce K, the maximal proper invariant subspace of the indecomposable left ideal I/J.

Lemma 2.7 *The quotient by K of the left ideal I/J has dimension at least $C_{n+1,d+s,r}$.*

Proof:

By establishing the existence of this number of basis vectors.

Under the quotient by J we can treat E^s as having the relevant properties of a well defined idempotent E, and build elements $(w,e)^s$ analogous to (w,e) (i.e. obeying

$$(e,w)(w,e) = (e,e) \quad)$$

using E^s in place of E in the definition.

Specifically, we will show that there is a finite $(w,e)^s$ or equivalent (i.e. a reflection $(w,e)^s_r$) for every sequence $\{w\} \in S(n+1, d+s, r)$.

Recall the definition of $S(n+1, d+s, r)$. Any $\{w\} \in S(n+1, d+s, r)$ has the property that there is no $w_j = d+r$ after the last $w_i = d$. Consider the sequence $\{w'\}$ given by

$$\{w'\} = (12)^t \;\; 1 \;\; d \;\; w_{i+1} \;\; w_{i+2} \; \ldots \; d+s$$

where $w_i = d$ is the last d in the sequence. Then $(w',e)^s$ is well defined and there exist $R, R' \in T_n(q)$ such that

$$RR' = R'R = 1$$

and

$$R(w',e)^s = (e,e)^s$$

by Proposition 21. On the other hand, there is a well defined basis vector for each modification of $\{w'\}$ to the *left* of $w'_i = w_i = d$ by Lemma 2.4.

This completes the proof of the Lemma 2.7.

Lemma 2.8 *The irreducible invariant subspace of I/J has dimension*

$$C'_{n+1,d+s,r}$$

and is isomorphic to K.

Proof:

Similar to above but with a 'twist' .....

We will show that there is an independent basis vector in the subspace for each sequence in $S'(n+1, d+s, r)$. Consider $\{t\} \in S'(n+1, d+s, r)$ and the finite vectors

$$\frac{1}{\sqrt{k_d}}\,(t, e)_r$$

Then consider the subset of sequences in $S'(n+1, d+s, r)$ labelled by integer j such that $t_j = d+r$ and there is no $t_{i>j} = d$. If one element of the subset gives rise to a basis vector for the subspace then they all do by lemmas 2.4 and equation 6.33. Thus we only need to find one vector for each possible j. We will do this by induction. Suppose an example exists for some j (it exists for the maximum possible value of j, specifically the vector

$$v' = \frac{1}{\sqrt{k_d}}\,(1212...1\; d+r\; d+s,\; e),$$

which is in the invariant subspace by Sublemma 2.4.2) then without loss of generality a vector of the form

$$v'' = \frac{1}{\sqrt{k_d}}\,(...\;\; d{+}r{-}2\;\; d{+}r{-}1\;\; d{+}r{-}2\;\; d{+}r{-}1\;\; \underbrace{d+r}_{s_j}\;\; d{+}r{-}1\;\; ...\; d{+}s\;, e_{d+s})$$

is present (Proposition 21 and equation 6.33). Now by Lemma 2.4 there exist operators $R, R' \in T_n(q)$ such that

$$RR' = R'R = 1$$

and, defining

$$f' = \frac{1}{\sqrt{k_d}}\,(...\;\; d{+}r{-}2\;\; d{+}r{-}1\;\; d{+}r\;\; d{+}r{+}1\;\; d{+}r\;\; d{+}r{-}1\;\; ...\; d{+}s\;, e_{d+s})_r$$

we have

$$f' = Rv''$$

and

$$v'' = R'f'.$$

To see this note that v'' and f' involve identical sequences after $s_j = d+r$, and that Lemma 2.4 therefore implies that they are each part of the basis for an irreducible representation of $T_{n+s-r}(q)$ (so in fact $R, R' \in T_{n+s-r}(q) \subset T_n(q)$).

Now consider a vector with subsequence of the form

$$w = \frac{1}{\sqrt{k_d}}(\ldots\ d+r-2\ \ d+r-1\ \ d+r\ \ d+r-1\ \ \underbrace{d+r-2}_{s_j}\ \ d+r-1\ \ldots, e)$$

Since the subsequence shown is just the reverse of that in v'', and the corresponding subsequence in f' is symmetrical under this reversal, then by symmetry there exist $P, P' \in T_n(q)$ such that

$$PP' = P'P = 1$$

and

$$f' = Pw$$

and

$$w = P'f'$$

Then

$$w = P'Rv''$$

and

$$v'' = R'Pw.$$

From Proposition 14 and Lemma 2.7 then both bounds on dimensions are saturated or I/J is irreducible.

In fact we can check the closure of the invariant subspace on the given set of $C'_{n+1,d+s,r}$ basis states directly. By definition 24 the only two situations in which it would be possible to get out of this set are when acting with U_i on vectors of the form

$$\frac{1}{\sqrt{k_d}}(\ldots(d+r-1\ \ d+r\ \ d+r-1)_i\ldots, e)$$

or

$$\frac{1}{\sqrt{k_d}}(\ldots(d+1\ \ d+2\ \ d+1)_i\ldots, e).$$

However, we see from the definition 24 that in both cases the *coefficient* of the vector outside the claimed basis set is zero when $r = r_c$.

For example,

$$U_i\ \frac{1}{\sqrt{k_d}}(\ldots(d+1\ \ d+2\ \ d+1)_i\ldots, e) =$$

$$-\underbrace{\sqrt{\frac{k_{d+1}}{k_{d+2}k_d}}}_{=0}\underbrace{(...(d+1\ d\ d+1)_i...,e)}_{not\ in\ S'}+\underbrace{\frac{1}{k_{d+2}}}_{=\sqrt{Q}}\frac{1}{\sqrt{k_d}}(...(d+1\ d+2\ d+1)_i...,e) \tag{7.23}$$

This completes the proof of Lemma 2.8.

Lemma 2.9 *The invariant subspace J has irreducible invariant subspace A of dimension $C'_{n+1,d-s,r}$. The quotient J/A is irreducible.*

Proof: Similar to Lemmas 2.6, 2.7 and 2.8.

Lemma 2.10 *The projective module I has an invariant subspace L generated by*

$$\frac{1}{\sqrt{k_d}}(g_{d+s}\ ,\ e)^s$$

with invariant subspace A and $L/A = K$.

Proof:
$\frac{1}{\sqrt{k_d}}(g_{d+s}\ ,\ e)^s$ is obtained from $(e_{d+s}, e_{d+s})^s$ by acting with $r-s$ 'strip' operators of the type $L_{i,j}(x)$. The effect on the J part of $(e_{d+s}, e_{d+s})^s$ is of the form

$$L_{i,j}(x)\ \ \frac{1}{k_d}(g_{d-s}, g_{d-s}) \propto \frac{1}{k_d}(1212...121\ 2\ 1\ d+1\ \ d-s, g)$$

(see Proposition 21 for details, and note that we also use Propostion 8). When the I/J part has reached $\frac{1}{\sqrt{k_d}}(g_{d+s}\ ,\ e)^s$ we have in the J part

$$\frac{1}{k_d}(1212...1\ \ \underbrace{d+r-s}_{s_i}\ \ d-s\ ,\ g).$$

Acting with U_i then gives

$$U_i\ \frac{1}{\sqrt{k_d}}(g_{d+s}\ ,\ e)^s = U_i\ \frac{1}{k_d}(1212...1\,d+r-s\ d-s, g)$$

$$= \Theta\,(1212...1\,d+r-s-1\;\underbrace{d+r-s-2}_{s_i}\;d+r-s-1\;\;d-s,g)$$

$$+\Theta'\,(1212...1\,d+r-s\;d-s,g)$$

(where Θ, Θ' are constants) by Proposition 8. These operators generate A.

This completes the proof of Lemma 2.10.

Sublemma 2.10.1 *The invariant subspace A is isomorphic to the irreducible quotient of I/J by K.*

Proof:

D is irreducible, and note that v is a homomorphic image of

$$((12)^p\;1\;\;d+s, (12)^p\;1\;\;d+s)^s$$

That is, we have

$$(e,e)^s \overset{v}{\mapsto} (e,e)^s v = v$$

by the definitions and the generic theorem. We may then build on either object from the right in the same way, since the required factors from definition 24 are either identical or else the same via

$$k_{d+s+1} = 1/k_{d-s}.$$

Note that the left ideal generated by v closes since U_i does not have an inverse (see equation 7.23 for an example), whilst $(I/K)/J$ closes at the corresponding point by virtue of the K quotient.

Altogether we see that our indecomposable projective I has Loewy structure

$$(I/K)/J \sim A$$

$$K/J \qquad\qquad\qquad\qquad J/A\;.$$

$$A$$

The projective I' generated by $(e_{d+2r-s}, e_{d+2r-s})^s$ has structure

$$(I'/K')/J' \sim J/A$$

$$K'/J' \sim A \sim (I/K)/J \qquad\qquad\qquad\qquad J'/A'$$

$$A' \sim J/A$$

- c.f. Theorem 2.

We are now in a position to identify the operators e α β and γ from the alternative statement of the theorem. Recall that with $\{g_{d-s}\} = 1212....12\,1\;d\;\;d-s$ we have

$$(e_{d+s}, e_{d+s})^s = (e_{d+s}, e_{d+s}) + (g_{d-s}, g_{d-s}).$$

In what follows (w,t) means $(w,t)^s_r$.

Let $Q_C(m)$ be a direct summand of $T_n(q)$ and, for $k = 1, 2, .., m$, $(e_k)_{ij} \in Q_C(m)$, where k indexes the indecomposable projectives in $Q_C(m)$. We put $k = 1$ for the projective generated by (e_s, e_s) with $s < r$, $k = 2$ for (e_p, e_p) with $p = 2r - s$, $k = 3$ for $p = 2r + s$, and so on (each new 'height' p a reflection of the previous one about the integer multiple of r immediately above it). Then with k corresponding to a typical $p = d + s$

$$(e_k)_{ij} = (x_i, e_{d+s})(e_{d+s}, x_j) \qquad x_i, x_j \in S(n+1, d+s, r)$$

$$(\alpha_k)_{aj} = \frac{1}{\sqrt{k_d}}(y_a, e_{d+s})(e_{d+s}, x_j) \qquad y_a \in S'(n+1, d+s, r)$$

$$(\beta_{k-1})_{zj} = \frac{1}{\sqrt{k_d}}(y_z, g_{d-s})(e_{d+s}, x_j) \qquad y_z \in S(n+1, d-s, r)$$

$$(\gamma_k)_{ij} = \frac{1}{k_d}(x_i, g_{d-s})(e_{d+s}, x_j) \qquad x_i \in S'(n+1, d-s, r) \sim S(n+1, d+s, r).$$

This completes the proof of the Main Theorem.

This has been quite a complicated proof, generating alot of extra information on the way. There are several relatively concise ways to check it, and we will describe a very powerful one in chapter 9.

Chapter 8

Graph Temperley-Lieb algebras

An enormous amount of work has been done on 2 dimensional statistical mechanics, considering that 3 and 4 dimensions are probably the most important physically! This is partly because models are more tractable in 2 dimensions. However, the imbalance must not persist indefinitely.

Graph Temperley-Lieb algebras are certain generalisations of Temperley-Lieb algebras, including those appropriate for building the transfer matrices for Potts models in arbitrary dimensions. In this chapter we construct various representations for these algebras, which are again parameterised by the scalar Q.

In particular we give the generically irreducible representations which are appropriate for building the transfer matrices of statistical mechanical models. These representations are well defined for all Q, whereas the representations coming directly from physical models are typically only defined for certain values of Q. We also give representations with bases derived from the partitions of n distinguishable objects (n is the number of nodes in the graph). These bases give representations for all possible spatial lattices! We will show how to compute the dimensions of these representations by a diagrammatic technique.

8.1 Introduction

The Temperley-Lieb algebras $T_n(Q)$ of chapter 6 may be defined as follows. Associate a generator U_i with each node i of an A_n graph (figure 8.1). The

Figure 8.1: The A_n graph for $n = 7$.

generators have relations

$$\begin{aligned} U_i U_i &= \sqrt{Q} U_i \\ U_i U_j U_i &= U_i \qquad \ldots \text{ if nodes } i \text{ and } j \text{ are connected by a bond,} \\ U_i U_j &= U_j U_i \qquad \ldots \text{ otherwise.} \end{aligned} \tag{8.1}$$

We now consider the effect of generalising the graph associated with the defining relations 8.1 to any graph constructed as follows. Form a 'pregraph' from a collection of nodes or 'sites' somehow connected by bonds, the graph is obtained by decorating the middle of every bond with a further site [1]. The algebras associated with the relations 8.1 in such cases will be called graph Temperley-Lieb (GTL) algebras. The algebra associated with a given pregraph G will be denoted $GT(Q)$ (so $T_{2n-1}(Q)$ becomes $A_nT(Q)$ and so on). We will examine the representation theory for these new algebras.

An immediate application of this work is in the 3 dimensional Potts spin and lattice gauge models, for which the algebra generators for an appropriate graph build the 2 dimensional layer transfer matrix. The identification of the irreducible representation associated with the free energy in these models simplifies the computation of the free energy on the finite lattice. The breakdown of the Q-generic structure of the algebra should signal a 3 dimensional series of critical field theory limits analogous to the central charge $c \leq 1$ conformal series ($Q = 4\cos^2(\pi/r); r \in \mathbf{Z}_+$) in 2 dimensions (c.f., for example, Cardy 1987).

In the next section we give the representations of graph Temperley-Lieb algebras (with Q positive integer) associated with Potts models, and describe the quotient relations obeyed by these representations In section 8.3 we give representations (with Q an indeterminate) associated with the set of partitions of n distinguishable objects. We identify generically irreducible subspaces and determine the corresponding quotient relations. In section 8.4 we discuss the physical consequences of these results.

---

[1] There are many further generalisations possible, involving directed bonds, multiple bonds and so on. Our choice is motivated by physical considerations (see later) given which such further complications are as yet superfluous.

8.2 The Potts Representation

8.2.1 Construction

For definiteness, then, let us start by considering the algebras appropriate for the 3d Potts models. The pregraphs here are square lattices, so the graphs are edge centred square lattices. The 3d Potts $l.m$-site layer transfer matrix may be written

$$T(v) = \prod_{\text{layer sites } i.} (v + Q^{1/2}U_{i.}) \prod_{\text{layer bonds } ij} (1 + vQ^{-1/2}U_{ij}) \qquad (8.2)$$

where $v = (\exp(\beta) - 1)$ and β is the coupling parameter, as in the 2 dimensional case (see, for example, Baxter 1982). The layer site label $(i.)$ may be regarded as a vector specifying a site position in the layer $i.=(i_1, i_2)$ $(i_1 \in 1,..,l; i_2 \in 1,..,m)$; and the bond label (ij) as a pair of vectors specifying adjacent (i.e. bond connected) sites in the layer.

For more general pregraphs the site label is not naturally specified as a 2 dimensional vector. We will, however, retain the distinction of indexing pregraph sites with a dotted letter, and bonds with the associated pair of sites. This pregraph notation seems to be more convenient in general than labelling generators with a node of the derived graph (as is done in the relations (1)). Henceforward we will always specify algebras by their pregraph, and the relations (1) (which apply to the *graph*) should be interpretted accordingly.

The matrices $U_{i.}$ and U_{ij} are given as follows:

$$U_{i.}, U_{ij} \in End(\bigotimes_{i_1=1,..,l; i_2=1,..,m} V^{(i.)}) \qquad (8.3)$$

where $V^{(i.)}$ is the Q dimensional vector space with basis the possible Potts spin variable values on site $(i.)$, say $s_{i.} \in \{1, 2, .., Q\}$. Then:

$U_{i.}$ acts trivially on $V^{(j.)}$ unless $i. = j.$ (i.e. $i_1 = j_1$ and $i_2 = j_2$);
$U_{i.}$ on $V^{(i.)}$ is $Q^{-1/2}\ M$, where M is the Q-by-Q matrix with all entries unity.

Meanwhile:

U_{ij} acts trivially on $V^{(k.)}$ unless $i. = k.$ or $j. = k.$;
U_{ij} on $V^{(i.)} \otimes V^{(j.)}$ is diagonal$(Q^{1/2}\ \delta(s_{i.}, s_{j.}))$.

Note that the generalisation of this definition to arbitrary graphs is automatic.

8.2.2 The Potts quotient algebra

We may think of these matrices as *defining* a finite dimensional 'Potts' algebra for each integer Q (the dimension of such an algebra is $\leq Q^{lm}$) or we may think of them as giving a representation of the abstract algebra associated with the original graph. The quotient relations describing the place of the Potts algebra in the abstract algebra are not obvious, but note, for instance, that the Potts representation is real and symmetric, while representations need not be unitarisable in general. The abstract algebra can be finite dimensional only if the pregraph has no closed loops and no nodes with coordination number greater than 3 (generically it should be a Coxeter-Dynkin diagram for finiteness). Finite dimension of the abstract algebra is not particularly significant physically, so not too much should be read into this.

In any case the Potts representation must obey at least one set of quotient relation on physical grounds, if they are not already a consequence of the relations (1). That is, $T(0)$ should be a primitive idempotent (at least up to normalisation). For the $A_nT(Q)$ algebras this is already a consequence of the original relations, but in other cases it simply amounts to the statement that at high temperatures the Potts model is disordered. The explicit relations may be determined by reference to inhomogeneous Potts model partition functions with free boundary conditions in the layering direction.

By similar arguments a minimal list of sets of quotient relations appropriate for such physical models (and all of which are trivial in the A_n case) is indicated as follows. For W any word of length $O(W)$ in the generators $\{U_{i.}, U_{ij}\}$ and

$$R_{b^{(N)}} = \prod_{i.\not\in b^{(N)}} (U_{i.}/Q^{1/2}) \tag{8.4}$$

where $b^{(N)}$ is any N element subset of nodes in the pregraph, so $0 \leq N \leq n$, and, for example,

$$R_0 = R_{b^{(0)}} = \prod_{\text{all pregraph nodes } i.} (U_{i.}/Q^{1/2}), \tag{8.5}$$

then:

$$R_{b^{(N)}} W R_{b^{(N)}} = \chi_{b^{(N)}}(W) R_{b^{(N)}} \quad (mod.\ R_{b^{(M)}}\ \forall\ b^{(M)} \subset b^{(N)}) \tag{8.6}$$

where $\chi_{b^{(N)}}(W)$ is a scalar, and in particular, generically $\chi_{b^{(n)}}(W) = 0$ $(O(W) > 0)$. The proof is outlined in section 8.3.6.

This means that we may, in principle, associate not necessarily distinct indecomposable representations with the left ideals generated from each $R_{b^{(N)}}$ (mod. $R_{b^{(M)}}$).

8.2.3 Reducibility of the Potts representation

To see that the Potts representation for $Q = 2, 3, 4, \ldots$ is reducible consider the similarity transformation by the matrix S defined by

$$\begin{aligned} S_{ij} &= (1/(Q+1-i)) \quad \ldots j \geq i \\ S_{ii-1} &= -1 \\ S_{ij} &= 0 \ \ldots \text{ otherwise.} \end{aligned} \tag{8.7}$$

on every subspace $V^{(k.)}$. The site operators then become diagonal $(Q^{1/2}, 0, 0, ..)$ and the bond operators do not mix between states with odd and even numbers of spins in state $s_{k.} = Q$.

We know from the A_n case that the decomposition described here for $Q = 2$ is complete. This is a remarkable result. It means that the $Q = 2$ A_n Potts algebra is isomorphic to every other $Q = 2$ n-node pregraph Potts algebra. In other words the operators for building arbitrary dimensional interactions may be built from just one local operator and a set of spatially translating and rotating conjugations, all of which already exist in the A_n (i.e. 2 dimensional Potts model) algebra.

For $Q = 3$ we have shown that the Potts representation contains at least 2 irreducible components. On the other hand we know from the A_n case that there are at most 3. We also know on physical grounds that the multiplicities of the first 2 irreducibles are 1 (from the uniqueness of the free energy) and 2 (from the degeneracy of the ordered state) respectively. The lower bounds for their dimensions (and that of the third, if it exists) from the A_n case, together with the total dimension of the Potts representation and the dimensions of the blocks we exhibited above, are given in the following table for enough cases to show the pattern:

n	$d1$	$d2$	$d3$	$dPotts$	$b1$	$b2$
1	1	1	0	3	2	1
2	2	3	1	9	5	4
3	5	9	4	27	14	13
4	14	27	13	81	41	40

We see that the only possibility is that, here also, the A_n Potts algebra is isomorphic to every other n-node pregraph algebra. The same remarkable consequences also pertain.

In fact the same is trivially true for $Q = 0, 1$. On the other hand it is *not* true for Q large enough.

8.3 Partition representations

8.3.1 The bases

Note that for $Q = 0$ the Potts representation is undefined, and for $Q = 1$ it is 1 dimensional ($U_{i.} = U_{ij} = 1 \ \forall \ i, j$). Otherwise the representations grow rapidly with l, m and we have shown that they are reducible. Furthermore, it is not obvious how to generalise to non-integer Q, although the defining relations make no obvious distinction. Let us proceed by constructing 2 further types of representation from a different (although related) source. One of these types is generically irreducible and both are defined for all Q. The *idea* for these representations comes from the Whitney polynomial (Baxter 1982) for certain non-planar graphs which we will describe later. However, the basis is most naturally described in a more general framework.

Consider the set of partitions of n distinguishable objects. For example with $n = 1, 2, 3$ we have

$$\{(1)\}, \{(1)(2), (12)\}, \{(1)(2)(3), (12)(3), (13)(2), (23)(1), (123)\}.$$

Another useful notation here is to replace the list of all partitions in each case with just the list of all partition shapes (as in Young diagrams (Robinson 1961)) preceded by their corresponding multiplicities (if other than unity). In this scheme the $n = 3$ set becomes

$$\{(1^3), 3.(21), (3)\}.$$

We will define $\mathcal{S}_n$ to be the number of partitions in each case. Then for $n = 1, 2, 3, 4, 5, 6, 7, 8$ we have

$$\mathcal{S}_n = 1, 2, 5, 15, 52, 203, 877, 4140,$$

and so on. We will show how to compute these numbers below.

If 2 objects (a, b say) are clustered together in a partition we say that they are connected, and write $a \sim b$. Of course $a \sim b$ and $b \sim c$ implies $a \sim c$.

Now associate to each partition the set of possible partitions of its connected clusters into 2 'teams' (including the possibility of empty teams). If a partition has c clusters then the set has 2^c elements. For example the set for (13)(2) is

$$\{((13)(2), \emptyset), ((13), (2)), ((2), (13)), (\emptyset, (13)(2))\}.$$

Here ((13)(2),$\emptyset$) means the element in which the clusters (13) and (2) are in the first team, and the second team is empty. The union of such sets over the partitions then provides a basis for a representation of any graph Temperley-Lieb algebra with an n node pregraph, as we will see shortly.

It is convenient to partially order the union by putting all the elements with i cluster '2nd teams' before all those with j cluster 2nd teams if $i < j$; and *within* this ordering by putting all elements with d clusters in total before those with e clusters in total if $d > e$. The number of elements in the union with i cluster 2nd teams, $\mathcal{S}_n(i)$, is computed below. Note that we have put the (1^n) partitions first in each fixed i subset.

8.3.2 Computation of $\mathcal{S}_n(i)$

To compute $\mathcal{S}_n(0)$ note that it is the sum of entries in the nth row of the following diagram:

$$\begin{array}{cccccc}
1 & & & & & \\
1 & 1 & & & & \\
1 & 3 & 1 & & & \\
1 & 7 & 6 & 1 & & \\
1 & 15 & 25 & 10 & 1 & \\
.. & .. & . & .. & .. & ..
\end{array}$$

where the number in the jth column of the nth row is the sum of the number in the $(j\text{-}1)^{th}$ column of the $(n\text{-}1)^{th}$ row and j times the number in the jth column of the $(n\text{-}1)^{th}$ row. To see this note that $\mathcal{S}_n(0)$ is also the number of sequences of length n, $\mathbf{p}=\{p_1, p_2, ..., p_n\}$, with the property that

$$1 \leq p_i \leq ((max\ p_j\ \forall\ 1 \leq j \leq i) + 1). \tag{8.8}$$

Proof: Associate p_i with node i and say nodes i, j are connected if $p_i = p_j$ (i.e. *label* the relevant cluster by p_i). Then the cluster containing node 1 is always cluster no.1 , that containing node 2 is cluster no.2 unless it is in cluster no.1, and so on. It is easy to see that this is a unique representation of partitions.

On the other hand the possibilities for $\mathbf{p}$ may be enumerated diagrammatically as follows:

$$\begin{array}{l|ll|lll|lll|lll|llll}
p_1: & & & & & & & 1 & & & & & & & & \\
p_2: & & & & 1 & & & & & & & 2 & & & & \\
p_3: & & 1 & & 2 & & & 1 & & & & 2 & & & 3 & \\
p_4: & 1 & 2 & 1 & 2 & 3 & 1 & 2 & 3 & 1 & 2 & 3 & 1 & 2 & 3 & 4
\end{array}$$

and so on. The claimed result follows.

Note also from this construction that the jth number in the nth row in the original diagram gives the number of partitions into exactly j non-empty clusters. Note that the maximal such contribution to any row comes from further to the right as n increases, so the asymptotic ratio is

$$\lim_{n\to\infty} \mathcal{S}_{n+1}/\mathcal{S}_n = \infty. \tag{8.9}$$

To compute $\mathcal{S}_n(i)$ note that the number of ways of choosing i clusters from j to go into the second team is $(j)!/(j-i)!(i)!$ if $j \geq i$ and zero otherwise. Thus the corresponding diagram for arbitrary i is obtained by multiplying each entry in the diagram above by the appropriate factor. For example for i=1,2,3 we have:

1				
1	2			
1	6	3		
1	14	18	4	
1	30	75	40	5

0				
0	1			
0	3	3		
0	7	18	6	
0	15	75	60	10

0				
0	0			
0	0	1		
0	0	6	4	
0	0	25	40	10

and so on.

8.3.3 Representation type 1

Assign one distinguishable object (from n) to each site of the pregraph. Then numbering elements in the union from

$$1, ..., \mathcal{S}_n(0), \mathcal{S}_n(0)+1, ..., \mathcal{S}_n(0)+\mathcal{S}_n(1), ..., \sum_i \mathcal{S}_n(i)$$

the representation is given by

$$(U_{i.})_{kl} \;=\; Q^{\delta_{kl}/2} \quad \ldots \text{ if completely disconnecting the object}$$
at site $i.$ of the pregraph and putting it in

$$
\begin{aligned}
& \qquad \text{the } 1^{st} \text{ team takes element } k \text{ to } l, \\
&= 0 \quad \ldots \text{ otherwise;} \\
& \qquad\qquad\qquad\qquad\qquad\qquad\qquad\qquad\qquad\qquad (8.10) \\
(U_{ij})_{kl} &= Q^{\delta_{kl}/2} \quad \ldots \text{ if connecting the objects} \\
& \qquad \text{at sites } i. \text{ and } j. \text{ of the pregraph, and putting} \\
& \qquad \text{the resultant cluster in the } 2^{nd} \text{ team if } i \text{ and or} \\
& \qquad j \text{ is in the second team, takes element } k \text{ to } l, \\
&= 0 \quad \ldots \text{ otherwise.}
\end{aligned}
$$

To see that this defines a representation note (c.f. the defining relations):

i) that the repeated application of $U_{i.}$ or U_{ij} simply results in the appearance of a factor of $Q^{1/2}$;

ii) that the order of application of any 2 operators is unimportant if they are not of the form $U_{i.}$ and U_{jk} with either $i. = j.$ or $i. = k.$;

iii) that disconnecting an object, connecting it and then disconnecting it again is equivalent to disconnecting it;

and iv) that connecting 2 objects, disconnecting one and then reconnecting it is equivalent to connecting the 2 objects.QED.

Note that no operation increases the number of clusters in the second team, so we have a sequence of invariant subspaces filtered by this number. Quotienting by elements with second team occupancy $< i$ we get representations for each fixed i. We call the fixed i basis the 'full' partition basis for each i.

Note that all these bases depend on the pregraph only in as much as they depend on n.

8.3.4 Representation type 2

The above representations are not, in general, irreducible. Fixing i, there is an invariant subspace associated with the sub-basis of elements with the property that they can be realised as 'boundary states' of walks on an extended pregraph (e.p.graph) constructed as follows (see also figure 8.2).

Step (1): To each node of the pregraph associate an infinite tower of nodes connected by a linear chain of bonds. The original node is the top level node of the tower (call it level 0).

Step (2): Considering the original pregraph, join the nodes at each given level in the resultant set of towers in the same way as the original nodes at level 0 are joined to construct the pregraph.

Now consider again the set of partitions of n distinguishable objects into clusters, with i clusters in the second team. The subset of elements we want

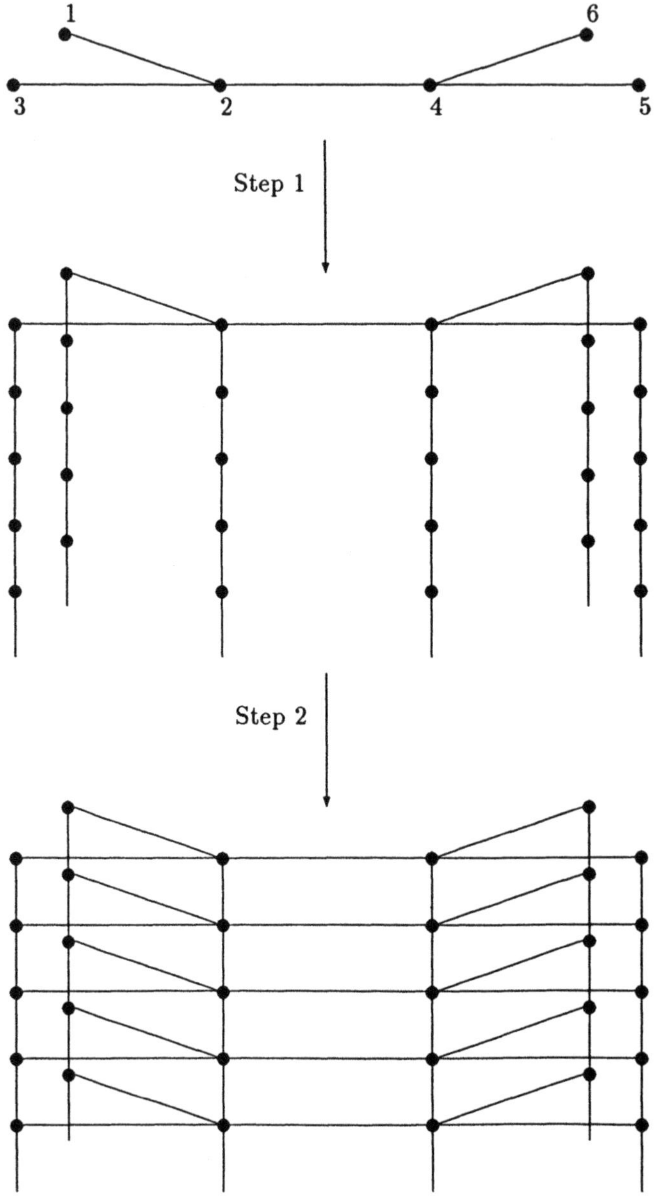

Figure 8.2: Constructing the e.p.graph for the pregraph $D_5^{(1)}$.

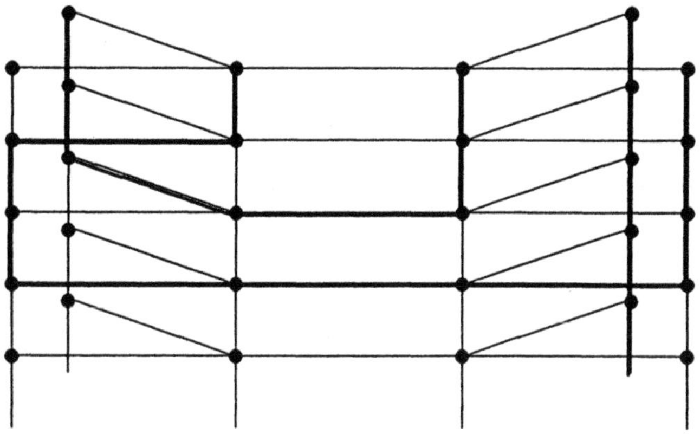

Figure 8.3: Realisation of the basis state ((14)(25)(3),(6)) for the pregraph $D_5^{(1)}$. Note that node 6 at level 0 is connected to the corresponding node in the bottom level shown, which is otherwise totally disconnected.

for an invariant subspace is that set for which it is possible to construct non-intersecting paths along bonds of the extended pregraph such that each set of connected nodes in the partition at level 0 is joined by a path, and that each cluster in the second team is joined by a path to a level with all nodes otherwise pathless.

To see this note from the definition 8.10 and the subsequent quotienting procedure that each $U_{i.}$ or U_{ij} has at most one non-zero entry per column. Thus starting from any basis state corresponding to all nodes disconnected, and acting with some $U_{i.}$ or U_{ij}, the effect is just to take us to another (or possibly the same) basis state. The new state is obtained from the old one by either connecting 2 nodes (acting with U_{ij}) or disconnecting one ($U_{i.}$). Any word in the U-operators (W, say) takes us from the original basis state to another by a *sequence* of moves corresponding to connecting nodes or disconnecting a node. Writing the disconnected basis state as $|0>$ we can express the effect of the sequence of moves by

$$W|0>= k_W|w>,$$

where $|w>$ is the basis state reached by the action of W and k_W is some scalar function of $Q^{1/2}$ (see below). Note that we have distinguished between the labels W and w since, in general, more than one W will produce a given $|w>$.

Now each of the moves *in turn* may be represented diagramatically by adding a 'layer' (consisting of a new level 0 pregraph and bonds connecting each node to the corresponding node in the next level) to the top of the e.p.graph. The layer must contain appropriate sections of paths, or 'steps', on some bonds. If a path is present between two nodes then they are connected, otherwise they are disconnected. Thus the required layer for a $U_{i.}$ contains steps on every bond between levels except at node i, and no paths between nodes within a level; while the layer for a U_{ij} contains steps on every bond between levels and a step between nodes i and j within a level. The effect of a $U_{i.}$ or U_{ij} on the present state depends only on the resultant connectivity of nodes at level 0 (although in general this connectivity will be achieved by paths passing through other levels in the e.p.graph), so any paths or parts of paths irrelevant to this connectivity, such as ***cul-de-sacs***, may be ignored. This is why we have described this realisation of basis states as 'boundary states' of walks on the e.p.graph.

We note immediately that, for an arbitrary pregraph, not all connectivities can be realised in this way, thus not all basis states are in the invariant subspace. We can determine the invariant subspace by carrying out a programme of adding layers with steps to the e.p.graph with no paths (i.e. acting with $U_{i.}$ and U_{ij} on some totally disconnected basis state) and noting the accessible connectivities, until the set of connectivities so produced becomes fixed.

For example, in the case shown in figure 8.2 (i.e. for $n = 6$ with the pregraph $D_5^{(1)}$), the basis state ((14)(25)(36),$\emptyset$) is not accessible, since any connection between 1 and 4 leaves room for only one non-intersecting path from 2,3 to 5,6 , while we require two. By the same measure ((14)(25)(3)(6),$\emptyset$) is possible. Similarly, in the case $i = 1$ the state ((14)(25)(6),(3)) is not accessible, while the state ((14)(25)(3),(6)) is accessible. An appropriate arrangement of paths in this case is shown by the thick lines in figure 8.3 (note that since we need only keep information about the connectivity of the top level it is often possible, and convenient, to compose the effects of several layers as defined above into one layer, which may then isolate or connect several nodes at once).

Carrying out the programme indicated above for the $D_5^{(1)}$ pregraph shown here we find that 198 out of 203 $i = 0$ basis states are accessible. In fact the $i = 0$ accessible subspace has been computed for many pregraphs by this straightforward procedure. The available results show no simple patterns to suggest a more sophisticated algorithm in general, and are unpublished (but easily reproducible).

For another, well understood example, if the pregraph is A_4 then the extended pregraph is a 4 site wide square lattice. Numbering the nodes of the pregraph (at level 0 on the e.p.graph) from one end to the other we see that we have excluded the basis state ((13)(24),$\emptyset$) from the subspace, since there is no path on the extended pregraph from 1 to 3 which does not cross any path from 2 to 4. Similarly ((13)(4),(2)) is excluded, since any path from 1 to 3 isolates 2 from all pathless levels. The case of pregraph A_k is precisely that of the Whitney diagram representation discussed in chapter 6.

In general the invariant subspaces correspond to the left ideals generated from $R_{b^{(i)}}$ (mod $R_{b^{(j)}}$ $b^{(j)} \subset b^{(i)}$) in the quotient algebra in which $R_{b^{(j)}}$ ($\forall b^{(j)}$; $j = 0, .., i$) is a primitive idempotent, as follows. The basis element with all sites isolated, and nodes in $b^{(i)}$ in the second team, corresponds to $R_{b^{(i)}}$. Multiplying (on the left) by any pre-graph bond U_{ij} which does not change the second team occupancy gives the partition with the appropriate pair of sites connected, and so on. This process just corresponds to reproducing the abovementioned paths at the operator level. An example of an excluded element, at second team occupancy $i = 2$ in the A_4 case, is $U_{12}U_3U_4$, since $U_1U_3U_4 = 0$ (mod $R_{b^{(1)}}, b^{(1)} = \{i. = 2\}$) implies $U_{12}U_1U_3U_4U_{12} = 0$.

For completeness we note that, up to overall factors of $Q^{1/2}$, the element of the left ideal generated from $U_1U_2U_3U_4U_5$ represented in figure 8.3 may be written

$$U_3U_{23}U_2U_{12}U_{24}(U_1)U_2U_4U_{23}U_{24}U_{45}(U_1U_2U_3U_4U_5)U_1U_2U_3U_4U_5.$$

This product, taken from left to right, can be extracted from the figure by reading from top to bottom. The bracketed factors each reduce to $Q^{1/2}$ on application of the defining relations (from equation 8.1).

In general if we replace every factor of the form $U_{i.}$ by $(U_{i.}Q^{1/2})$ and every factor of the form U_{ij} by $(U_{ij}Q^{-1/2})$ in a word W then the factor $k_W(Q^{1/2})$ in

$$W|0> = k_W(Q^{1/2})|w>$$

becomes just Q^C. Here C is the number of isolated paths, including single nodes, in the interior of the e.p.graph (i.e. those which do not affect the level 0 connectivity). In our example, for instance, the above renormalisation changes each of the bracketed factors to Q, corresponding to the presence of the isolated node in position 1 at level 3 (that is the fourth level down) and the isolated nodes in positions 1,2,3,4 and 5 at level 4 (the bottom level shown). We will prove the general result in section 8.3.6.

In the corresponding representations $T^k(v)$ is, up to similarity transformations, the transfer matrix for the dichromatic polynomial (chapter 1) (Q is the other parameter) associated with an extended pregraph of just k levels. This is, perhaps, a strange object to consider. Nonetheless, it provided the stimulation for these ideas! The extended pregraph of k levels is a useful notion for later on. We will adopt the notation G^k for such an extension of the graph G.

8.3.5 Fixed n pregraph dependence of type 2 bases

Note, from the generalised version of the definition 8.1, that if an algebra has a pregraph G , and another algebra has a pregraph H, then $GT(Q)$ is a subalgebra of $HT(Q)$ if H contains G as a subgraph. The 'maximal' pregraph with n nodes is the one in which every site is connected to every other by a bond. We sometimes call this the infinite dimensional case, in reference to the associated physical models. At the other extreme, a graph with more than one connected component has an algebra corresponding to the direct product of (commuting) algebras associated with each connected component.

In the case $i = 0$, note that the algebras for the pregraphs in which at least one site is connected by a bond to every other have an irreducible representation given by the full partition basis ($i = 0$), since in these cases every partition may be realised from the (1^n) partition by some sequence of moves in which sites are connected along bonds or sites are isolated. The instances in which such pregraphs are also tree graphs are unique for each n (up to permutations of sites), having one coordination number $(n-1)$ site and $(n-1)$ coordination number one sites. These cases are called 'daisy' graphs.

For general i only the algebra for the maximal pregraph always has an irreducible representation associated with the full fixed i basis.

As bonds are removed from the maximal pregraph, then, the subspaces accessed from (1^n) (i.e. the (1^n) elements in any fixed i basis) in the way described above occasionally shrink. One possible limit of this procedure is the case in which all site coordination numbers are 2 (except for a pair of coordination number 1 sites at the ends of the chain of bonds). We sometimes call this A_n case the 'linear' pregraph. This case is the original Temperley-Lieb algebra for which such a sub-basis was discussed in Martin 1986b. Here the number of accessible partitions is $C_n = 1, 2, 5, 14, 42, 132, ...$ with

$$\lim_{n\to\infty} C_{n+1}/C_n = 4 \tag{8.11}$$

(Blote and Nightingale 1982) [2]. Of course the linear pregraph is not the unique connected endpoint of such a bond dissolution process (but that's

---

[2] The number of accessible basis elements for linear pregraphs and *general* i is, by continuity (see chapter 6), the dimension of the irreducible representation of the symmetric group S_{2n} corresponding to the 2 row tableau shape $(n+i, n-i)$.

another story). The assymptotic growth rate above tells us that every positive integer Q value below 4 has an exceptional (i.e. non-generic) Temperley-Lieb algebra associated for large enough n, as we will see shortly.

8.3.6 Generic irreducibility of type 2 representations

First let us show that $R_{b^{(i)}}$ is, at least up to quotients, a primitive idempotent, and hence that (provided the algebra is semi-simple) the representation associated with the left ideal is irreducible (otherwise it is indecomposable).

For given n, consider the tree of pregraphs related by bond dissolutions starting from the maximal case and ending with tree graphs (which become disconnected if any further bonds are dissolved). For $n = 1, 2$ the endpoints coincide. For $n = 3$ we have just 2 possible connected graphs, figure 8.4. In both cases the $i = 0$ irreducible subspace coincides with the original 5 dimensional space, but for $A_2^{(1)}$ we must impose the non-trivial quotient relation

$$U_0U_1U_2U_{01}U_{12}U_{20}U_0U_1U_2 = y(Q^{1/2})U_0U_1U_2 \tag{8.12}$$

(where $y(Q^{1/2})$ is known - see later) on the abstract algebra 8.1 before

$$R_{b^{(0)}} = Q^{-3/2}U_0U_1U_2$$

becomes primitive.

For $n = 5$ the daisy graph endpoint is shown in figure 8.5. For the corresponding algebra $R_{b^{(j)}}$ is a primitive idempotent modulo quotient relations of the form

$$R_{b^{(i)}}U_{01}U_{03}U_{0.}U_{02}U_{04}U_{0.}U_{01}U_{03}U_{0.}U_{02}U_{04}R_{b^{(i)}} = x(Q^{1/2})R_{b^{(i)}} \tag{8.13}$$

To determine the list of such relations in general is difficult and, in fact, unnecessary.

To see that $R_{b^{(i)}}$ is primitive (modulo such relations) in general, i.e. for any graph, note that all the $U_{j.}$ are lower triangular in any representation with fixed second team occupancy i, so that $(R_{b^{(i)}})_{kk}$ is zero unless all the $(U_{j.})_{kk}$ $(j. \not\in b^{(i)})$ are non-zero. Each $(U_{j.})_{kk}$ is non-zero only if disconnecting a site leaves the basis element k alone, so for any partition other than of (1^n) type some such matrix element must be zero. Similarly, for (1^n) type elements of the form

$$((m_1)(m_2)...(m_{n-i}), (m_{n-i+1})...(m_n))$$

where any $m_{p>n-i} \not\in b^{(i)}$, the action of $R_{b^{(i)}}$ is to reduce the number of second team clusters. Conversely for the unique (1^n) type element of the

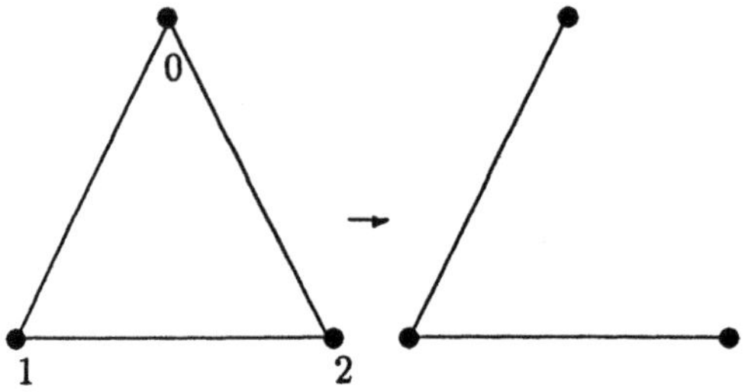

Figure 8.4: The maximal ($A_2^{(1)}$)and daisy/linear (A_3) pregraphs for n=3.

Figure 8.5: The daisy pregraph ($D_4^{(1)}$) for n=5.

form

$$((m_1)(m_2)...(m_{n-i}),(m_{n-i+1})...(m_n))$$

where $b^{(i)} = \{m_{n-i+1},...,m_n\}$ (numbered $k = 1$, say) , no $(U_j.)_{11}$ ($j. \notin b^{(i)}$) is zero. QED. Alternatively, note that the action of $R_{b^{(i)}}$ on any element of the fixed second team occupancy i subset is either to take it to "zero" (i.e. an element of occupancy$< i$) or to take it to the element $k = 1$ above, since $R_{b(i)}$ disconnects and puts into the first team all nodes $\notin b^{(i)}$. Thus either the result has less than i clusters in the second team or it is precisely $k = 1$.

This argument also allows a straightforward determination of the required quotient relations in each case, e.g. the form of $x(Q^{1/2})$ in equation 8.13.

The representations on the accessible subspaces for any pregraph are thus irreducible provided the algebra is semi-simple.

8.3.7 Quotient relations for type 2 representations

It follows from the treatment of the *Potts* representation in section 8.5, and in particular the e.p.graph representation of words $W(Q)$, together with the discussion in section 8.3.3, that the quotient relations generically coincide (when $Q \in Z_+$, and formally regarding Q as an indeterminate integer in 8.5).

To see this note the following. Forming the word $W'(Q)$ (where $W' = R_{b(M)}WR_{b(M)}$) in the e.p.graph, it describes some connectivity from one end of the e.p.graph to the other, and some connectivity of nodes at each end of the e.p.graph, together with some isolated clusters. From the definition of type 2 representations we will show that here, as in the Potts case, either the number of distinct lines passing through the diagram decreases or $\chi_{b(M)}(W)$ (defined by analogy with equation 8.6) just picks up a factor of Q for each isolated cluster.

Specifically, note that all the $U_{i.}$ matrices in equation 8.10 may be arranged to be lower triangular; and the U_{ij} matrices upper triangular. There is, therefore, an equivalent representation (for $Q \neq 0$) with the ***non-vanishing*** matrix elements in equation 8.10 replaced by

$$(U_{i.})_{kl} = Q^{1/2} \quad and \quad (U_{ij})_{kl} = Q^{\delta_{kl}-1/2}.$$

Recall that the quotient relations are basis independent. Then note that:

(i) introducing $U_{ij}Q^{-1/2}$ into $W(Q)$ cannot change the number of isolated clusters if it does not change the connectivity; but if it does change the connectivity then it must do so by connecting 2 isolated clusters, and so reduce the number of isolated clusters by 1. Meanwhile, in our equivalent representation, the appropriate matrix element is such as to introduce a factor of precisely 1 in the former situation and Q^{-1} in the latter.

(ii) On the other hand, $U_i.Q^{1/2}$ increases the number of isolated clusters by 1 whether or not it changes the connectivity (since either way it isolates a node). But from the definition of the equivalent representation the matrix element of $U_i.Q^{1/2}$ is precisley Q in either situation.

This result implies that the Potts representation is, generically, just a direct sum of type 2 representations. The multiplicities have not been determined in general.

8.4 Physical consequences

We can show that the type 2 representation associated with R_0 is the representation associated with the free energy in statisitical mechanical models. To see this note that the free energy is the largest magnitude eigenvalue of the transfer matrix when the coupling is real. In this case the transfer matrix is positive, i.e. all the elements are positive, and by Perron's theorem it has a unique largest magnitude eigenvalue, which is positive and has a 'positivisable' (i.e. positive up to an overall phase) eigenvector. From the definition it is easy to see that R_0 is the matrix with all entries 1 (up to an overall factor) in the Potts representation. Since it is a primitive idempotent this implies that the Potts representation contains the irreducible representation discussed above exactly once as an irreducible component (and also as an indecomposable module). Again from the Potts representation it is easy to see that R_0 is not orthogonal to the positivisable eigenvector associated with the free energy in this basis. Therefore the irreducible representation associated with the primitive idempotent R_0 is the one responsible for the part of the transfer matrix spectrum containing the free energy.

Note that the Potts representation has asymptotic growth rate of dimension d_n (n sites) of

$$\lim_{n \to \infty} d_{n+1}/d_n = Q \tag{8.14}$$

This is independent of the graph. If we take linear pregraphs then the irreducible from R_0 , contained in the Potts representation, has generic asymptotic growth rate 4 (see above). This means that for large enough n and $Q < 4$ the generic 'irreducible' representation contained in the Potts representation has dimension greater than the Potts representation. In such cases the true irreducible must be smaller than the generic irreducible, i.e. the left ideal generated from the primitive idempotent is no longer the basis for an irreducible representation. This means that the abstract algebra is no longer semi-simple. In other words the Temperley-Lieb algebra has an exceptional structure for $Q = 0, 1, 2, 3$ (with $Q = 4$ a limiting case of the generic structure). This observation has been the subject of much interest in 2 dimensional physics (see, for example, Cardy 1986).

Asymptotic growth rates for arbitrary sequences of pregraphs are not known at present. The most interesting question for physics is perhaps the asymptotic growth rate for pregraphs associated with 3 dimensional models. The rate for coordination number n pregraphs (i.e. essentially infinite dimensional models) is infinite. This means that here all integer Q values are exceptional for large enough n.

It is intriguing that the three dimensional limiting case seems to be at $Q > 4$, while the crossover between second and first order phase transitions for Potts models seems to be closer to $Q = 3$ than $Q = 4$ (from numerical and strong coupling eveidence). This is contrary to the pattern in two dimensions.

8.5 Quotient relations: the Potts representation

We will now outline a demonstration that the Potts representation obeys the quotient relations 8.6.

(N=0) In the Potts representation $(R_0)_{ij} = Q^{-n} \quad \forall \quad i, j$. QED.

(N=1) $R_{b^{(1)}}$ takes the form $M_{n-1} \otimes 1_Q$ where M_{n-1} is R_0 for a graph with the node $b^{(1)}$ deleted and 1_Q is the Q dimensional unit matrix. But the Potts generators are all invariant under a global redefinition of site variables so, with E_{ij} a Q dimensional elementary matrix,

$$R_{b^{(1)}} W R_{b^{(1)}} = \sum_{i,j=1,..,Q} C_W(i,j) M_{n-1} \otimes E_{ij} \tag{8.15}$$

where $C_W(i,i)$ is independent of i and $C_W(i,j)$ $i \neq j$ is independent of i and j. Now quotient by R_0 from the (N=0) case above. QED.

(N> 1) It is useful to define an inhomogeneous transfer matrix, generalising equation 8.2, by

$$T(\{v_{i.}, v_{ij}\}) =$$
$$\prod_{\text{pregraph nodes } i.} \frac{(v_{i.} + Q^{1/2} U_{i.})}{(1+v_{i.})} \prod_{\text{pregraph bonds } ij} \frac{(1 + v_{ij} Q^{-1/2} U_{ij})}{(1+v_{ij})}$$

Clearly any word W can then be written as a product of such transfer matrices, since any generator can. The physical picture is of a Potts system with some couplings at zero temperature ($\beta = \infty, v = \infty$) so that the connected spins are frozen together; and some couplings at high temperature ($\beta = 0, v = 0$) so that the connected spins are effectively decoupled. We see from the transfer matrix that whether or not a generator is included in each case depends on whether it is associated with a node or a bond. From the definition of the Potts generators we confirm that the presence of a bond generator freezes the relevant spins together, while a node generator decouples them. It is then helpful to write out W on G^k [3] (k as large as necessary to contain W, i.e. $k \leq O(W)$). The nodes of G^k can be thought of as the sites of a Potts lattice. The transfer matrix transfers attention from one level to the next. If a bond is *frozen* in the transfer matrix then it is marked in the e.p.graph.

For example, a marked e.p.graph for

$$\begin{aligned} W &= U_{1.}U_{3.}U_{5.}U_{6.}U_{12}U_{23}(U_{56})U_{54}U_{5.}U_{4.}U_{54} \\ &\quad U_{56}U_{5.}U_{6.}U_{56}U_{3.}U_{1.}U_{34}U_{21}U_{2.}U_{4.}U_{34}U_{21}U_{1.}U_{3.}U_{5.}U_{6.} \end{aligned} \quad (8.16)$$

in $A_6T(Q)$ is shown in figure 8.6. The construction is precisely analogous to that described for the e.p.graph path realisation of basis states for partition representations in section 8.3.3, with path 'steps' replaced by marked bonds. Note, however, that the e.p.graph in figure 8.6 has been drawn on its side, compared to figure 8.2. In figure 8.6, in analogy to figure 8.3, we have, where possible, accounted for more than one factor in a layer - in general any number of successive factors of the form $U_{i.}$ followed by any number of successive factors of the form U_{ij} may clearly be incorporated in the same layer. In the diagram each $U_{i.}$ (i.e. *unmarked* horizontal bond) should be read as $Q^{1/2}U_{i.}$ and each U_{ij} (marked vertical bond) as $Q^{-1/2}U_{ij}$. We will denote by $W(Q)$ the product obtained from W by making these replacements.

Any W can be represented this way, and any such diagram corresponds to some W. In general the word may be reducible (i.e. $W \propto W'$ with $O(W) > O(W')$) by the relations 8.1. In our example $W(Q) = QW'(Q)$ where $W' = U_{1.}U_{3.}U_{5.}U_{6.}$. The factor of Q comes from the relations 8.1, but physically it corresponds to the fact that the isolated cluster in the top right of the diagram may take any of Q possible values for its spins (which are frozen together). The remaining spins are determined by the boundary conditions. For general pregraphs the word will not necessarily reduce using the relations in this way, but isolated clusters will clearly still give rise to factors of Q, from the physical picture, in the Potts representation.

[3] The extended pregraph (e.p.graph) G^k is defined in section 8.3.3.

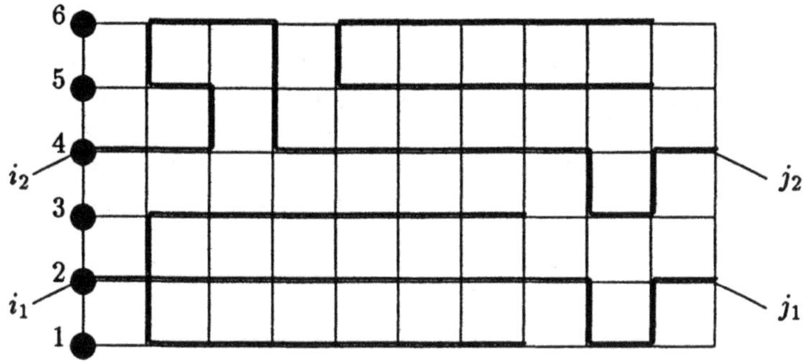

Figure 8.6: A word from $A_6T(Q)$ drawn on $(A_6)^{11}$.

Now $R_{b(N)}$ takes the form $M_{n-N} \otimes 1_{Q^N}$, so

$$R_{b(N)} W R_{b(N)} = \sum_{i_1,j_1,\ldots,i_N,j_N=1}^{Q} C_W(i_1,j_1,\ldots,i_N,j_N)\ M_{n-N} \otimes E_{i_1,j_1} \otimes \ldots \otimes E_{i_N,j_N} \tag{8.17}$$

This corresponds to imposing free boundary conditions at the 'external' points $i_a, j_a \notin b^{(N)}$ (in the pregraphs at levels 0 and $(k-1)$ of the e.p.graph, respectively) and then specifying completely the boundary states of $i_a, j_a \in b^{(N)}$ to obtain a matrix element C_W. By writing the *whole* of the left hand side as a word (for example, in the figure we have $R_{b(2)} W' R_{b(2)}$ where $b^{(2)} = \{2,4\}$) we see that, for large enough Q, C_W just depends, up to overall factors, on the connectivity of the 'external' nodes i_a and j_a ($i_a, j_a \in b^{(N)}$) of G^k. If the boundary conditions are consistent with the connectivity (i.e. if $i_a \sim j_b$ then the spins $i_a = j_b$; if $i_a \sim i_b$ then $i_a = i_b$; and if $j_a \sim j_b$ then $j_a = j_b$) the result is just Q^c where c is the number of isolated connected clusters. Otherwise it is zero. We thus see that, up to overall factors, any $R_{b(N)} W R_{b(N)}$ diagram may be contracted to its 'basic' connections before computing C_W. Because of the $R_{b(N)}$ factors at each end, the only case not covered by the quotienting is $i_a \sim j_b$ if and only if $a = b$. All other connectivities involve fewer distinct (non-interconnected) lines passing through the diagram, and can thus be represented by elements of ideals already quotiented out (recall equation 8.6).

To see this in an example, consider the case in which W from equation 8.16 is replaced by the same thing but with the bracketed factor (U_{56}) missing. Call this word X. Then only one line passes through the diagram, and $i_2 \not\sim j_2$. In this case the relations simplify the word to $Q^{(n-1)/2}R_{b^{(1)}}$ with $b^{(1)} = \{2\}$. Specifically, $X(Q) = QQ^{(n-1)}R_{b^{(1)}}$. Again the frozen bond picture ensures that such a simplification occurs on physical grounds in the Potts representation, even if the relations 8.1 do not imply it.

In general, if i distinct lines cross the diagram, then the corresponding word is in the double sided ideal generated from some $R_{b^{(i)}}$ (since adding generators to a word never increases the number of distinct lines across its diagram). Note that all cases in which $i_a = j_a$ are covered by the quotient relations or by $R_{b^{(N)}}$, and all cases in which $i_a \neq j_a$ for some $i_a \in b^{(N)}$ are covered by the quotient relations. That is, such boundary conditions imply a *maximum* amount of connectedness across the diagram.

We have shown that the left hand side of equation 8.17 can be replaced by the left hand side of equation 8.6.

For example, with N=2, if all 4 external nodes are disconnected then the LHS of equation 8.17 is in the double sided ideal generated from R_0. If all 4 are connected then it is in the ideal generated from $R_{b(1)}$, and so on.

Finally, note that for sufficiently small Q there are more connectivities possible than states of the boundary nodes, so the connectivities cannot be treated independently. This is a signal of the breakdown of the generic structure of the algebra.

Chapter 9

Hecke Algebras

9.1 Review

As we have already seen, the n-site layer transfer matrices for a wide range of statistical mechanical models may be written in the form

$$T = \left(\prod_{i=1,..,n} (1 + xU_{2i-1}) \right) \left(\prod_{i=1,..,n-1} (x + U_{2i}) \right) . \tag{9.1}$$

In a huge class of models (see e.g. Date *et al* 1987, Pasquier 1988) the matrices $\{U_i \; : \; i = 1, ..., 2n-1\}$ give a representation for the generators of a Hecke algebra $H_k(q)$, where $k = 2n-1$ and

$$q = \frac{\sqrt{Q} \pm \sqrt{Q-4}}{2}$$

i.e.

$$q + q^{-1} = \sqrt{Q}.$$

This algebra is abstractly defined by the k generators $\{U_i\}$ and relations

$$\begin{aligned} U_i U_i &= \sqrt{Q} U_i & (9.2) \\ U_i U_{i\pm1} U_i - U_i &= U_{i\pm1} U_i U_{i\pm1} - U_{i\pm1} & (9.3) \\ U_i U_{i+j} &= U_{i+j} U_i \qquad (j \neq 1) & (9.4) \end{aligned}$$

which generalise automatically to even k. In as much as this algebra is nothing more than the quotient of the braid group algebra by quadratic local relations (see section 9.1.1), it is the most general algebra appropriate for building a transfer matrix of the form of equation 9.1. In most statistical mechanical contexts the variable q can be an arbitrary scalar parameter, although this parameter is fixed for a given model. On the other hand the variable x in equation 9.1 is a function of the temperature (see chapter 2). The precise representation of the algebra involved in the transfer matrix depends again on the model, *and* on the boundary conditions.

As generalisations of the Temperley-Lieb algebras, interest in these algebras is similarly widespread. As with Temperley-Lieb, in addition to their relevance for statistical mechanics and conformal field theory they have an intimate connection with knots and braids (see e.g. Deguchi *et al* 1988, Lickorish 1988, and references therein), and string field theory. In this chapter we show how to determine the structure of these algebras. The unitarisable part, in particular, may then be read off as a quotient.

9.1.1 Technical notes

The dimension of the algebra, or equivalently the number of words in the generators (including the unit) which are linearly independent under the relations, is clearly independent of q.

The elements of the algebra

$$t_i^{\pm 1} = 1 - q^{\pm 1} U_i \tag{9.5}$$

may be used as alternative generators. They obey the braid relations

$$t_i t_{i\pm 1} t_i = t_{i\pm 1} t_i t_{i\pm 1} \tag{9.6}$$

$$t_i t_{i+j} = t_{i+j} t_i \qquad (j \neq 1) \tag{9.7}$$

plus

$$(t_i - 1)(t_i + q^2) = 0. \tag{9.8}$$

When $q = 1$ the relations 9.6, 9.7 and 9.8 define the group algebra for the symmetric group on $k+1$ objects, S_{k+1}. As we will see, there is a readily accessible hierarchy of quotient algebras $NH_n(q)$ corresponding to N row Young tableau.

The algebra $H_k(q)$ has an involution

$$D : U_i \mapsto \sqrt{Q} - U_i$$

corresponding to reflecting tableau in a diagonal line. As with all quotients of the braid group algebra it has another involution

$$R : U_i \mapsto M U_i M^{-1} = U_{k-i+1}$$

corresponding to the reflection symmetry on the physical lattice. We desribe the construction of an operator M such that

$$MU_iM^{-1} = U_{k-i+1}$$

in the more general context in chapter 3.

Similarly, let us define $U_{k+1} = U_0$ as a certain special element of $H_k(q)$ which obeys the relations as if the indices are defined mod $2n$, but which is not a new generator. Then the algebra has further automorphisms

$$t : U_i \mapsto GU_iG^{-1} = U_{i+1} \tag{9.9}$$

(corresponding to duality/translation on the lattice) where a convenient, but not unique, realisation for G is given by

$$G = \prod_{i=1}^{k} t_i. \tag{9.10}$$

In particular, $U_0 = GU_kG^{-1}$, and $U_1 = GU_0G^{-1}$.

Note that any $H_k(q)$ may, therefore, be alternatively generated by a *pair* of generators $\{U_1, G\}$. We will return to these points later.

We will use the notation W^T for the element of the algebra obtained by writing the generators in element W in reverse order. Note that the defining relations are invariant under the reversal $W \rightarrow W^T$.

Note that $H_k(q) \subset H_m(q)$ if and only if $k < m$. The subalgebra can be realised in various ways, the most obvious being identification of the first k generators. We will occasionally discuss constructions involving some set of generators in an abstract sense, without specifying k. Such constructions apply to any algebra containing sufficient generators.

9.2 On the structure of $H_k(q)$

Preview

The main *generic* result of the next three sections is to be found on page 253. For n a positive integer, r a complex number *excluding* the rationals,

$$q = e^{i\pi/r}$$

and hence

$$Q = 4\cos^2(\pi/r) \tag{9.11}$$

this is a theorem giving the structure of the algebra $H_n(q)$ defined by the generators $\{U_i : i = 1, 2, .., n\}$ and relations 9.2, 9.3 and 9.4.

After defining two primitive central idempotents for each algebra, and introducing some notation, we define a set of orthogonal operators for each algebra before arriving at our theorem.

9.2.1 A primitive central idempotent in $H_n(q)$

1) A sequence of idempotents in $H_n(q)$

Definition 44 *For $m = 1, 2, 3, .., n+2$ define $E_m \in H_n(q)$ by*

$$E_1 = E_2 = 1$$

and then

$$E_m \in H_{m-2}(q) \subset H_n(q)$$

and

$$E_m E_m = E_m$$

and for $i = 1, 2, .., m-2$

$$E_m U_i = U_i E_m = 0.$$

Let us consider the existence and uniqueness of such an element.

Definition 45 *Define $I[m-2] \in H_{m-2}(q)$ by $I[0] = 1$ and*

$$I[m-2] = I[m-3](1 - k_{m-1}U_{m-2})I[m-3]$$

The existence of $I[m-2]$ for a given value of r is guaranteed unless some k_n required in its construction has a pole at that point.

Proposition 25 *If $I[m-2]$ exists then*

$$E_m = I[m-2]$$

Proof:

The proof is almost the same as for proposition 6:

From the definition E_m is a primitive central idempotent in $H_{m-2}(q)$ containing the unit. Any two such objects must be identical. On the other hand assume that for some non-negative integer b we have

$$E_{b+2} = I[b]$$

and

$$(I[b]U_{b+1})^2 = k_{b+2}^{-1} \, I[b]U_{b+1}$$

(both are clearly true for $b = 0$). Then

$$\begin{aligned} I[b+1]I[b+1] &= (I[b] - k_{b+2}I[b]U_{b+1}I[b])^2 \\ &= I[b] - 2k_{b+2}I[b]U_{b+1}I[b] + k_{b+2}^2(I[b]U_{b+1})^2 I[b] \\ &= I[b+1] \end{aligned} \tag{9.12}$$

and

$$U_i I[b+1] = U_i I[b] I[b+1] = 0$$

for $(1 \leq i \leq b)$ and

$$U_{b+1}I[b+1] = U_{b+1}I[b] - k_{b+2}(U_{b+1}I[b])^2 = 0. \tag{9.13}$$

Furthermore, we have

$$\begin{aligned} (I[b+1]U_{b+2})^2 &= I[b+1]U_{b+2}(I[b] - k_{b+2}I[b]U_{b+1}I[b])U_{b+2} \\ &= I[b+1]U_{b+2}(\sqrt{Q} - k_{b+2}) \\ &\quad -k_{b+2}I[b+1](U_{b+1}U_{b+2}U_{b+1} - U_{b+1})I[b] \\ &= k_{b+3}^{-1}I[b+1]U_{b+2}. \end{aligned} \tag{9.14}$$

The last term in the penultimate expression here vanishes identically in Temperley-Lieb, while we have had to use the (already established) orthogonality property of $I[b+1]$.

Note that we have used the symmetry, by construction, of $I[b]$ under reversal of operator order

$$I[b] = I[b]^T.$$

This completes the proof of proposition 25.

Translation/reflection notation

We will take across in its entirety the translation/reflection notation of section 6.3.4.

Examples

The operator $E_m^{(t)}$ may be obtained from the recursive formula for E_m by translation. The first couple are:

$$E_3^{(t-1)} = 1 - Q^{-1/2} U_t \tag{9.15}$$

$$E_4^{(t-1)} = 1 + (U_t U_{t+1} + U_{t+1} U_t - \sqrt{Q}(U_t + U_{t+1}))/(Q-1) \tag{9.16}$$
$$+ \frac{1}{\sqrt{Q}(Q-1)} (U_t - U_t U_{t+1} U_t).$$

Note the symmetry under $U_i \rightarrow U_{b+1-i}$ in E_{b+2}, which is a general consequence of its uniqueness.

These examples illustrate the point that the construction procedure fails for some Q values, $Q = Q_c$, say, where the formal idempotent has terms whose coefficients have a pole. In these cases the appropriate idempotent to provide a basis for the trivial representation is obtained by simply subtracting the potentially divergent part before taking $Q \rightarrow Q_c$ (see the argument in section 6.3.4).

9.2.2 Another primitive central idempotent

Under the automorphism D defined by

$$U_i \mapsto \sqrt{Q} - U_i$$

we have

$$D(E_m) = F_m.$$

Definition 46 *For $n \geq b \geq a > 0$ define $F_{ab} \in H_n(q)$ by*

$$F_{ab} = F_{3-a+b}^{(a-1)}.$$

Consequently, if $a \leq i \leq b$ then

$$U_i \, F_{ab} = F_{ab} \, U_i = \sqrt{Q} \, F_{ab}.$$

For example,

$$F_{11} = \frac{U_1}{\sqrt{Q}}$$

$$F_{12} = \frac{U_1 U_2 U_1 - U_1}{\sqrt{Q}(Q-1)}.$$

We similarly define

$$E_{ab} = E_{3-a+b}^{(a-1)}$$

so that

$$U_i\ E_{ab} = E_{ab}\ U_i = 0$$

if $a \leq i \leq b$. For example $E_{1\ -1} = 1$.

Note that, as with E_{ij}, F_{ij} may not be well defined for all Q (consider our examples). However, it follows from the definition that for each problematic Q there exists a well defined constant k such that kF_{ij} is finite, although nilpotent. Then clearly

$$U_i\ \ kF_{ij} = \sqrt{Q}\ \ kF_{ij}.$$

9.2.3 The quotient algebras $NH_n(q)$

We define a sequence of quotient algebras of $H_n(q)$ as follows. The quotient $NH_n(q)$ is obtained by imposing the quotient relations

$$F_{N+2} = 0.$$

Note that the case $N = 2$ corresponds to the Temperley-Lieb algebra.

9.2.4 The word problem

Suppose X, Y and Z are words in $H^{(1)}_{n-1}(q)$ (i.e. words in $H_n(q)$ containing no factors of U_1). A word $W \in H_n(q)$ of the form

$$W = X\ U_1 Y\ U_1\ Z$$

is said to have U_1 separation

$$U_1(W) = Length(Y).$$

For any other form $U_1(W)$ is undefined.

For example,

$$U_1(U_2U_1U_2U_3U_4U_1) = 3.$$

Note that $U_1(W)$ is not invariant under the relations (strictly speaking it is a function on a subset of words in the free algebra on n generators).

Proposition 26 *Let word W have $U_1(W) = d$, then either $d = 0$ or W may be rewritten as a linear combination of words W' with $U_1(W') < d$ plus words with no U_1.*

Proof:

Write W in the form

$$W = X \prod_{i=1}^{d+2} U_{x_i} \ Z.$$

Since $x_{d+2} = 1$ either $d = 0$ or there exists positive integer p such that

$$x_i = i \quad for \ all \quad 1 \leq i < p$$

$$x_p \neq p.$$

Now either $x_p > p$ in which case by relation 9.7

$$W \mapsto W' = X \ U_{x_p} U_1 \ ... \ U_1 \ Y$$

with

$$U_1(W') = d - 1;$$

or $a = x_p < p$ in which case we have

$$W = X \ U_1 U_2 U_3 ... U_a U_{a+1} ... U_{p-1} U_a ... U_1 \ Z$$

which, using relation 9.7 again, becomes

$$W' = X \ U_1 U_2 U_3 ... (U_a U_{a+1} U_a) .... U_1 \ Z.$$

Using relation 9.6 we may then rewrite this as

$$X \ U_1 U_2 U_3 ... (U_{a+1} U_a U_{a+1} - U_{a+1} + U_a) .... U_1 \ Z.$$

which becomes

$$X \ U_{a+1} \ U_1 U_2 U_3 ... U_a U_{a+1} .... U_1 \ Z$$

$$-X \ U_1 U_2 U_3 ... U_{a+1} .... U_1 \ Z + X \ U_1 U_2 U_3 ... U_a .... U_1 \ Z.$$

The first of these terms has U_1 separation $d - 1$, and the other two have separation $d - 2$.

This completes the proof of proposition 26 (c.f. Wenzl 1988 and, for the Temperley-Lieb case, Jones 1983).

Corollary 26.1 *Every word can be written as a linear combination of words containing at most one* U_1.

A truncation of a word W is any word obtained from W by omiting some factors (all words have 1 as a truncation, for example). It follows immediately that every word can be written as a linear combination of truncations of words of the form

$$X\ U_2U_1\ Y\ U_2\ Z$$

where X, Y, Z are words in $H^{(2)}_{n-2}(q)$.

Similarly the algebra $H_n(q)$ is spanned by words of the form

$$X\ U_3U_2U_1\ Y\ U_3U_2\ Z\ U_3\ T$$

where X, Y, Z, T are words in $H^{(3)}_{n-3}(q)$. Eventually we see that $H_n(q)$ is spanned by truncations of

$$\prod_{i=1}^{n}\left[\prod_{j=n}^{i}{}' U_j\right].$$

We then have the following

Proposition 27 *Let W be any word in $H_n(q)$ and N an integer $0 \leq N \leq n$. Then there exist elements $X, Y \in H_n(q)$ and $Z \in H^{(N+1)}_{n-N-1}(q)$ such that*

$$F_{1N}\ W\ F_{1N} = X\ \ F_{1N+1}\ Y + F_{1N}\ Z.$$

Proof:

If there is no U_{N+1} in W then the proposition follows from the definition of F_{1N}. Otherwise, since $F_{1N} = (U_i/\sqrt{Q})F_{1N}$ for $i = 1, 2, .., N$ we may, without loss of generality, consider the case

$$F_{1N}WF_{1N} = F_{1N}\ X\ \left[\prod_{i=N+1}^{1}{}'\ U_i\right]\ \ ...\ F_{1N}$$

where $X \in H^{(N+1)}_{n-N-1}(q)$ and hence commutes with F_{1N}.

From the definition we see that $F_{1N}\ \left[\prod_{i=N+1}^{1}{}' U_i\right]$ may be written as a linear combination of F_{1N+1} and F_{1N}.

Thus $F_{1N}\ W\ F_{1N}$ may be written as a part with the required form together with a part of the original form, but with one fewer factor of U_{N+1}. Since the number of such factors is finite the proposition follows by iteration.

9.2.5 Sequence notation

We now need to generalise the sequence notation of section 6.3.2. The case described there corresponds, in what follows, to $N = 2$. This notation initially amounts to the same thing as the partition and Young tableau notation which is standard in the $q = 1$ (permutation group) case (see e.g. Robinson 1961). However, it seems ultimately to be better suited to provide a description of the special cases.

Definition 47 *Define the set NR_n of N-tuples of non-negative integers*

$$\alpha = (\alpha_1, \alpha_2, .., \alpha_N)$$

with the properties

$$\alpha_i > \alpha_j \Rightarrow j > i$$

and

$$\sum_i \alpha_i = n + 1.$$

Define a total order on NR_n by $\alpha > \beta$ if there exists integer j such that

$$\alpha_i = \beta_i$$

for $N \geq i > j$, and

$$\alpha_j > \beta_j.$$

For example, the N-tuple $(1, 1, 1, .., 1)$ is the first element and $(N, 0, 0, .., 0)$ the last in this order in NR_{N-1}.

We will regard $MR_n \subset NR_n$ for $M < N$ by extending the M-tuples to N-tuples by adding zeros on the right.

Definition 48 *For given N define e_i as the N-tuple with j^{th} component*

$$(e_i)_j = \delta_{ij}.$$

Definition 49 *Define an N-sequence, or just a* sequence, *$\{s\}$ to be any ordered set of N-tuples of positive integers s_i indexed by $i = 0, 1, 2, 3, ...$*

$$\{s\} = s_0 s_1 s_2 \, .. \, s_i \, ..$$

with the properties

$$s_0 = (0, 0, ..., 0)$$

and, with addition and subtraction of N-tuples to be performed component-wise,

$$s_i - s_{i+1} = -e_j \tag{9.17}$$

for some j.

The length *of $\{s\}$ is the number of N-tuples in the set.*

Definition 50 *If $\{s\}$ is a sequence of length $n+1$ and $\{t\}$ a sequence of length $m+1$, then the composite $\{w\} = \{st\}$ is a sequence of length $n+m+1$ defined as follows:*

$$w_i = s_i \qquad for \quad 0 \leq i \leq n$$

$$w_{i+n} = t_i + s_n \quad for \quad 0 \leq i \leq m.$$

Definition 51 *If $\{s\}$ has a subsequence of the form*

$$g \; \overbrace{g + e_j}^{s_i} \; g + e_j + e_k$$

with k greater than (resp. less than) j, then s_i is said to be a maximum (resp. minimum) of $\{s\}$.

Definition 52 *If s_i is a minimum of $\{s\}$ as above then the sequence $\{s^i\}$ is defined by*

$$\begin{aligned} s^i_l &= s_l \qquad (l \neq i) \\ s^i_i &= s_i + e_k - e_j. \end{aligned}$$

Note that we can think of N-sequences as increasing walks from the origin on N-dimensional hypercubical lattices (in the positive coordinate sector). It will be helpful to introduce this language to make contact with physics later on.

Definition 53 *For $n \in \mathbf{Z}_+$ define $NS(n)$ as the set of all N-sequences $\{s\}$ having length $n+1$, i.e.*

$$\{s\} = s_0 s_1 s_2 \ldots s_i \ldots s_n$$

and with final entry $s_n \in NR_n$.

Definition 54 *For $\alpha \in NR_n$ define $S(\alpha)$ as the subset of $NS(n)$ consisting of sequences with final entry $s_n = \alpha$.*

Definition 55 *Define the subset $R(\alpha) \in S(\alpha)$ as the set of N-sequences with the property that every element of every sequence is an element of some set NR_m with $m \leq n$.*

Definition 56 *Define a partial order on sequences $\{s\}$ in $S(\alpha)$ (and hence on sequences in $R(\alpha)$) by*

$$\{s\} \leq \{t\}$$

if and only if $\{t\}$ can be obtained from $\{s\}$ by a sequence of moves of the form

$$\{s\} \to \ldots \to \{v\} \to \{v^i\} \to \ldots \to \{t\}$$

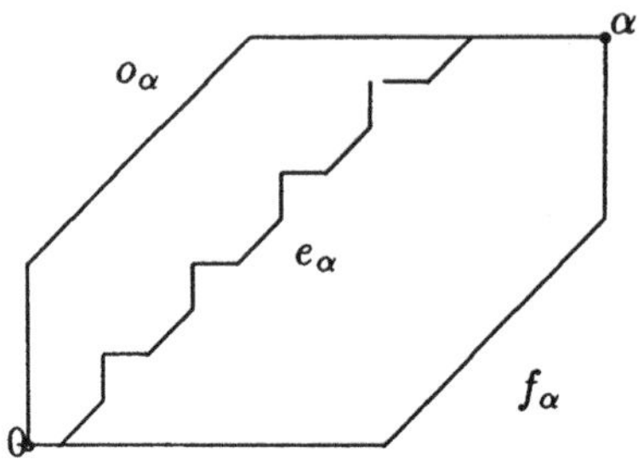

Figure 9.1: The highest (f_α), lowest (o_α), and e_α walks for $\alpha = (8,5,4)$.

The lowest sequence in the partial order in $S(\alpha)$, call it $\{o_\alpha\}$, is the unique sequence passing through the points

$$(0,0,..,0),(0,0,..,0,\alpha_N),(0,0,..,\alpha_{N-1},\alpha_N),..,(0,\alpha_2,..,\alpha_{N-1},\alpha_N),\alpha.$$

The poset is a lattice.

Let us define the sequence $\{e_\alpha\}$ as the lowest sequence in $R(\alpha)$, i.e. the sequence in which, at each step, the j coordinate is raised in preference to the k coordinate if $j > k$, as long as this is consistent with the property

$$(s_i)_j > (s_i)_k \Rightarrow j < k$$

for all i, and is consistent with α. For example, with $\alpha = (4,3,1)$

$$(0,0,0)(1,0,0)(1,1,0)(1,1,1)(2,1,1)(2,2,1)(3,2,1)(3,3,1)(4,3,1).$$

Walk diagrams

It is quite useful to have a diagrammatic version of these sequences. We achieve this for $N = 2,3,4,..$ by thinking of the elements of each sequence as points on a square (resp. cubical, hypercubical) lattice, and then projecting down to 2 dimensions, as in figure 9.1. The figure exemplifies the diagram for the lowest sequence $\{o_\alpha\}$ for $\alpha = (8,5,4)$

$$(0,0,0)(0,0,1)(0,0,2)..(0,0,4)(0,1,4)..(0,5,4)(1,5,4)..(8,5,4)$$

and some descendants, including $\{e_\alpha\}$.

Definition 57 *Define a partial order on sequences in $NS(n)$ by $\{s\} > \{t\}$ if $s_n > t_n$ under the total order of NR_n, and by definition 56 if $s_n = t_n$.*

Definition 58 *Define $P(\alpha)$ as the set of pairs of sequences $(\{s\},\{t\})$ for all $\{s\},\{t\} \in R(\alpha)$.*

Definition 59 *Define $S_{\{t\}}(\alpha)$ as the subset of $P(\alpha)$ obtained by taking all elements of $P(\alpha)$ with the same second sequence, $\{t\}$.*

Definition 60 *Define $NP(n)$ as the disjoint union of sets $P(\alpha)$ over all values of $\alpha \in NR_n$.*

9.3 On bases for $H_n(q)$

Definition 61 *Let us define the N-tuple ν_N by*

$$\nu_N = (1, 1, 1, ..., 1).$$

For given α we then define integers $d_M(\alpha)$ by writing

$$\alpha = \sum_{M=1}^{N} d_M(\alpha)\, \nu_M.$$

We further define integers

$$\alpha(j) = \sum_{i=j}^{N} \alpha_i.$$

Recall that $\{o_\alpha\}$ is the lowest sequence in $S(\alpha)$. Then

Definition 62 *Define idempotents $(o_\alpha, o_\alpha) \in H_n(q)$ by*

$$(o_\alpha, o_\alpha) = E_{1+\alpha_N} \left(\prod_{i=1}^{N-1} E_{1+\alpha_{N-i}}^{(\alpha(N-i+1))} \right). \tag{9.18}$$

For example,

$$(o_{(1,1,1)}, o_{(1,1,1)}) = 1.$$

The idea of this construction is that (o_α, o_α) is a product of commuting idempotents arranged so that

$$(o_\alpha, o_\alpha)U_i = U_i(o_\alpha, o_\alpha) = 0$$

unless i is a minimum of the sequence $\{o_\alpha\}$.

By analogy with the permutation group we call the left sided ideal generated by (o_α, o_α) the α-permutation module. In other words, it is the left ideal induced from the trivial module for the subalgebra

$$\otimes_{i=0}^{N-1} H_{\alpha_{N-i}-1}^{(\alpha(N-i+1))}(q).$$

9.3.1 A basis for α-permutation representations

Definition 63 *If $\alpha \in NR_n$, x is an N-tuple of complex numbers and $\{s\}, \{t\}$ are sequences in $S(\alpha)$, then $(s,t)_{(x)}$ is a function of x taking values in $H_n(q)$ and obtained iteratively from $(o_\alpha, o_\alpha)_{(x)} = (o_\alpha, o_\alpha)$ as follows:*

Suppose $s_i = s_{i-1} + e_j = s_{i+1} - e_k$ a minimum of $\{s\}$ (i.e. $j > k$), then with

$$g = (s_{i-1})_k - (s_{i-1})_j + (x_k - x_j)$$

we have

$$(s^i\ ,\ t)_{(x)} = \sqrt{k_g k_{g+1}}\left(1 - \frac{U_i}{k_g}\right)(s\ ,\ t)_{(x)} \tag{9.19}$$

and in general

$$(t,s)_{(x)} = (s,t)^T_{(x)}. \tag{9.20}$$

It follows from Definition 56 that every pair in $P(\alpha)$ is covered by this construction. The order in which the identities may be applied in moving from the initially defined operator to $(s,t)_{(x)}$ is not unique. A direct calcualtion using relation 9.4 reveals, however, that the definition of $(s,t)_{(x)}$ is unique, i.e. independent of the choice of order.

Note the following useful identity

Proposition 28 *With $\{s\}, \{s^i\}$ defined as in Definition 63 we have*

$$U_i\ (s^i,t)_{(x)} = \sqrt{k_g k_{g+1}}\left(1 - \frac{\sqrt{Q}}{k_g}\right) U_i(s,t)_{(x)}$$

$$= \frac{1}{k_{g+1}}(s^i,t)_{(x)} - \sqrt{\frac{k_g}{k_{g+1}}}\ (s,t)_{(x)}$$

by direct calculation, whereupon

$$(s,t)_{(x)} = \sqrt{\frac{k_{g+1}}{k_g}}\left(\frac{1}{k_{g+1}} - U_i\right)(s^i,t)_{(x)} = \sqrt{\frac{1}{k_{g+1}k_g}}\,(1 - k_{g+1}U_i)\,(s^i,t)_{(x)} \tag{9.21}$$

Corollary 28.1 *For each x, the set of order $N!$ obtained by acting on the left of (o_{ν_N}, o_{ν_N}) (i.e. $\{(s, o_{\nu_N})_{(x)}$ for all $s \in S(\nu_N)\}$) gives a basis for $H_{N-1}(q)$.*

Consider figure 9.2, which illustrates the sequences for some of this set (i.e. some of $S(\nu_N)$) in the case $N = 5$. It is straightforward to deduce from this a minimal set of the truncations defined in section 9.2.4.

We also have

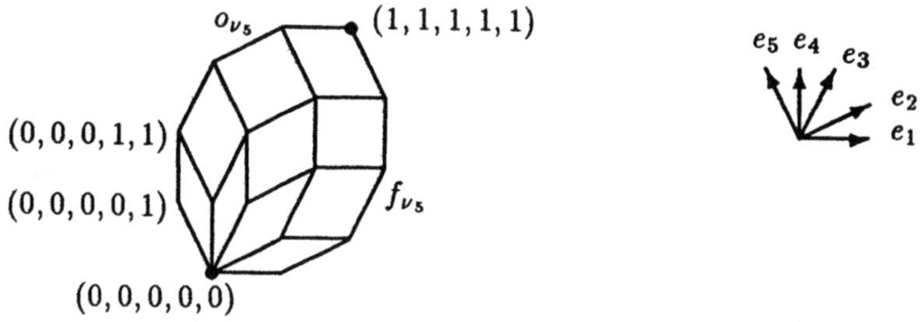

Figure 9.2: A projection, defined by the images of the 5 orthonormal vectors e_i shown, of a view of a 5 dimensional unit cube.

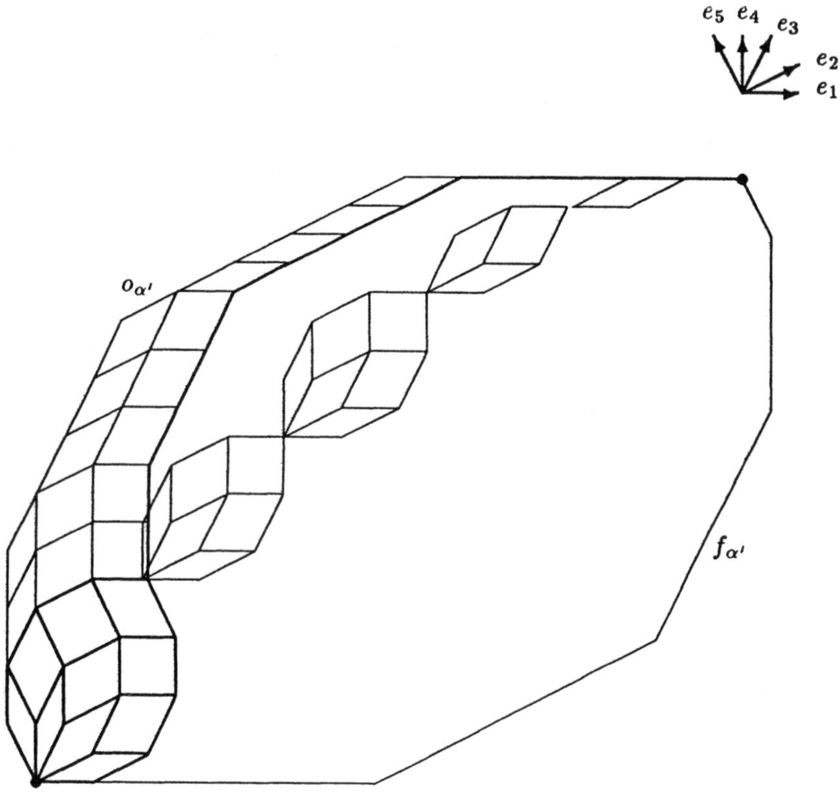

Figure 9.3: A projection, defined by the images of the 5 orthonormal vectors e_i shown, of a square patch surface on the $6 \times 5 \times 4 \times 3 \times 1$ hypercubical lattice, illustrating an iterative step in the text.

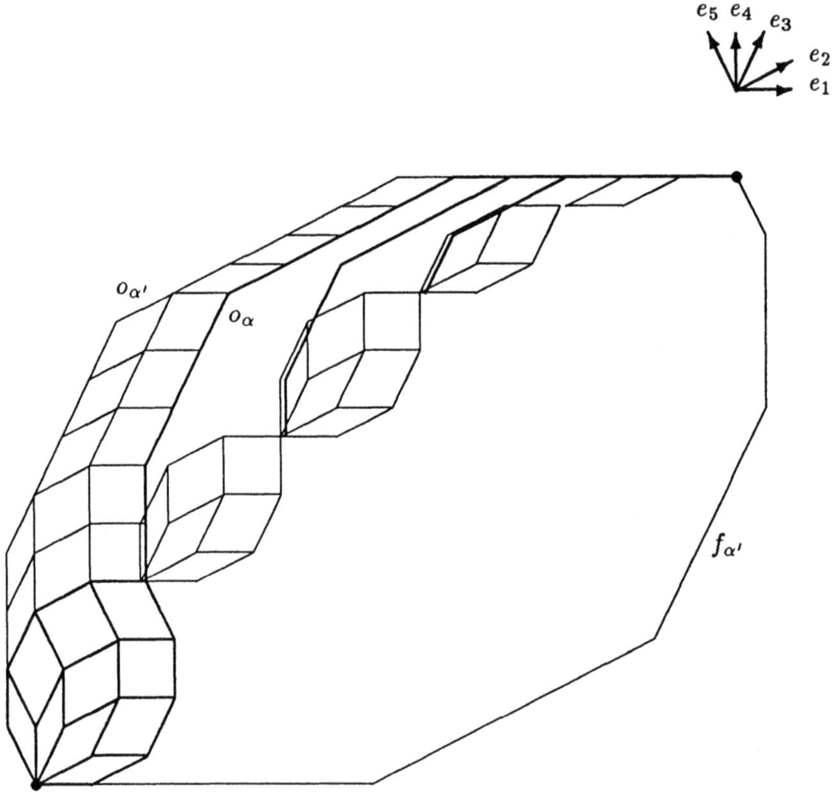

Figure 9.4: A projection, defined by the images of the 5 orthonormal vectors e_i shown, of a square patch surface on the $6 \times 5 \times 4 \times 3 \times 1$ hypercubical lattice, illustrating an iterative step in the text.

Proposition 29

$$U_i(...(g\ g+e_j\ g+2e_j)_i\ ...,\ t)_{(x)} = 0$$

Proof: Similar to proposition 8, i.e. by induction:

Assume the proposition is true for all occurences of the given subsequences in sequences below a certain sequence $\{w\}$ in the partial order (it is true for all occurences in $\{o_\alpha\}$ by construction). By the defining relations it is sufficient to prove the proposition in the case

$$\{w\} = ..., w_{i-2}, w_{i-2}+e_j, \overbrace{w_{i-2}+e_j+e_k}^{w_i}, w_{i-2}+e_j+2e_k, ...,$$

i.e. to prove that here $U_i(w,t)_{(x)} = 0$.

Let

$$\{z\} = ..., w_{i-2}, w_{i-2}+e_k, \overbrace{w_{i-2}+e_k+e_j}^{z_i}, w_{i-2}+e_j+2e_k, ...$$

and

$$\{y\} = ..., w_{i-2}, w_{i-2}+e_k, \overbrace{w_{i-2}+2e_k}^{y_i}, w_{i-2}+e_j+2e_k, ...$$

(where all the unspecified steps are taken to be the same). Since $\{y\}$ is below $\{z\}$ is below $\{w\}$ we have that $U_{i-1}\ (y,t)_{(x)} = 0$ by assumption. Then for some scalar g we have

$$\begin{aligned} U_i(w,t)_{(x)} &= U_i\sqrt{k_g k_{g+1}}(1-U_{i-1}/k_g)\sqrt{k_{g-1}k_g}(1-U_i/k_{g-1})(y,t)_{(x)} \\ &= U_i k_g\sqrt{k_{g-1}k_{g+1}}(1-U_i/k_{g-1}+U_{i-1}U_i/(k_g k_{g-1}))(y,t)_{(x)} \\ &= k_g\sqrt{k_{g-1}k_{g+1}}(1-Q^{1/2}/k_{g-1}+1/(k_g k_{g-1}))U_i(y,t)_{(x)} \\ &= 0 \end{aligned} \tag{9.22}$$

where we have used the Hecke relation in the antipenultimate step.

This completes the proof of proposition 29.

Remark 4 *The proof of this identity requires of the E_m factors appearing in Definition 63 only that they have the property $U_iE_m = 0$ for $i = 1, 2, .., m-2$.*

Corollary 29.1 *If $\{t\} \in S(\alpha)$ then the left ideal generated by (o_α, t) is spanned by $S_t(\alpha)$, i.e. by $\{(v,t)$ for all $\{v\} \in S(\alpha)\}$.*

Proof: The action of U_i on (v,t) with $\{v\}$ any sequence is covered by propositions 28 and 29.

9.3.2 On irreducible representations

In general each element of $H_k(q)$ in definition 63 corresponds to a pair of walks on a hypercubical lattice, each begining at the origin and ending at some point α. We will see that for certain values of x subsets of $S_t(\alpha)$ give rise to irreducible representations.

The case $x_i = -i$

Note from Definition 63 that the operators $(s,t)_{(x)}$ are well defined for all irrational r and irrational entries in x. When the components of x take values from the integers it can happen that some of the operators cease to be well defined.

In the case $x_i = -i$ definition 63 fails because of zeros appearing in the denominator in equation 9.19.

For example, consider the case $\alpha = (1,1,1)$. The sequence $\{o_\alpha\}$ is

$$(0,0,0),(0,0,1),(0,1,1),(1,1,1)$$

then among the remaining sequences in $S(\alpha)$ are

$$\{s\} = (0,0,0),(0,1,0),(0,1,1),(1,1,1)$$
$$\{s^2\} = (0,0,0),(0,1,0),(1,1,0),(1,1,1)$$

and

$$\{f\} = (0,0,0),(1,0,0),(1,1,0),(1,1,1).$$

We have, for example,

$$(s,o_\alpha)_{(x)} = \sqrt{k_{x_2-x_3}k_{x_2-x_3+1}}(1-U_1/k_{x_2-x_3})\,(o_\alpha,o_\alpha)$$

which is not defined at $x = (-1,-2,-3)$. However, consider

$$\sqrt{k_{x_2-x_3}}(s,o_\alpha)_{(x)}\Big|_{x=(-1,-2,-3)} = -\sqrt{k_2}\,U_1\,(o_\alpha,o_\alpha).$$

The next operator in our example then becomes

$$\sqrt{k_1}\,(s^2,o_\alpha)_{(x)} = \sqrt{k_3k_2}(1-U_2/k_2)\,[-\sqrt{k_2}\,U_1]\,(o_\alpha,o_\alpha)$$

which is alright, and finally

$$\begin{aligned} k_1\,(f,o_\alpha) &= k_2\,U_1\,\sqrt{k_3k_2}(1-U_2/k_2)\;U_1\;(o_\alpha,o_\alpha) \\ &= \sqrt{k_3k_2}\,Q^{1/2}\,(Q-1)\,F_{12}\;\underbrace{(o_\alpha,o_\alpha)}_{=1}. \end{aligned}$$

More generally we have

Proposition 30 *For y_ϵ an $N+1$-tuple with elements $(y_\epsilon)_i = -i(1+\epsilon)$ there exists a positive constant δ such that*

$$k_{1+\epsilon}^{N/2}\,(e_{\nu_{N+1}}, o_{\nu_{N+1}})_{(y_\epsilon)}$$

is finite for all $|\epsilon| \leq \delta$, and in particular

$$k_1^{N/2}\,(e_{\nu_{N+1}}, o_{\nu_{N+1}})_{(y_0)} = f_N\; F_{1N}$$

where f_N is a finite constant.

Proof: by direct calculation from the definitions and an induction on N.

By definition 63 and the inductive assumption (true for $N=1$)

$$\sqrt{k_1}\,(e_\alpha, o_\alpha)_{(y_0)} = \left[-\sqrt{k_2}\,U_1\left[\prod_{i=2}^{N}\sqrt{k_i k_{i+1}}\,(1 - U_i/k_i)\right]\right] f_{N-1}\; F_{2\,N}.$$

The proposition then follows by a rearrangement precisely analogous to that used for E_m in proposition 18.

It follows that if $(e_\alpha, o_\alpha)_{(y_0)}$ is well defined, or may be made well defined by renormalisation by a well defined constant as above, then the well defined version generates an invariant subspace of the left sided ideal generated by (o_α, o_α), the invariant subspace being spanned by $\{(s, o_\alpha)_{(y_0)}$ for all $s \in R(\alpha)\}$ (by the definition of F_{ij} and E_m).

Proposition 31 *There exists a non-negative integer $g(\alpha)$ such that*

$$k_1^{g(\alpha)/2}\,(e_{\alpha'}, o_{\alpha'})_{(y_0)}$$

is well defined.

Proof:

Note that in general if $\alpha' = \alpha + \nu_N$ for some N then by proposition 29

$$(e_{\alpha'}, o_{\alpha'})_{(x)} = (e_\alpha, o_\alpha)_{(x)}\,(e_{\nu_N} o_\alpha, o_{\alpha'})_{(x)}$$

and that by the definition

$$(e_{\nu_N} o_\alpha, o_{\alpha'})_{(x)} =$$

$$(e_{\nu_N}, e_{\nu_N})_{(x)}\left[{\prod_{i=N}^{2}}'\left[\prod_{j=1}^{\alpha_i}\left[\prod_{l=1+j}^{j+i-1}\left[\sqrt{k_{i-l}k_{i-l+1}}\left(1 - \frac{U_{l+\alpha'(i+1)}}{k_{i-l}}\right)\right]\right]\right]\right]$$

$$.(o_{\alpha'}, o_{\alpha'}).$$

The idea is to make repeated use of these two expansions and hence write $(e_{\alpha'}, o_{\alpha'})_{(x)}$ as an explicit product. We can then use

Proposition 32

$$F_{1\,N-1}\ \ (e_{\nu_N} o_\alpha, o_{\alpha'})_{(x)} =$$

$$F_{1\,N-1}\ \ (e_{\nu_N}, e_{\nu_N})_{(x)}$$

$$\cdot \left[\prod_{i=N}^{2}{}' \left[\prod_{j=1}^{\alpha_i} \left[\prod_{l=1+j}^{j+i-1} \left[\sqrt{k_{i-l}k_{i-l+1}} \left(-\frac{U_{l+\alpha'(i+1)}}{k_{i-l}}\right)\right]\right]\right]\right] (o_{\alpha'}, o_{\alpha'}).$$

Proof:
For this we will need

Proposition 33 *For χ_{1N} denoting any operator which, for some χ, may be written in the form*

$$\chi_{1N} = U_i\ \chi$$

for all $1 \le i \le N$, then for any complex functions $C(i), D(i)$

$$E_{11} \left[\prod_{i=2}^{N+1} (C(i) + D(i)\ U_i)\right] \chi_{1N}$$

$$= E_{11} \left[\prod_{i=2}^{N+1} (D(i)\ U_i)\right] \chi_{1N}$$

$$= E_{11}\ \chi_{2\,N+1}.$$

Proof:
First identity: suppose we expand the brackets and take the coefficient of $C(2)$, then U_1 can commute through to E_{11} from χ_{1N}, so only the $D(2)\ U_2$ part survives. But then if any subsequent U_i from the product is missing, all the $U_{j<i}$ commute through to, and may be absorbed in χ_{1N}, again releasing U_1. QED.

Second identity: even easier!
Proposition 32 then follows by iterating on j and then i. The organisation of the argument is most easily seen by reference to figure 9.3.

Using Proposition 33 we may then apply Proposition 32 directly to $(e_{\alpha'}, o_{\alpha'})_{(y_0)}$. The organisation of the argument is most easily seen by reference to figure 9.4, which illustrates the occurence of a nested sequence of operators of the form simplified in Proposition 32. The troublesome factors in Proposition 32 at $x = y_0$ are of the forms k_1 and k_0. These appear when $i - l = 1, 0, -1$. By direct calculation we find that, including the factors

from Proposition 33, then the accumulated k factors in Proposition 32 at $x = y_0$, altogether X, say, may be characterised as follows:
if $\alpha_N = 0$ then $\sqrt{k_1}\, X$ is finite;
if $\alpha_N = N-1$ then $\sqrt{k_0}\, X$ is finite;
otherwise, X is finite.
Since $\alpha_N = 0$ always occurs before $\alpha_N = N-1$ in the build up of any α, then there exists non-negative integer $g(\alpha)$ such that the product in Proposition 31 is finite.

This completes the proof of Proposition 31.

By considering the rest of the basis generated by $k_1^{g(\alpha)/2}\,(e_\alpha, o_\alpha)_{(y_0)}$ we find

Proposition 34 *There exist finite constants $h(\alpha)$ such that the operators*

$$(e_\alpha, e_\alpha) = \frac{k_1^{g(\alpha)}}{h(\alpha)}\,(e_\alpha, e_\alpha)_{(y_0)}$$

obey

$$(e_\alpha, e_\alpha)(e_{\alpha'}, e_{\alpha'}) = \delta_{\alpha,\alpha'}(e_\alpha, e_\alpha)$$

for all α, α'.

9.4 The generic structure of $H_n(q)$

Definition 64 *For $\{s\}, \{t\} \in R(\alpha)$ define (s,t) to be the element of $H_n(q)$ obtained as in definition 63 but using (e_α, e_α) as the initial idempotent.*

We then have (c.f. Hoefsmit 1974, Wenzl 1988)

Theorem 3 (Generic Theorem) .
(i) If the operators (u,s) and (t,v) are well defined then:

$$(u,s)(t,v) = \delta_{st}(u,v) \tag{9.23}$$

(ii) When all defined, the set of operators (t,v) for all $(\{t\},\{v\}) \in NP(n+1)$ with $N = n+1$ are a basis for, and as elementary operators exhibit the entire (multi-matrix) structure of, the algebra $H_n(q)$.

Proof :

Similar to the proof of theorem 1:

part(i): by induction.

Introduce a partial order on the set of pairs of sequences $(\{s\},\{t\})$ in equation 9.23 (i.e. not necessarily in $NP(n+1)$) by

$$(\{s\},\{t\}) < (\{u\},\{v\}) \; if \; \{t\} < \{v\}$$

and

$$(\{s\},\{t\}) < (\{u\},\{t\}) \; if \; \{s\} < \{u\}.$$

Assume that the identity is proved for all pairs of sequences below $(\{s\},\{t\})$, and all $\{u\},\{v\}$. It is true by proposition 34 if $\{s\}$ and $\{t\}$ are the lowest sequences in their respective partial orders. We stress that the assumption implies that if $\{w\}$ is below $\{t\}$ then the identity holds for pairs $(\{x\},\{w\})$ for *all* $\{x\}$.

Consider $\{w\}$ to be a sequence immediately below $\{t\}$, i.e. $\{t\} = \{w^h\}$ for some h, then by definition 63

$$(u,s)(t,v) = (u,s)\sqrt{k_g k_{g+1}}(1 - U_h/k_g)(w,v) \tag{9.24}$$

for some g. Pictorially this means that t and w differ only in the neighbourhood of $(..., w_{h-1}, w_h, w_{h+1}, ...)$, where they have the walk shapes $\wedge$ and $\vee$ respectively in some lattice plane. There are then three possibilities. The first is that

$$(u,s)U_h = 0 \tag{9.25}$$

in which case s, v and t are all necessarily distinct and the product $(u,s)(w,v)$, and hence $(u,s)(t,v)$, vanishes by assumption.

In the two remaining possibilities $(u,s)U_h$ is expressible as a linear combination of (u,s) and (u,a) (for some sequence $\{a\}$), using

$$(u,a) = (u,s)\sqrt{k_b k_{b+1}}(1 - U_h/k_b) \tag{9.26}$$

for some b, if $\{s\}$ precedes $\{a\}$ in the partial order, and

$$(u,s) = (u,a)\sqrt{k_b k_{b+1}}(1 - U_h/k_b) \tag{9.27}$$

if $\{a\}$ precedes $\{s\}$.

In the former case the shapes of $\{s\}$ and $\{a\}$ in the neighbourhood of h are $\vee$ and $\wedge$ respectively (in the latter, the roles are reversed). In the former case then, s and t are distinct, a and w are distinct, and $a = t$ if and only if $s = w$, whereupon $g = b$. In general then

$$(u,s)(t,v) \;\; = \;\; (u,s)\sqrt{k_g k_{g+1}}(1 - U_h/k_g)(w,v)$$

$$
\begin{aligned}
&= \sqrt{k_g k_{g+1}}(u,s)(w,v) \\
&\quad -(\sqrt{k_g k_{g+1}}/k_g)(k_b(u,s) - \sqrt{k_b/k_{b+1}}(u,a)\,)(w,v) \\
&= \sqrt{k_g k_{g+1}}(1-(k_b/k_g))(u,s)(w,v) \qquad (9.28)
\end{aligned}
$$

which vanishes either identically or by assumption.

In the latter case w and s are distinct, and t and a are distinct, and $s = t$ if and only if $a = w$, whereupon $b = g$. In either situation here we have

$$
\begin{aligned}
(u,s)(t,v) &= (u,s)\sqrt{k_g k_{g+1}}(1-U_h/k_g)(w,v) \\
&= \sqrt{k_g k_{g+1}}(u,s)(w,v) \\
&\quad - \left(\frac{\sqrt{k_g k_{g+1}}}{k_g}\right)\left(1-\frac{\sqrt{Q}}{k_b}\right)\sqrt{k_b k_{b+1}} \\
&\quad \cdot \left(k_b\,(u,a) - \sqrt{\frac{k_b}{k_{b+1}}}\,(u,s)\right)(w,v) \\
&= -\sqrt{\frac{k_{g+1}}{k_g}}\left(k_b - \sqrt{Q}\right)\sqrt{k_b k_{b+1}}(u,a)(w,v). \qquad (9.29)
\end{aligned}
$$

This vanishes by assumption unless $a = w$, in which case $b = g$ (so the coefficient is unity) and

$$(u,a)(a,v) = (u,v)$$

by assumption, since a preceeds s.

Thus, if the identities hold for all pairs of walks below $\{s\}, \{t\}$ they also hold for $\{s\}, \{t\}$. They therefore hold for all $\{s\}$. If they hold for all $\{s\}$ with $\{t\}$ then we can replace $\{t\}$ with $\{w\}$ immediately above $\{t\}$ and $\{v\}$ with $\{t\}$ in the above cases and the same arguments apply.

This completes the proof of theorem 3(i).

Corollary 3.1 *If the operators (w,s) are well defined for all $\{w\}, \{s\} \in S(\alpha)$ then the left ideal generated by any (w,s) is a simple module.*

Corollary 3.2 *If all the operators (s,t) for a given $H_n(q)$ are well defined then they span a (not necessarily proper) subalgebra with multi-matrix structure.*

Proof: These operators satisfy the definition of elementary operators for an algebra with such a structure.

part (ii):

Let us adopt the convention that F_{ij} may be regarded as a word in the generators of $H_n(q)$. We need the following definitions:

Let us define idempotents $(e_\alpha \circ e_\alpha) \in H_n(q)$ by

$$(e_\alpha \circ e_\alpha) = {\prod_{j=N}^{2}}' \left(\left(\prod_{i=1}^{d_j(\alpha)} F_{1\,j-1}^{((i-1)j)} \right)^{\left({\sum_{k=N}^{j+1}}' k d_k(\alpha)\right)} \right)$$

or iteratively by

$$(e_\alpha \circ e_\alpha) = 1$$

if $\alpha = (n+1)$, and then for $\alpha' = \alpha + \nu_{N+1}$ by

$$(e_{\alpha'} \circ e_{\alpha'}) = F_{1N} \quad (e_\alpha \circ e_\alpha)^{\overbrace{(N+1)}^{translation}}.$$

This means that

$$U_i(e_\alpha \circ e_\alpha) = \sqrt{Q}(e_\alpha \circ e_\alpha)$$

if i is a maximum of e_α. For example,

$$(e_{(5,3,1)} \circ e_{(5,3,1)}) = F_{12}\, F_{44}\, F_{66}$$

$$= \frac{1}{Q\sqrt{Q}(Q-1)}\,(U_1U_2U_1 - U_1)\, U_4\, U_6\,.$$

Definition 65 *If $(\{s\},\{t\})$ is in $NP(n+1)$ then $(s \circ t)$ is a word in the generators of $H_n(q)$ (counting F_{ij} as a word) obtained iteratively from $(e_\alpha \circ e_\alpha)$ as follows:*

$$(s^i \circ t) = U_i\,(s \circ t) \tag{9.30}$$

and

$$(t \circ s) = (s \circ t)^T. \tag{9.31}$$

Proposition 35 *The words $(v \circ w)$ $\forall$ $(\{v\},\{w\}) \in NP(n+1)$ span the same vector space as the elements (s,t) $\forall(\{s\},\{t\}) \in NP(n+1)$ where these latter are defined.*

Corollary 35.1 *Words corresponding to distinct elements of $NP(n+1)$ are linearly independent.*

Corollary 35.2 *The words $(v \circ w)$ $\forall$ $(\{v\}, \{w\}) \in NP(\nu_N)$ span the same space as the elements (s,t) $\forall(\{s\},\{t\}) \in NP(\nu_N)$ where these latter are defined.*

Proof of Proposition 35:
First note that $(e_{\nu_N} \circ e_{\nu_N}) = (e_{\nu_N}, e_{\nu_N})$ by definition. Let us abreviate $H_n(q)$ to H. Then more generally

Proposition 36 *For each N-tuple $\alpha \neq \nu_N$, and with α' the next element down in the α order, there exists*

$$X \in H\ (e_{\alpha'} \circ e_{\alpha'})\ H$$

such that $(e_\alpha, e_\alpha) = (e_\alpha \circ e_\alpha) - X$.

Proof:
(e_α, e_α) may be written in the form

$$F_{1N} W F_{1N}$$

which, for some X involving F_{1N+1} and hence as defined above is

$$X + F_{1N}\ Z$$

where $Z \in H^{(N+1)}_{n-N-1}(q)$, by proposition 27. Again from the definition of (e_α, e_α) the operator Z may in turn be written in the form FWF, and the same argument applies. We then iterate to achieve the required result.

Corollary 36.1 *The left module $H(e_\alpha \circ e_\alpha)/H(e_{\alpha'} \circ e_{\alpha'})H$ has dimension equal to the order of $R(\alpha)$, and is isomorphic to $H(e_\alpha, e_\alpha)$ where the latter is defined.*

Then from the definitions 63 and 65 each (s,t) is a linear combination of words $(p \circ v)$ such that $\{p\} \leq \{s\}$ and $\{v\} \leq \{t\}$.
There is a bijection which takes

$$(s,t) \leftrightarrow (s \circ t).$$

From theorem 3 (i) the elements (s,t) are linearly independent when defined. The reduction of words by applying the defining relations does not depend on q, except in an overall power of $\sqrt{Q}$.
This completes the proof of proposition 35.

Proposition 37 *The set of 'words' in definition 65 is a basis for the regular representation of $H_n(q)$.*

Outline Proof:

By a suitably organised application of the defining relations. For given α consider the double sided ideal generated by

$$(e_\alpha, e_\alpha) = (e_\alpha \circ e_\alpha) + X$$

where, with α' the next element down in the total order of NR_n, X is an element of the double sided ideal generated by $(e_{\alpha'} \circ e_{\alpha'})$ by proposition 36. If at any stage in this procedure we act with U_i on the left, say, on the current word (regarding (e_α, e_α) as a word, for convenience), we obtain a new word which appears in the iterative definition above *unless* the current subsequence around s_i is of the form

$$(g\ g + e_j\ g + e_j + e_k)_i \qquad (j < k)$$

or

$$(g\ g + e_j\ g + 2e_j)_i.$$

However, in the first of these cases we do not get a new reduced word because of relation 1, and in the second case because of relation 2 and or proposition 29. By starting with the lowest value of α in the partial order and increasing we note, from the definition of $(e_{(n+1)} \circ e_{(n+1)})$, that the double sided ideal generated by the unit is finally included in this scheme.

This completes the proof of proposition 37 and of theorem 3 (ii).

9.5 On the non-generic structure of $H_n(q)$

The orders of the sets $R(\alpha)$ are readily computed, so Theorem 3 gives the generic structure of $H_n(q)$.

What happens when r is rational? Well, the generically simple modules $H(e_\alpha, e_\alpha)/H(e_{\alpha'}, e_{\alpha'})H$ are well defined for all q, provided the factors F_{ab} involved in the initial idempotents are well defined (or are made well defined, as they always can be). On the other hand, these modules do not necessarily remain simple. We need to determine their invariant subspaces or, equally well, their module morphisms.

This is still quite a complicated problem, and it is best split into more manageable pieces.

9.5.1 On the structure of $NH_n(q)$

Generically everything goes through as in the previous section, except that we must restrict attention to α higher in the total order than the last $N+1$-tuple (i.e only N-tuples and shorter). The α in which α_N takes its largest possible value, and then α_i for $i < N$ takes its largest possible value consistent with α_{i+1} taking its largest possible value, is now the lowest in the order. The associated idempotent $(e_\alpha \circ e_\alpha)$ gives rise to an indecomposable projective module, since the usual quotient is then by elements associated with $N+1$-tuples, and this is now automatic.

For given N let us introduce the shorthand H_n for $NH_n(q)$. We will define

$$F = F_{n+2\, n+N} \in H_{n+N}$$

and note that

$$F_{n+1\, n+N} = 0$$

in H_{n+N}.

Note that $H_{n+N} F H_{n+N}$ is the subspace of H_{n+N} excluding the parts associated with $N-1$-tuples and higher in the total order (and hence excluding the unit). For example, with $N=3$ it excludes the Temperley-Lieb quotient algebra. With $N=2$ it excludes only the unit.

Note that (e_α, e_α) is idempotent so that, for example, any generically simple module $H_n(e_\alpha, e_\alpha)/H_n(e_{\alpha'}, e_{\alpha'})H_n$ is a quotient of H_n as a left H_n module. Note that H_n has a unit, and note from section 5.3 that the multiplication

$$H_{n+N} F H_{n+N} \times H_{n+N} F H_{n+N} \rightarrow H_{n+N} F H_{n+N}$$

is surjective. Then from sections 5.3 and 5.4.1 we have the following

Theorem 4 *H_{n+N} F H_{n+N} and H_n are Morita equivalent.*

Proof:

Note that $H_{n+N} \supset H_n$ and

$$[H_n, F] = 0$$

so FH_{n+N} and $H_{n+N}F$ are (projective) bimodules. By proposition 27 and the NH quotient

$$FH_{n+N}F \cong H_n$$

and then since F is idempotent we have

$$FH_{n+N} \otimes_{H_{n+N}FH_{n+N}} H_{n+N}F = F \otimes_{H_{n+N}FH_{n+N}} FH_{n+N}F \cong FH_{n+N}F.$$

Finally

$$H_{n+N}F \otimes_{H_n} FH_{n+N} \cong H_{n+N}FH_{n+N}$$

which fulfils the requirements of section 5.4.1.

Provided that we can determine directly the morphisms between modules in the HFH and H/HFH sectors this theorem gives us, by iteration on n and N, the structure of H_{n+N}.

The remaining problems are partly organisational. The equivalence has the bonus of suggesting a natural way to display the morphisms of modules of $NH_n(q)$ for all n. For each n we need a node for each generically simple module, labelled by α. But since α and $\alpha + \nu_N$ (the H_{n+N} module) have equivalent morphisms (up to morphisms with modules whose images do not exist in H_n), they might as well sit on the same spot.

For example, consider the case $N = 3$. All the modules for this case may be arranged onto a triangular lattice, as shown in figure 9.5. Various α values, each for a different n, are overlaid at each node. The first few lying over the origin have been written out explicitly. Full circles on the nodes of the diagram represent generically simple (g.s.) modules for $[n = 0]_{mod\ 3}$, empty circles represent $[n = 1]_{mod\ 3}$ and no circles $[n = 2]_{mod\ 3}$. The arrows may be translated to show moves $n \rightarrow n + 1$ (and hence the underlying Bratelli diagram).

In the figure the Temperley-Lieb (TL) edge is the upper boundary of the diagram for a given n, and is, in general, parallel to the example shown. At the given n value all the modules on the edge are of the form $(a+b, b, 0)$, i.e. of Temperley-Lieb type. The node at the right hand end of this line $((n + 1, 0, 0))$ corresponds to the trivial representation.

Before addressing the problem of morphisms between modules in different sectors in general, let us illustrate the whole programme with an example. The simplest non-trivial one is $N = 2$. We are now in a position to develop a much slicker alternative proof of Theorem 2.....

9.5.2 On $T_k(q)$ modules

In the boundary diagrammatic representation of $T_k(q)$ (chapter 6) we note that, as we multiply an existing word by further generators the number of lines which travel right through the diagram cannot increase. Consider, for example, the diagram for the word

$$R = U_1 U_3 U_5 ... U_k$$

(k odd), which is shown in figure 9.6.

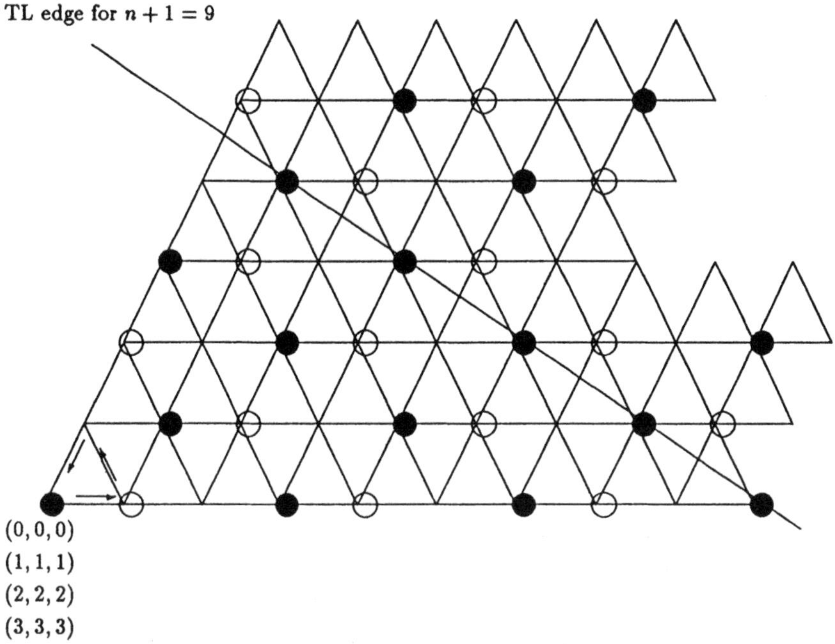

Figure 9.5: The projected generic Bratelli diagram for the tower of algebras $3H(q)$.

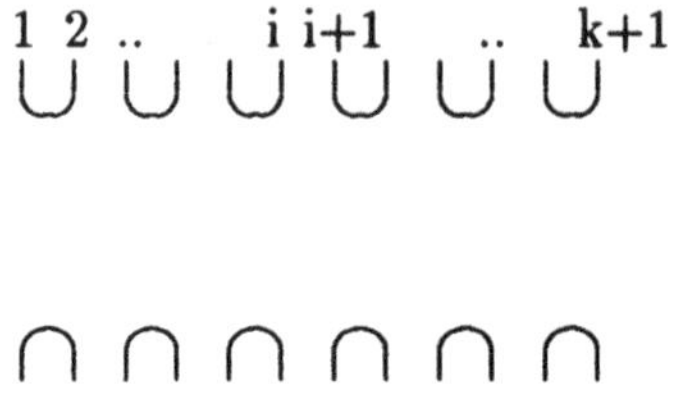

Figure 9.6: Diagram for the word $U_1U_3U_5...U_k$ (k odd).

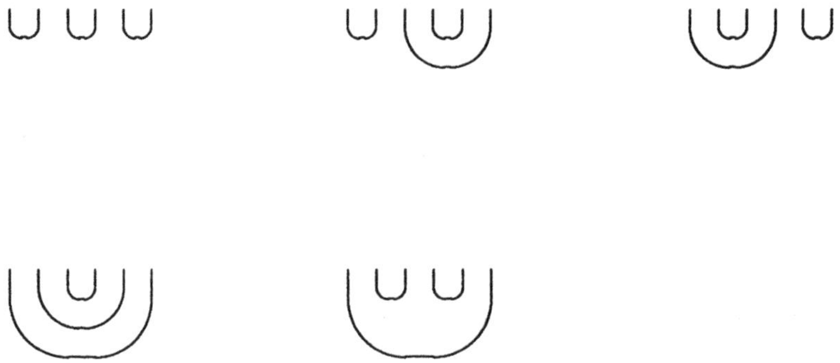

Figure 9.7: A basis for a right T_5 module.

The left sided ideal generated from this word forms a left T_k module. It follows incidentally that the *top halves* of the diagrams generated from the diagram in figure 9.6 form a basis for this module (here 'left' multiplying operators act from the top), since the bottom half is not affected by the process of generating the diagrammatic version of the left ideal. For example, with $k = 5$ we generate the top half diagrams shown in figure 9.7. The action of a generator U_i on a basis element is also indicated, by an example, in figure 9.8.

Considering the left ideal generated from an idempotent word with one fewer factors than R, and quotienting by the double sided ideal generated by R, we form another left T_k module of the g.s. type. The process of 'taking the top halves' in this case requires us to cut two lines. Shrinking

Figure 9.8: The action of U_2 on the second basis element in the previous figure.

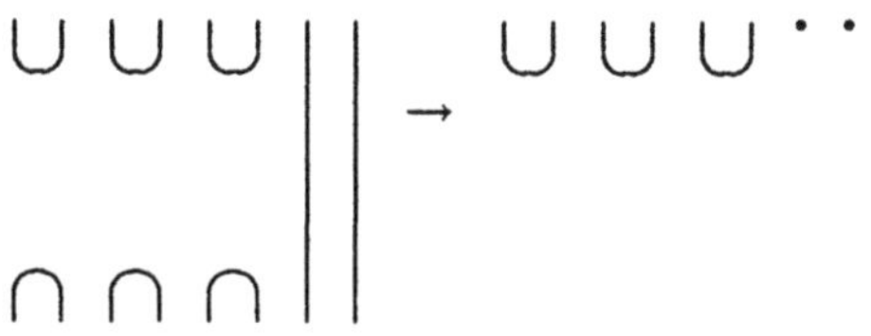

Figure 9.9: Cutting a diagram to form a half diagram.

the cut lines to dots and throwing the bottom halves away gives a basis for the new module. This process is illustrated in figure 9.9.

We can iterate this procedure, quotienting at each stage by the double sided ideal generated by the previous idempotent. Eventually we obtain a module with a single element basis given by the unique top half diagram of $(k+1)$ unconnected dots. Here, because of the quotienting, the action of any U_i is to take the basis state to zero.

Here the first module (that with 0 or 1 dots for k odd or even respectively) will be called T_k module .0 (no quotienting). The second module (2 or 3 dots) will be called T_k module .1, and so on. The last one (when k is odd) is T_k module $.(k+1)/2$. Collectively they will be called half-diagram or dot modules.

It is easy to see by induction (or as a special case of the construction in chapter 13) that the dimensions of these modules for odd k are the same as the numbers given in the even rows of the truncated Pascal's triangle. The order in the row is the same as the order of construction of the modules above (i.e. $0, 1, ..., (k+1)/2$). The procedure may be generalised to even values of k (the odd rows).

The complete set of *full* diagrams forms a basis for a bimodule for the regular representation of T_k. Alternatively, we could reconstruct the same set of diagrams from the half diagrams in the following way. Form the set of basis states for the right T_k modules corresponding to the left modules constructed above. Form the set of all possible pairs of right and left module basis elements, with the right module basis element arranged immediately below the left module one. If the number of dots in each element of the pair is different, discard the pair (this means we keep only pairs from corresponding modules). If the number of dots agrees there is exactly one way of forming a diagram from the pair by joining up the dots, since lines cannot cross. Each resultant diagram *is* a diagram in the original sense, and is clearly distinct since its top and bottom half combination is unique. Furthermore, if we cut this diagram into a top and bottom half we retrieve the right-left pair. Thus any full diagram may be thought of as a right-left pair, a diadic. Since we know how to compose full diagrams this implicitly defines an inner and outer product for composition of left and right basis vectors.

It is useful to match the diagrammatic representation with a diadic notation. First we will give a notation for vectors (left or right). We number every node distinctly, we then assign one of the two numbers to both nodes in each connected pair. Finally we renumber, so that, reading across the diagram, the first number is 1, and the first occurence of $n > 1$ is preceded by at least one occurence of $n - 1$. This renumbering is only necessary to determine equivalence of vectors. The outer product is given by generalising the matching of paired nodes to include those paired down the diagram. Thus for example the diagram for U_1 in T_4 becomes $|11234 >< 11234|$. The inner product is given by identifying numbers in corresponding positions, and then applying the appropriate quotient. If two numbers are already identical then a factor of $\sqrt{Q}$ is introduced, since this corresponds to closing a loop. Here are a couple of examples:

$$|11234 >< 11234||55667 >= \sqrt{Q}|11223 >$$

$$|11234 >< 11234||56778 >= |11223 > \stackrel{quot.}{\rightarrow} 0$$

where, at the last stage, we have applied the quotient. The quotient applies here since $|11223 >$ was implicitly quotiented out to make the module containing $|12334 > \equiv |56778 >$ a generically simple T_4 module.

Modules and functors

Suppose we form certain bimodules $(k, k-2)$ which are left T_k modules and right T_{k-2} modules, in this way. The resulting bimodule vectors consist of

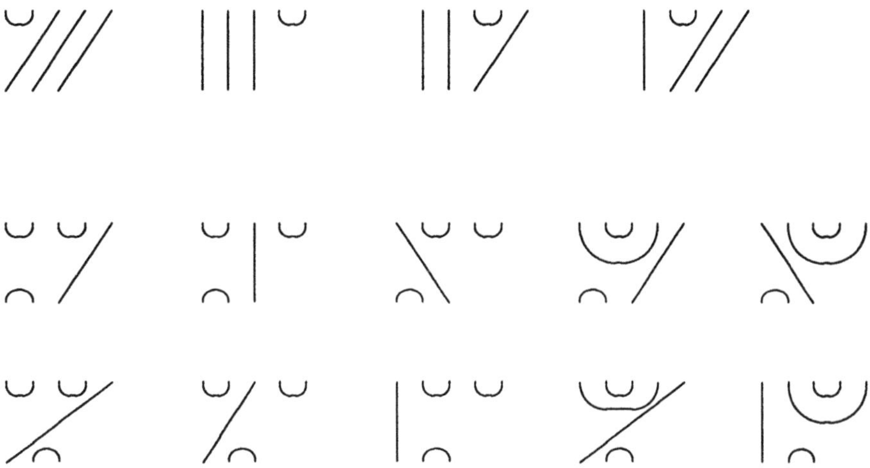

Figure 9.10: Basis vectors for the (4,2) bimodule.

the compositions of T_k module m elements with T_{k-2} module m elements only (for all possible m). This is the effect of matching the number of dots. In particular we do not form any bimodules which contain submodules acting as left T_k module $(k+1)/2$ in this way, since this requires more then the maximum available number of dots in any right T_{k-2} module. The bimodules we will consider are taken to contain the compositions of all vectors with all vectors, so they are $(k, k-2)$ bimodules without requiring any quotienting.

We can repeat this construction to form the dual bimodule $(k-2, k)$. For example, the vectors shown in figure 9.10 give a basis for the bimodule $(4,2)$. The diagrams are the same for (2,4), but upside down!

If we compose the bimodule and the dual bimodule by a tensor product

$$(k, k-2) \otimes_{T_{k-2}} (k-2, k),$$

i.e. we compose every vector in $(k, k-2)$ (as an element of a right T_{k-2} module) with every vector in $(k-2, k)$ (as an element of a left T_{k-2} module), then the resulting independent vectors span the regular representation T_k bimodule, except that the unit is missing. It follows form figure 9.10 that this is isomorphic to $T_k F T_k$ with $F = U_k$, c.f. our general construction of section 9.5.1. This means that the surjectiveness properties of section 5.3

apply. Then if we compose the dual bimodule with the bimodule

$$(k-2,k)\otimes_{T_k}(k,k-2)$$

we form the regular representation T_{k-2} bimodule.

We can use the $(k,k-2)$ and $(k-2,k)$ bimodules to define set maps between the sets of left T_k and left T_{k-2} modules as follows. Let us consider the left T_k modules M and N, and the left T_{k-2} modules M' and N'. Then

$$F: M \mapsto (k-2,k)\otimes_{T_k} M$$

where the tensor product $\otimes$ signifies the composition (under the quotient) of every vector in $(k-2,k)$, regarded as a right T_k module, with every vector in M. Note that, because of the quotient, if M is a T_k module $.m$, then the composition of a vector in M with a vector in $(k-2,k)$ will only be non-vanishing if the bimodule vector is in a right T_k submodule $.p \leq m$ (i.e. its bottom half is in a right T_k module $.p \leq m$). The map is defined similarly on N. Note that F maps T_k module m to T_{k-2} module $.m$. Thus the images of M and N are distinct if and only if M and N are distinct.

The map defined using $(k,k-2)$ is

$$G: M' \mapsto (k,k-2)\otimes_{T_{k-2}} M'.$$

For example, suppose M' is the left T_2 module .1 (i.e. the single half diagram of three dots). Then composing it with each bimodule vector from figure 9.10 we find that compositions with vectors from the second and third rows are all quotiented to zero. Compositions with the first row produce the vectors for the left T_4 module .1.

We can extend F to define a map of algebra homomorphisms complimentary to the set map. Suppose there is an algebra homomorphism $\psi: M \to N$. Let us consider $x\otimes y \in F: M$, i.e. $x \in (k-2,k)$ and $y \in M$, such that

$$\psi: y \mapsto \psi(y),$$

then

$$F(\psi): x\otimes y \mapsto x \otimes \psi(y).$$

A T_{k-2} module homomorphism map extending G may be similarly defined. The direction of homomorphisms is preserved (with respect to the set map) by construction in either case, as is the identity map. Furthermore, suppose we have two composable algebra homomorphisms ψ and ψ' such that

$$y \overset{\psi}{\mapsto} \psi(y) \overset{\psi'}{\mapsto} \psi'(\psi(y)) = (\psi'\psi)(y),$$

then the composite of morphisms maps under F to $F(\psi')F(\psi) = F((\psi'\psi))$, again by construction.

Recall that the joint map of modules and morphisms is called a functor (see e.g. Maclane and Birkoff 1979). Composing the functor maps F and G we find, using the surjectivity properties and section 5.3, that both FG and GF act as the identity on the appropriate modular categories. For example,

$$M \overset{F}{\mapsto} (k-2,k) \otimes M \overset{G}{\mapsto} (k,k-2) \otimes ((k-2,k) \otimes M)$$

$$= ((k,k-2) \otimes (k-2,k)) \otimes M \quad = \quad M$$

since the product of bimodules is the regular representation bimodule (except for the identity in the example above, but we have already excluded the case in which M is the corresponding left T_k module). Applying all full diagrams to a given module clearly just reproduces the same module. Applying all diagrams except the identity reproduces the same module unless it is the identity (i.e. all dots) module.

Any module homomorphism among the T_k modules (excluding the last module) thus implies a matching homomorphism among the T_{k-2} modules, and *vice versa*.

We showed in chapter 6 that the modules with basis the ket vectors are generically simple. This means that they have no interesting homomorphisms. In order to get an *interesting* example we must consider a nongeneric case.

The example is as follows. The $T_2(e^{i\pi/3})$ module .1 maps onto an invariant subspace of the $T_2(e^{i\pi/3})$ module .0 (we stress that we have now taken $Q = 1$). To see this note that the action of $U_1 = |112><112|$ and $U_2 = |122><122|$ on the vector $(|112> - |122>)$:

$$U_1(|112> - |122>) \quad = \quad \sqrt{Q}|112> - |112>$$

$$\overset{Q=1}{\equiv} \quad U_2(|112> - |122>) = 0$$

is the same as the action on $|123>$. The vector $(|112> - |122>)$ spans an invariant subspace of the module spanned by $|112>$ and $|122>$. We write

$$|123> \overset{\phi}{\mapsto} |112> - |122> .$$

The functor F pulls this through to a homomorphism $F(\phi)$ of $T_4(e^{i\pi/3})$

modules via

$$\begin{array}{rcl}
|11234><234||567> & = & |11234> \\
& \stackrel{F(\phi)}{\mapsto} & \\
|11234><234|(|556>-|566>) & = & |11223>-|11233> \\
|12234><134||567> & = & |12234> \\
& \stackrel{F(\phi)}{\mapsto} & \\
|12234><134|(|556>-|566>) & = & |12213>-|12233> \\
|12334><124||567> & = & |12334> \\
& \stackrel{F(\phi)}{\mapsto} & \\
|12334><124|(|556>-|566>) & = & |11223>-|12332> \\
|12344><123||567> & = & |12344> \\
& \stackrel{F(\phi)}{\mapsto} & \\
|12344><123|(|556>-|566>) & = & |11233>-|12233>
\end{array}$$

with all the other bimodule compositions going to zero. It is easy to check that the right hand sides span the module mapped onto by $F: |123>$ and that the left hand sides span an isomorphic invariant subspace of another module, that mapped onto by $F: \{|112>, |122>\}$.

Conversely, considering the homomorphism of T_4 modules shown above as the starting point, we can recover the original T_2 homomorphism by applying G. For example:

$$\begin{array}{rcl}
|123><14423||55678> & = & |123> \\
& \mapsto & \\
|123><14423|(|55667>-|55677>) & = & |112>-|122>.
\end{array}$$

9.5.3 The structure of $2H_k(q)$

More generally with $N = 2$, then, we need only determine the morphisms associated with the trivial module to completely determine the structure of the algebra. These follow immediately from Proposition 7 via the arguments of section 5.3. We leave it as an exercise to check Theorem 2 from this perspective.

9.5.4 The structure of $3H_k(q)$ and higher N

A complete analysis is beyond the scope of this book. However, there are a number of aspects which can be brought out fairly simply with the machinery we have already developed. Firstly we can determine the morphisms associated with the trivial representation. Specifically we can identify invariant subspaces of various modules which are isomorphic to the trivial module.

It follows from the definition that every (s,t) is either finite or may be made finite by multiplying by a well defined constant k. On the other hand, many sequences $\{s\}$ have the property that they have no minima and all their maxima occur with $g = (s_i)_k - (s_i)_j - k + j$ an integer multiple of r. Then from the definition we have, provided (s,t) is not itself finite, that

$$U_i \; k(s,t) = 0$$

for all i. Such a vector therefore spans an invariant subspace isomorphic to the trivial module. To see where these vectors lie for given r it is convenient to draw out our triangular lattice marking in the lines for which g takes the critical values. Then any trajectories whose only changes in direction are reflections (giving maxima) at these critical lines correspond to sequences giving vectors of this type!

For example, consider the case $r = 4$ shown in figure 9.11. The formal definition of the element indicated gives at worst a $k_0^{3/2}$ divergence (a factor $k_0^{1/2}$ for each reflection, plus one other). However, the terms at this level cancel each other, and the true divergence is $k_0^{1/2}$. Deleting this factor to obtain a finite element of the module we find that this spans a one dimensional invariant subspace, as expected. The general case proceeds similarly.

Figure 9.12 indicates how to establish, following an argument almost identical to the proof of proposition 32, that the operator associated with the final sequence $\{f_\alpha\}$ may be written in the form

$$(f_\alpha, e_\alpha) = k^{(\alpha)}(o_\alpha, o_\alpha)^{(-)}(f_\alpha \circ e_\alpha)(e_\alpha, e_\alpha)$$

where $k^{(\alpha)}$ is another overall scalar factor.

Armed with this result, the other invariant subspaces of g.s. modules may be exhibited by means analogous to the reflection arguments of section 7.4.

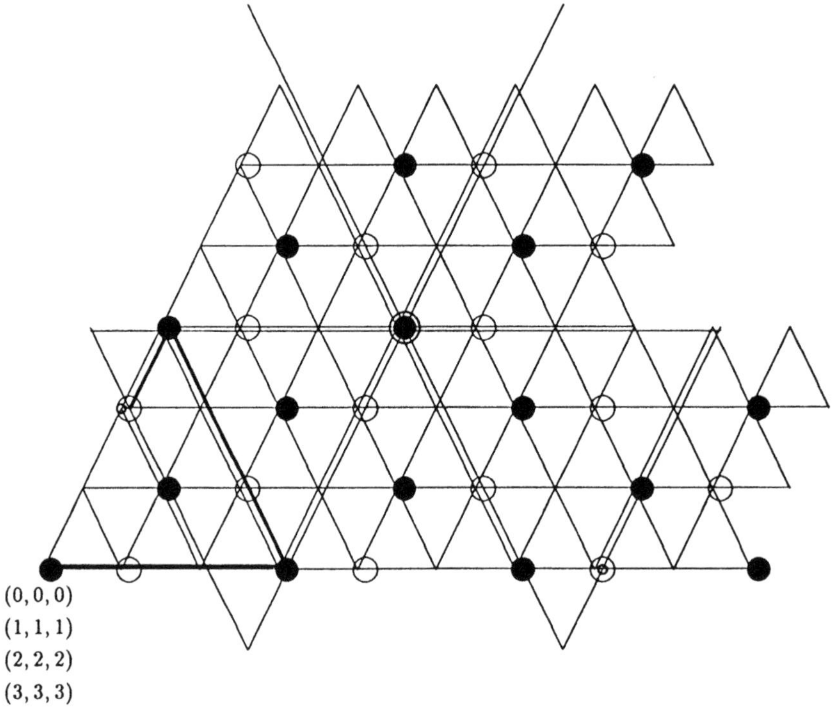

Figure 9.11: The projected diagram for the tower of algebras $3H(q)$ with $r = 4$. Critical lines have been translated slightly to make them visible against the lattice background. The solid trajectory from the origin to $\alpha = (3,3,1)$ exhibits an invariant subspace of this module isomorphic to $\alpha = (7,0,0)$, which is also marked.

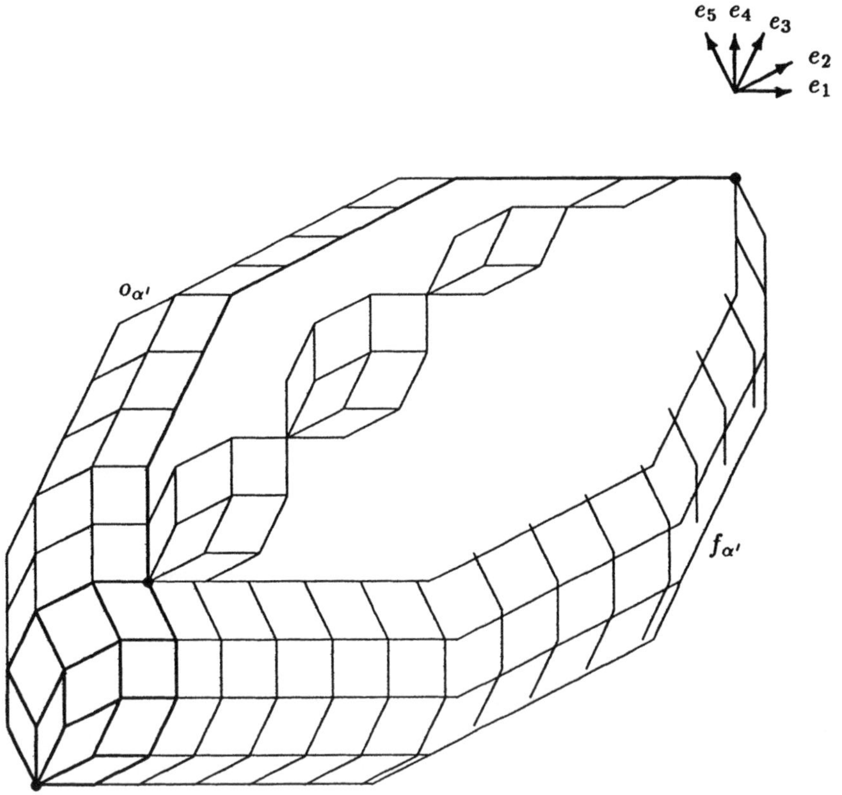

Figure 9.12: A projection, defined by the images of the 5 orthonormal vectors e_i shown, of a square patch surface on the $6\times5\times4\times3\times1$ hypercubical lattice, illustrating an iterative step in the text.

9.6 The defining representation of $NH_k(q)$

The defining representation of $T_k(q)$ appears in statistical mechanics in the diagonal row transfer matrix for the asymmetric 6-vertex models. These were the models discussed by Temperley and Lieb (1971) - see chapter 12. In this statistical mechanical interpretation of the representation the basis space is the set of all possible configurations of $k+1$ bond variables, each taking 2 possible values. In other words, in this representation $U_i \in End(V_2^{\otimes k+1})$. Ordering the product of vector spaces (and labeling them in order, to identify the corresponding bond variables in the transfer layer) as $V_2^{(1)} \otimes V_2^{(2)} \otimes V_2^{(3)} \otimes ... \otimes V_2^{(k+1)}$ we have

$$U_i = 1_2 \otimes 1_2 \otimes 1_2 \otimes ... \otimes \begin{pmatrix} 0 & 0 & 0 & 0 \\ 0 & q & 1 & 0 \\ 0 & 1 & q^{-1} & 0 \\ 0 & 0 & 0 & 0 \end{pmatrix} \otimes ... \otimes 1_2 \otimes 1_2 \qquad (9.32)$$

where the 4×4 matrix acts on $V_2^{(i)} \otimes V_2^{(i+1)}$. Note that in the abstract representation theory sense the ordering of vector spaces is arbitrary. We could define further matrices in which the 4×4 factor acted on any pair $V_2^{(i)} \otimes V_2^{(j)}$ (c.f. chapters 4 and 8).

It follows (after some work) from the formalism in section 6.3.2 and the results in section 6.4 that this representation is faithful for all q, and hence from chapter 5 that $T_k(q) \cong End_{U_qSl(2)}(\otimes^{k+1} V_2)$.

The representation immediately breaks up into blocks as follows. Writing $V_2 = \{1, 2\}$ we see that non-zero off-diagonal entries in U_i just take $1 \otimes 2$ $(\in V_2^{(i)} \otimes V_2^{(i+1)}) \rightarrow 2 \otimes 1$. Writing an arbitrary basis state as $a_1 a_2 a_3 ... a_{k+1}$ $(a_i \in 1, 2)$ we then note that only states with the same number of 1's are mixed. Let us write $n(1)$ for the number of 1's. It follows from the definition of the representation that these fixed $n(1)$ blocks are generalisations of permutation representations of the symmetric group. The generic multiplicities of irreducible representations in the full representation, and hence the generic structure of $U_qSl(2)$, follow immediately (see e.g. Robinson 1961). The non-generic structure may also be deduced, but more work is required.

In the 6-vertex model context $n(1) - n(2)$ is the surfiet of up arrows over down arrows in the 6-vertex model lattice layer (i.e. the Bethe ansatz parameter, see e.g. Baxter 1982).

The generalisation of the results noted in this section to other Hecke quotients $NH_k(q)$ (R-matrix representation in $End(V_N^{\otimes k+1})$, and so on) is straightforward. We leave it as an exercise (but see also chapter 12).

Chapter 10

Algebraic formalism for Z_Q symmetry

Subtitle: Operator formalism for locally and globally Z_Q symmetric models, duality and lattice gauge theory representations.

In this chapter we will define generalised Clifford algebras, and simplicial Clifford algebras, which are certain subalgebras of generalised Clifford algebras. These subalgebras include the algebras appropriate for building the transfer matrices for general Z_Q symmetric lattice models in arbitrary dimensions. We define canonical representations of these algebras. We use these representations to construct representations of graph Temperley-Lieb algebras appropriate for the 3 dimensional Potts lattice gauge theories.

10.1 Introduction

The operator formalism we have used in building transfer matrices for 2 dimensional Potts and equivalent models, and central in the construction of solutions to the Yang-Baxter equations, can be extended on the one hand to Z_Q symmetric spin models in 2 dimensions, and on the other hand to Potts models in arbitrary dimensions.

In general these extensions do not appear to shed much new light on the search for thermodynamic limit *solutions*, either to models with more physically realistic Hamiltonians, or to models in higher dimensions. They do, however, provide a systematic framework for the exploitation of symmetry in the computation of observables, at least on the finite lattice. The results of such computations, together with a physical picture of Z_Q symmetric spin models in two dimensions, are discussed in chapter 11. In the present chapter we will forget about star triangle relations and so forth for a while and describe the *general* framework for the operator formulation of Z_Q symmetric models (including locally symmetric cases, i.e. lattice gauge models) in arbitrary dimensions.

Because of the duality properties of lattice models in 3 dimensions (see, for example, Savit 1980) we are able to exhibit two graph Temperley-Lieb subalgebras appropriate for 3 dimensional models in the generalised Clifford algebra. These correspond to the Potts spin and gauge models respectively. The gauge model case enables us to construct new representations for the graph Temperley-Lieb algebra.

Conversely, since many of the generic irreducible representations of this algebra are already known (and are much smaller than the lattice gauge model representation) we can greatly facilitate the computation of finite lattice results for 3 dimensional Potts gauge models by block diagonalising the transfer matrix into its irreducible components.

10.1.1 Notation

In this chapter we deal with models having 'near neighbour' interactions. Before we can give a working definition of this restriction it will be useful to introduce some extra notation.

Consider an ordered set $A \subset \mathbf{Z}^d$ of points spanning a c dimensional subspace of $\mathbf{Z}^d \subset \mathbf{R}^d$. If no point in $\mathbf{Z}^d$ is an interior point of any segment in the convex hull of $A \subset \mathbf{R}^d$ then the convex hull is called a c dimensional (lattice) simplex in $\mathbf{Z}^d$. The minimum order of A is $c+1$, the maximum order is 2^c. If the order is 2^c then the simplex is called a hypercubical simplex.

Consider a d dimensional hypercubical lattice model [1], with Z_Q valued variables sitting on sites, or oriented bonds, or some other set of identical oriented hypercubical simplices s_c (with dimension $c < d$). We will refer to these variables *ensemble* as the field. It is useful to denote configurations of the field by $f \in Hom(s_c, Z_Q)$, and of an individual variable $x \in s_c$ by $f(x) \in \{0, 1, 2, ...Q-1\}$. We further extend this definition so that

$$f(ax+by) = (af(x)+bf(y))_{mod\ Q} \qquad (x, y \in s_c;\ a, b \in Z). \tag{10.1}$$

The interactions sit on the set of $c+1$ dimensional hypercubical simplices s_{c+1} which are bounded by the simplices s_c, or on simplices s_{c+2} which are bounded by s_1, and so on. For definiteness we will take the former case.

[1] The restriction to hypercubical lattices is really for notational convenience.

If $Y \in s_{c+1}$ we write BY for the set of $x \in s_c$ bounding Y, and ∂Y for the usual boundary operator on Y. For example, with the sites $(i) \in s_0$ round a plaquette $(ijkl) \in s_2$ numbered anticlockwise (1),(2),(3),(4), and the bonds oriented (12), (23), (43), (14), we have

$$B(1234) = \{(12), (23), (43), (14)\}$$

and

$$\partial(1234) = (12) + (23) - (43) - (14).$$

10.2 Z_Q symmetric models

In a Z_Q symmetric model the interactions on simplices $Y \in s_{c+1}$ are character functions of the Z_Q valued variables on their boundaries. The variable on simplex $x \in s_c$ may be represented in the form $f(x)$, or $F(x)$ defined by

$$F(x) = \exp(2\pi i f(x)/Q).$$

Suppose that for each $x \in BY$ we have a function

$$\epsilon_x : Y \to \pm 1$$

such that

$$\sum_{x \in BY} \epsilon_x(Y) = 0.$$

We then define

$$p_f(Y) = \sum_{x \in BY} \epsilon_x(Y) f(x), \tag{10.2}$$

and the interaction on simplex Y is written in the form

$$\chi_{(\beta_r)}(p_f(Y)) = \sum_{r=0,..Q-1} \beta_r \exp(2\pi i p_f(Y) r/Q). \tag{10.3}$$

For example we could take

$$p_f(Y) = f(\partial Y), \tag{10.4}$$

ie. $\epsilon_x = \pm 1$ depending on orientation. In general, however, we just require the overall condition accompanying equation 10.2, so that the interaction is invariant under a global Z_Q phase redefinition. In the case $c = 0$ there is no distinction, and we have the usual Z_Q symmetric spin models. For local gauge symmetry, however, the interaction must be of the form of equation 10.4. Then, for instance, $c = 1$ gives a Z_Q lattice gauge theory. That is, $p_f(Y)$ is invariant under the local gauge transformation

$$f(x) \to f(x) + f_0(\partial x)$$

where $f_0 \in Hom(s_0, Z_Q)$, $f_0(x \in s_0) \in \{0, 1, 2, ..., Q-1\}$, and so on. This old result ($\partial^2 = 0$) has some interesting consequences at the operator level, as we will see!

The partition function for such models is

$$Z = \sum_f \exp(\beta \sum_{Y \in s_{c+1}} \chi_{(\beta_r)}(p_f(Y))) \tag{10.5}$$

where the first sum is over all possible field configurations.

In order to define a model with a unique coupling constant β we must specify the interaction parameters β_r. For example,

$$\beta_r = (\delta_{r,1} + \delta_{r,Q-1})/2$$

together with equation 10.4 gives the *gauge clock model* interaction

$$\chi(p_f(Y = (1234))) = \left[F((12))F((23))F^{-1}((43))F^{-1}((14))\right] + [..]^\dagger \tag{10.6}$$

when $c = 1$, and the *standard clock model* when $c = 0$. The spin and gauge Potts models have $\beta_r = 1/Q$. Note that $\beta_r = \beta_{Q-r}$ for unitarity.

10.2.1 Transfer matrices

Let us define the notion of a single $d-1$ dimensional layer of the lattice. The defining constituents of such a layer, as far as any given model is concerned, are the variables (sorted into two sets, as we will see) and interactions associated with the layer. These objects are collected from those in the region of general physical location of the layer in a way which depends on the model. Loosely speaking, that region is a slice of the lattice of thickness one lattice spacing in the layering direction.

For $d = 2$ and $c = 0$ the layer consists of the variables on two parallel adjacent linear chains of sites, and the bond interactions within one chain and those between the two chains. For $d = 3$ and $c = 0$ it is two parallel adjacent square lattices of site variables, the bond interactions within one square lattice and between the two square lattices.

We will call the perpendicular direction the 'time' direction, without loss of generality. We can then call variables or interactions 'spacelike' if their simplices are orthogonal to the time direction, and 'timelike' otherwise.

The set of variables on simplices within the 'earlier' of the two parallel adjacent $d-1$ dimensional sublattices, together with those between the two sublattices, are called the leading edge of the layer.

The forward direction of time takes us from the leading edge to the trailing edge. All the variables associated with the layer are either associated with the leading edge (if spacelike and in the leading edge or if timelike connecting the two edges) or the trailing edge (if spacelike and in the trailing edge or timelike and connecting to a new edge).

For the purposes of identifying *variables* in the layer the simplices included in the trailing edge are just the time translations of those appearing in the leading edge, and each edge is common with another layer. On the other hand, each *interaction* in the full lattice may be uniquely associated with a given layer. The interactions associated with a layer are all those depending only on variables in the leading edge and all those depending on variables in both edges.

We will write just $x \in edge$ for the restriction of $x \in s_c$ to a single edge, and $Y \in layer$ for $Y \in s_{c+1}$ in a single layer.

Recall that the transfer matrix is a matrix whose ij^{th} element is the partition function for a layer, where the leading and trailing layer edge field configurations are completely specified by i and j respectively. In other words

$$i, j \in \bigotimes_{x \in edge} V^{(x)} \tag{10.7}$$

where $V^{(x)} \in \{0, 1, ..., Q-1\}$ is the space of states of $f(x)$. In the present case 'near neighbour interactions' means that the partition function for a layer *can* be determined with only this information, and that each interaction involves at most one variable from the leading edge.

We will call interactions which are completely determined by the configuration of the trailing edge 'type b', and the rest 'type a'. Type a interactions include one pair of spacelike variables which are related by time translation, and all the rest are timelike. In this case the relevant x uniquely locates Y, and it is then convenient to use this x as a label for Y. The transfer matrix may then be written in the following form

$$T = \prod_{Y:type\ a} (M_Y) \prod_{Y:type\ b} (M_Y) \tag{10.8}$$

where the products are over $Y \in layer$ and M_Y is the single interaction transfer matrix.

Generalised Clifford algebras

For S some set of order n and Q as above we define a generalised Clifford algebra $G_{2n}(Q)$ by generators $\{a_x, b_x; x \in S\}$ and relations

$$\begin{aligned} a_x^Q &= b_x^Q = 1 \\ a_x b_x &= W b_x a_x \\ W &= exp(2\pi i/Q) \\ a_x a_w &= a_w a_x \\ b_x b_w &= b_w b_x \\ a_x b_w &= b_w a_x \quad w \neq x. \end{aligned} \tag{10.9}$$

For example, Q=2 gives the usual Clifford algebra.

As we will demonstrate shortly by giving an explicit representation, for $S = s_c$ the single interaction transfer matrix M_Y may then be written as follows:

$$Y : type\ a \qquad M_Y = \sum_{j=0,..Q-1} \alpha_{Yj}(a_x \prod_{w \in BY \setminus x} b_w^{\epsilon_w})^j \tag{10.10}$$

$$Y : type\ b \qquad M_Y = \sum_{j=0,..Q-1} \alpha_{Yj}(\prod_{w \in BY} b_w^{\epsilon_w})^j \tag{10.11}$$

Here $BY \setminus x$ means the set of simplices bounding Y not including x and its time translation. The signs ϵ_w coincide with those in equation 10.3, and scalar α_{Yj} is a readily determined function of $\{\beta, \beta_r\}$ in any given model (see later).

The dimension of $G_{2n}(Q)$ is clearly Q^{2n}, since any word may be written (up to scalars) in the form

$$W(\{\alpha_x, \beta_x\}) \propto \prod_{x \in edge} a_x^{\alpha_x} b_x^{\beta_x}$$

where $\alpha_x, \beta_x \in \{0, 1, .., Q-1\}$ and $W = W'$ if and only if $\alpha_x = \alpha_x'$ and $\beta_x = \beta_x'$ $\forall x \in edge$.

Note that this algebra carries no information about the spacial structure of the model. We get the same algebra in any dimension for any c, provided that the value of n is the same.

10.2.2 Simplicial Clifford algebras

For a given physical model the transfer matrix can be built with a sub algebra generated by operators associated with interactions in the layer.

We introduce a set of operators indexed by these interactions as follows

$$\{S_x, S_Y \mid x \in spacelike\ s_c\ in\ edge;\ Y \in type\ b\ in\ layer\}$$

where $x \in$ *spacelike* s_c *in edge* are taken to label the interactions of type a in the layer. The operators are defined as follows

$$\begin{aligned} type\ a: \quad S_x &= a_x \prod_{w \in BY \setminus x} b_w^{\epsilon_w} \qquad (10.12)\\ type\ b: \quad S_Y &= \prod_{w \in BY} b_w^{\epsilon_w} \end{aligned}$$

These objects obey the following relations:

$$\begin{aligned} S_x^Q &= 1 \\ S_Y^Q &= 1 \qquad (10.13)\\ S_x S_Y &= W^{e(x,Y)} S_Y S_x \end{aligned}$$

where $e(x, Y)$ is the coefficient of x in ∂Y, or more generally the coefficient of $f(x)$ in $p_f(Y)$, (i.e. $\pm 1, or\ 0$).

In 2 dimensional spin models the algebra generated by these objects is isomorphic to $G_{2n}(Q)$ (at least with suitable boundary conditions, see Martin and Launer 1989), but in higher dimensions it defines a simplicial Clifford algebra $G_{2n}(Q, p_f)$ which is generally a proper subalgebra.

Note that equation 10.2 implies that

$$\prod_{spacelike\ x \in edge} S_x$$

is a central element of order Q.

In the case of spin models, and with the $c = 0$ notation ($x = site\ i.$) and ($Y = bond\ ij$), where i and j are layer edge sites, the presentation is

$$\begin{aligned} S_{i.}^Q &= S_{ij}^Q = 1 \qquad (10.14)\\ S_{i.} S_{ij} &= W S_{ij} S_{i.} \end{aligned}$$

with all other operators commuting. We will return to this case later.

10.2.3 Canonical representations

The appropriate representations of a_x, b_x are as follows:

$$a_x, b_x \in End(\bigotimes_{x \in edge} V^{(x)}) \qquad (10.15)$$

a_x acts trivially on $V^{(w)}$ if $w \neq x$, on $V^{(x)}$ it acts as the matrix N given by

$$N = \begin{pmatrix} 0 & 1 & 0 & .. & .. & .. \\ 0 & 0 & 1 & 0 & .. & .. \\ .. & 0 & 0 & 1 & 0 & .. \\ & & & & . & \\ & & & & & 1 \\ 1 & 0 & .. & & .. & 0 \end{pmatrix} \tag{10.16}$$

while b_x acts trivially on $V^{(w)}$ if $w \neq x$, on $V^{(x)}$ it acts as the matrix M given by

$$M = \begin{pmatrix} 1 & 0 & 0 & .. & .. & .. \\ 0 & W & 0 & 0 & .. & .. \\ .. & 0 & W^2 & & 0 & .. \\ & & & . & & \\ & & & & . & 0 \\ 0 & 0 & .. & & .. & W^{Q-1} \end{pmatrix} \tag{10.17}$$

Note that these matrices define a manifestly irreducible Q^n dimensional representation of $G_{2n}(Q)$ (equation 10.9). On the other hand we have already seen that the dimension of the algebra is Q^{2n}. We thus note that $G_{2n}(Q)$ is isomorphic to the Q^n dimensional complex matrix algebra $M_{Q^n}(C)$ (i.e. $G_{2n}(Q)$ has trivial center). For this reason we refer to the above representation as the *canonical* representation.

10.3 Temperley-Lieb subalgebras

10.3.1 Lattice gauge model representations

Note that for *any* gauge model we can use the gauge symmetry to fix the gauge and take all timelike (longitudinal) bond variables to $f(x) = 0$. This means that we can drop the factors of b_w in equations 10.10 and 10.11. In 3 dimensions all the variables in a transfer matrix edge are then the bonds of a square lattice. For the 3 dimensional Potts gauge model defined above, and indeed for any model without factors of b_w appearing in equation 10.10, the coefficients α are just

$$\alpha_{xj} = \exp(\beta\chi(j)) \tag{10.18}$$

(10.19)

$$\alpha_{Yj} = \sum_{r=0,..,Q-1} (W^{-jr}\exp(\beta\chi(r)))/Q.$$

In the Potts case $\chi(j) = \delta_{j,0}$, and putting equation 10.18 into equation 10.10 we find that the matrices

$$U_x = Q^{-1/2} \sum_{j=0,..,Q-1} (a_x)^j \tag{10.20}$$

$$U_Y = Q^{-1/2} \sum_{j=0,..,Q-1} (\prod_{w \in BY} b_w^{\epsilon_w})^j \tag{10.21}$$

build the layer transfer matrix

$$T(v) = \prod_{spacelike\ x \in edge} (v + Q^{1/2} U_x) \prod_{spacelike\ Y \in layer} (1 + v Q^{-1/2} U_Y) \tag{10.22}$$

where $v = (\exp(\beta) - 1)$. These matrices obey the following relations

$$\begin{aligned} U_x U_x &= \sqrt{Q} U_x & & x \in \ spacelike\ bonds\ in\ edge \\ U_Y U_Y &= \sqrt{Q} U_Y & & Y \in \ spacelike\ plaquettes\ in\ layer \\ U_x U_Y U_x &= U_x & & ...if\ x \in BY \\ U_Y U_x U_Y &= U_Y & & ...if\ x \in BY \\ U_x U_Y &= U_Y U_x & & ...if\ x \notin BY \end{aligned} \tag{10.23}$$

for a certain graph Temperley-Lieb algebra. The graph is again just the square lattice. To see this first note that fixing any 3 of the four states in the basis space for the non-trivial part of the action of U_Y (i.e. $\otimes_{x \in BY} V^{(x)}$ where BY consists of the 4 bond variables round a plaquette) always leaves a submatrix ($\in End(V^{(x)})$) of the form

$$diagonal(c_1, c_2, c_3, ..., c_Q),$$

where $c_i = Q^{1/2}$ for some unique i and $c_i = 0$ otherwise. Then note that a duality transformation takes plaquettes to bonds (and so spacelike plaquettes to timelike bonds, and hence sites, after projecting down onto the square lattice).

More explicitly, with

$$U_x, U_Y \in End(\bigotimes_{bonds\ x \in edge} V^{(x)}), \tag{10.24}$$

then:

U_x acts trivially on $V^{(w)}$ unless $x = w$;

U_x on $V^{(x)}$ is $Q^{-1/2}$M , where M is the Q-by-Q matrix with all entries unity.

Meanwhile:

U_Y acts trivially on $V^{(x)}$ unless $x \in BY$;
U_Y on $\otimes_{x \in BY} V^{(x)}$ is $diagonal(\sqrt{Q}\delta_{f(\partial Y),0})$.

On the other hand, the Z_Q spin model has the Q-state Potts model as a special case, and so always provides a representation of the appropriate Temperley-Lieb algebra by this specialisation. The matrices $U_{i.}$ and U_{ij}, in the notation of eqn(14), are given by:

$$U_{i.}, U_{ij} \in End(\bigotimes_{sites\ i \in edge} V^{(i.)}) \tag{10.25}$$

where $V^{(i.)}$ is the Q dimensional vector space with basis the possible Potts spin variable values on site $(i.)$, say $s_{i.} \in \{0, 1, .., Q-1\}$. Then:

$U_{i.}$ acts trivially on $V^{(j.)}$ unless $i. = j.$;
$U_{i.}$ on $V^{(i.)}$ is $Q^{-1/2}$M.

Meanwhile:

U_{ij} acts trivially on $V^{(k.)}$ unless $i. = k.$ or $j. = k.$;
U_{ij} on $V^{(i.)} \otimes V_{j.}$ is diagonal$(Q^{1/2}\delta(s_{i.}, s_{j.}))$.

Duality in 3 dimensions then takes

$$(vQ^{-1/2}) \rightarrow (Q^{1/2}/v),$$

and

$$\{U_{i.}\} \leftrightarrow \{U_Y\}$$

and

$$\{U_{ij}\} \leftrightarrow \{U_x\},$$

but the representations are clearly not isomorphic (see later).

Note that the relations 10.23 hold for arbitrary dimensional Potts lattice gauge models, but they are not dual to a *graph* TL algebra in higher dimensions than 3, as we can see by looking at the coordination numbers (i.e. the number of other generators with which each generator has non-trivial commutation relations). In d dimensions each Y is surrounded by 4 x's, and each x by $2(d-2)$ Y's, while each bond is bounded by 2 nodes and each node surrounded by $2(d-1)$ bonds in the corresponding graph algebras.

It is interesting to recall that, in 2 dimensions, the local duality transformation is just

$$U_i \rightarrow U_{i+1}$$

(at the level of operators). Note in particular that 2 duality transformations 'equals' 1 translation (of course the whole system is translation invariant, so duality becomes involutive). In 3 dimensions the situation appears much more complicated. The relationship between the 2 representations above is not obvious. In the case $Q = 2$ we can build a gauge representation out of the spin model representation as follows:

Look at the 2 dimensional layer edge. Consider the medial lattice of this square lattice, which is also a square lattice. Then each $U_{i.}$ is associated with a medial bond x, and we may put $U_x = U_{i.}$. Inside each medial plaquette Y are 4 original lattice bonds, take any parallel pair of these $(ij), (kl)$ (say) and form

$$U_Y = ((1 - 2U_{ij})(1 - 2U_{kl}) + 1)/2.$$

This gives the gauge representation for the medial lattice. But the transformation to the medial lattice is not a duality transformation, and infact involves a scale change! Clearly the procedure is not trivial, and for $Q \neq 2$ it is even more complicated.

In higher dimensions the models dual to the spin model are more complicated still, but the generalisation to *construct* representations of higher dimensional Temperley-Lieb algebras is straightforward. Specifically, in d dimensions we take $c = d - 2$.

The breakdown of the Q-generic structure of the graph Temperley-Lieb algebra (with graph the square lattice) should signal a 3 dimensional series of critical field theory limits analogous to the central charge $c \leq 1$ conformal series ($Q = 4\cos^2(\pi/r), r \in Z$) in 2 dimensions (see Cardy 1986 and chapter 4). Thus it is particularly interesting to investigate the asymptotic rate of growth of dimensions of this algebra with lattice size, which gives the breakdown point. With this end in mind we will now use the gauge model picture to study the algebra in a bit more detail.

10.3.2 Quotient relations for the gauge representation

The quotient relations describing the place of the 'Potts algebra' (defined by the representation following equation 10.25), and the 'gauge algebra' (equation 10.24), in the abstract graph Temperley-Lieb algebra are not obvious. But note , for instance, that the Potts and gauge representations are real and symmetric, while representations need not be unitarisable in general.

Recall that a minimal (that is to say not necessarily complete) list of sets of quotient relations obeyed by the Potts representation may be indicated as follows. For W any word of length $O(W)$ in the generators $\{U_{i.}, U_{ij}\}$ and

$$R_{b(N)} = \prod_{i. \notin b(N)} (U_{i.}/Q^{1/2}) \tag{10.26}$$

where $b^{(N)}$ is any N element subset of the n nodes (say) in the layer edge, so $0 \leq N \leq n$, and, for example,

$$R_0 = R_{b(0)} = \prod_{\text{all nodes } i.} (U_{i.}/Q^{1/2}), \tag{10.27}$$

then:

$$R_{b(N)} W R_{b(N)} = \chi_{b(N)}(W) R_{b(N)} \quad (mod.\ R_{b(M)}\ \forall\ b^{(M)} \subset b^{(N)}) \tag{10.28}$$

where $\chi_{b(N)}(W)$ is a readily determined scalar. The proof is in chapter 8.

The R_0 relation, with R_0 redefined by duality to

$$R_0^* = \prod_x Q^{-1/2} U_x$$

(remember R_0 is to be thought of as a high temperature transfer matrix), still holds for the gauge representation. Infact R_0^* is clearly not the same idempotent as R_0 in the Temperley-Lieb algebra, but it is an idempotent corresponding to the same irreducible representation. It thus gives the same free energy. To prove this we will construct the irreducible representation explicitly.

Consider the set $\{v_i\}$ of possible non-contractible 'boundaries' drawn on the bonds of the square lattice, accessible from the empty case (no boundaries) by arbitrary sequences of operations consisting of:

i) drawing a boundary around a single plaquette; and

ii) putting the 2 plaquettes adjacent to a single bond inside the same boundary (i.e. removing any intermediate segments).

For example, operation (ii) on bond x is illustrated by the move from v_4 to v_9 in Figure 1, which also gives an example of a complete set $\{v_i\}$. In other words we have the set of possible *minimal* distributions of lines drawn on bonds sorting the plaquettes into isolated clusters in the plane by operations (i) and (ii) (note that not all sortings are realised by this procedure). This set gives a basis for representations of the algebra defined by the relations 10.23 as follows.

$$(U_Y)_{ij} = \begin{cases} Q^{\delta_{ij}/2} \text{ if building a boundary around plaquette } Y \text{ takes distribution } i \text{ to } j; \\ = 0 \text{ otherwise;} \end{cases}$$

$$(U_x)_{ij} = \begin{cases} Q^{\delta_{ij}/2} \text{ if cutting and contracting away any boundary passing through bond } x \text{ takes } i \text{ to } j; \\ = 0 \text{ otherwise;} \end{cases}$$

In our example, with the plaquettes and basis elements numbered as indicated,

$$U_{Y=1}\ U_{Y=2}\ v_1 = Q^{1/2}\ v_4;$$

and with x the bond marked in v_4,

$$U_x\ \ v_4 = v_9;$$

and so on.

The basis space is essentially dual [2] to the partition basis discussed in chapter 8. The correspondence is that every node in the same cluster becomes a plaquette *inside* some bounded region, and travelling to a node in a different cluster necessitates crossing a boundary. The difference at the operator level is that the new representation includes bond generators at the edges which are not bounded at one end by a node generator. In other words, if we build a representation corresponding to the set of bonds and plaquettes shown in Figure 2 (for example) then the corresponding graph is as shown in Figure 3, where white nodes are not included.

Note in particular that both $\prod_x U_x$ and $\prod_Y U_Y$ are manifestly primitive idempotents (since each projects the entire basis space into a different 1 dimensional subspace) and that the set of possible boundaries corresponds to the left ideal generated from either of these. Specifically $\prod_x U_x$ corresponds to the state with no boundaries, and $\prod_Y U_Y$ the case where every bond is a boundary.

For the Potts model $\prod_Y U_Y$ is the primitive idempotent corresponding to high temperature (zero coupling) boundary conditions (see Baxter 1982), and guaranteeing, by Perron's theorem, a unique largest eigenvalue when couplings are real in the bulk of the system. For the gauge model $\prod_x U_x$ is the corresponding idempotent.

[2] i.e. square lattice goes to square lattice, but the edges of the dual systems are slightly different, in the usual fashion of duality (see Savit 1980), as exemplified in figures 10.2 and 10.3.

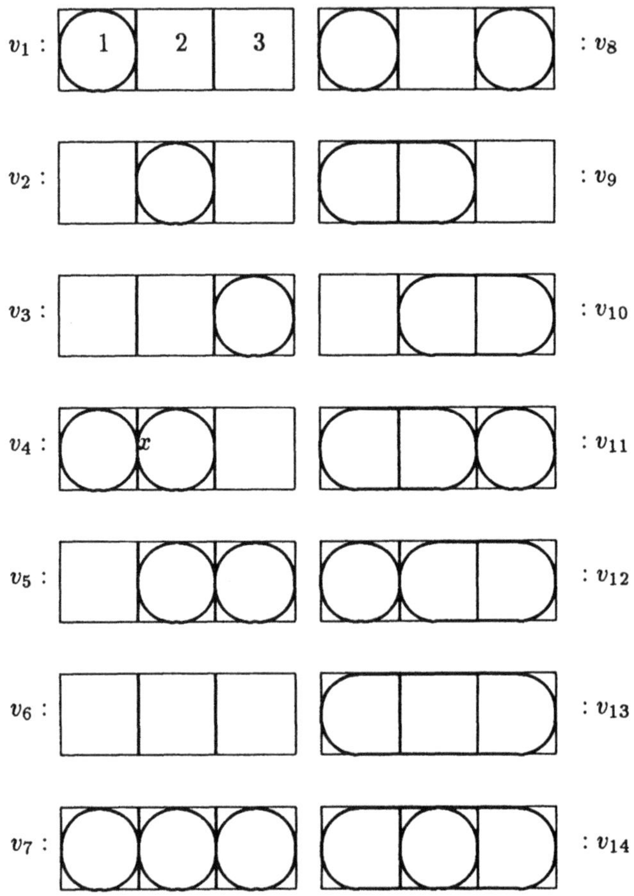

Figure 10.1: Non-contractible boundaries (indicated by curved lines) on a 3 by 1 lattice (straight lines). Note that the outside is regarded as a plaquette so, for example, v_6 and v_{13} are distinct.

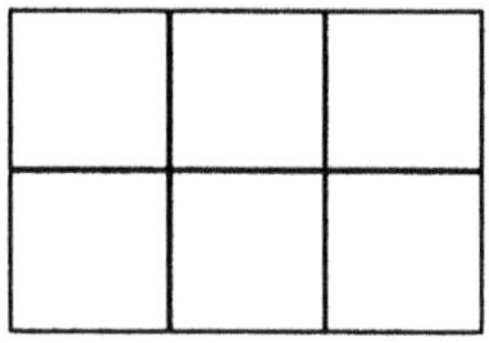

Figure 10.2: A 3 by 2 lattice of bonds and plaquettes.

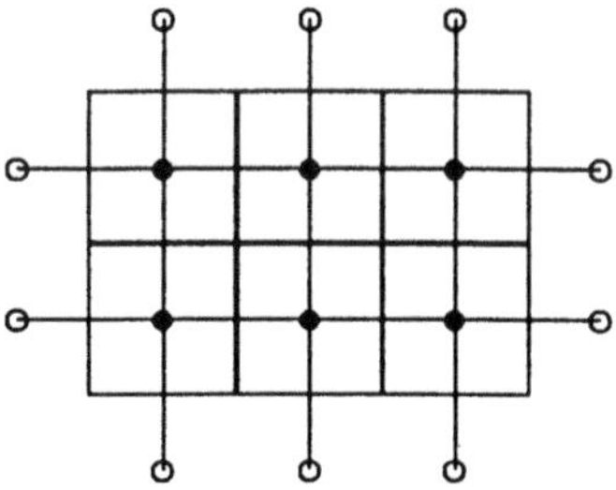

Figure 10.3: A 3 by 2 lattice and its dual lattice.

Remarks

It is possible to determine the structures of the simplicial Clifford algebras for arbitrary simplicial dimension c (which we will leave as an excercise). The full multiplicities of irreducible representations in the 3 dimensional gauge model representations of Temperley-Lieb algebras had not been determined at time of going to press.

Note that the gauge representation has assymptotic growth rate of dimension d_n (n sites) of

$$\lim_{n\to\infty} d_{n+1}/d_n = Q^2 \tag{10.29}$$

in 3 dimensions, after temporal gauge fixing. There is still scope for some more gauge fixing after this.

In many of the algebras we have defined the relations are associated with some regular graph or lattice. The automorphisms of the lattice then give rise to automorphisms of the algebra. It would be interesting to see how these automorphisms preserve the irreducible structure of the algebras in general. In the 2 dimensional Hecke cases the similarity transformations which realise the automorphisms play important roles in the inversion symmetry and braid group representations (chapters 2 and 3)!

Chapter 11

The modelling of phase transitions

Subtitle: Interpreting exact results for finite lattices.

11.1 Zeros of the partition function

In statistical mechanics, phase transitions correspond to thermodynamic limit singularities of derivatives of the free energy $f(\beta)$. In fact we might as well take any non-analytical point of $f(\beta)$ on the β real axis as a statistical mechanical critical point. Of course in practice we would not identify a phase transition purely on this basis, but expect abrupt changes in the behaviour of order parameters, and some supporting 'physical' (i.e. energy versus entropy) explanation as well, as we will illustrate in this chapter.

On the other hand, the partition function Z for a model with variables taking a finite range of values on a finite lattice can often be written as a polynomial in some coupling dependent variable $x(\beta)$. For example, if the Hamiltonian can be normalised to take integer values, as in the case of the Potts model, then such a variable is $x = \exp(\beta)$. Zeros are therefore the only analytical feature of the partition function. Furthermore, these zeros will lie off the real positive x axis, since all the coefficients, each of which corresponds to the number of configurations with a given energy, are positive. Consequently all observables vary smoothly with x in the region of real β.

The zeros of the partition function can only possibly approach the real axis in the thermodynamic limit. A point x_c where a continuous distribution of zeros cuts the axis in this limit would give a critical point. Such a distribution of zeros would in particular provide a mechanism to allow a different function representing an observable on either side of x_c. The exact solutions we describe in chapters 3 and 4 bear out this picture.

In the literature it is common to find discussions of the zero distribution of the partition function in the complex external magnetic field variable (at fixed β). Many exact results, stemming from the original Yang-Lee circle theorem (see Lee and Yang 1952), pertain to this bulk property of the system (see, for example, Suzuki and Fisher 1971). The more fundamental coupling parameter problem has turned out to be relatively intransigent, and consequently less exposed in the literature. Only relatively recently have exact solutions and related algebraic approaches to finite lattices rendered this problem accessible. Finite β *and* finite magnetic field remains an enigma.

Let us consider a partition function which formally illustrates the proposed mechanism for phase transitions. Suppose that the partition function for some model on a lattice of size N takes the form

$$Z = x^{2N} + 1 \tag{11.1}$$

with $x = \exp(\beta)$ (the one dimensional $Q = 2$-state Potts model, for example, has

$$Z = (x-1)^N \left(\left(\frac{x+1}{x-1} \right)^N + 1 \right)$$

which differs from equation 11.1 in no significant way). In the form 11.1, the zeros are spaced at constant intervals around the unit circle in the complex x plane. As $N \to \infty$ the zeros close up to give a uniform line density. The free energy, as it were, is

$$f = N^{-1} \ln(Z) = \ln(2)/N + \ln(x) + \ln(\cosh(N\beta))/N$$

and the internal energy is

$$U = -\frac{\partial N^{-1} \ln(Z)}{\partial \beta} = -1 - \tanh(N\beta).$$

As we let N increase, the change in U close to $\beta = 0$ becomes more and more rapid, until in the limit it is discontinuous. This would be the statistical mechanical signal associated with a *first order* phase transition (section 1.4.2).

In the limit we could have written the free energy in such a way as to make the zero distribution of the partition function manifest:

$$\lim_{N\to\infty} f = \beta + (1/(2\pi)) \int_{-\infty}^{\infty} a(y) \ln(\beta + iy) dy$$

where the line density of zeros $a(y) = 1$. Suppose we could generalise this simple model to realise other line densities $a(y)$, analytic at $y = 0$. We have

$$U = -1 + (1/(2\pi i)) \int_{-\infty}^{\infty} \frac{a(y)dy}{y - i\beta} \qquad (11.2)$$

where, since the integrand has a simple pole at $y = i\beta$, the integral changes by $2\pi a(0)$ as β changes sign. Now since $a(y)$ is certainly real and positive with $a(y) = a(-y)$ for real y we will always get first order transitions this way.

On the other hand supposing that for small y

$$a(y) \sim |y|^{1-p} \qquad (0 \le p < 1),$$

for example, we find that the specific heat exponent α (from section 1.4.2) has a simple dependence on p. When $p = 0$ we see immediately that the resultant specific heat has a logarithmic divergence, which means a second order phase transition. More generally, the specific heat critical exponent for $0 < p < 1$ may be obtained by examining the asymptotic behaviour of $U(\beta)$ as $\beta \to 0$. One way of doing this is to note, changing the integration range to exclude a non-singular region, that it is dominated in the limit by a special case of a hypergeometric integral (see Abramowitz and Stegun 1965, for example) with well known asymptotic properties. We find the specific heat

$$\frac{\partial U}{\partial \beta} \overset{\beta \to 0}{\sim} \beta^{-p}$$

i.e. the specific heat exponent is $\alpha = p$.

We see that the transition remains second order for the range $0 \le p < 1$. Note from equation 11.2 that it is the behaviour of the line density of zeros *close to the critical point* which determines the exponent. This means that essentially the same picture will pertain when the partition function cannot be expressed as a polynomial, and we must work directly with β.

Returning to the finite lattice it is reasonable to suppose that, for a sufficiently large system, the distribution of zeros will reveal some of the limiting structure. In particular it may give clues as to the shape and density of the limiting distribution (it clearly does in our toy example, and it does in all the examples to follow).

Note that a finite lattice Potts model distribution is, in principle, always calculable by direct enumeration. On the other hand, direct enumeration of, say, 100^5 configurations is not a realistic task for any present or future computer. The algebraic techniques discussed in this book may or may not be of use in the exact solution of thermodynamic limit models, but they can *always* simplify finite lattice computations. The picture discussed above has been familiar for many years (see, for example, Fisher 1964, Abe 1967, Abe and Katsura 1970, Nilsen and Hemmer 1967) as a mechanism for phase transitions, but it is only in the last few years that algebraic computing has reached a level of sophistication sufficient to generate finite lattice results in which the kind of features described above can be identified with any confidence.

11.2 Semi-infinite systems

We can get a better picture of what is going on, and introduce some more realistic models, by considering the ramifications of the above discussion for the transfer matrix formalism.

We have already established that the transfer matrix may be block diagonalised by transformations corresponding to translational and internal symmetries of the system (and therefore independent of the coupling parameters). Let us be more specific about this. Suppose we restrict attention from the full transfer matrix algebra H, say, as defined in chapter 2, to the subalgebra A generated by the periodic transfer matrices $\{T_{(N)}(x_1, x_2)$ for all $x_1, x_2\}$. In the Potts model case we are interested in the A module given by $V = \otimes^N V_Q$. We can decompose V into simple $H = T_k(q)$ modules using chapter 6 (the multiplicities are given in Martin 1988). If we wish to compute the free energy then we can restrict attention to the unique appropriate H submodule $S^{(0)}$, say. For any H module W let us introduce $D = End_A(W)$. If we can decompose W as a direct sum of D modules, then from chapter 5 elements of A only mix these submodules if they are isomorphic.

For example, note that

$$[G^2, T_{(N)}(x_1, x_2)] = 0$$

and

$$[M, T_{(N)}(x_1, x_2)] = 0$$

from chapter 2 (remember G^2 is just a particular transfer matrix, in the staircase formulation), so the appropriate representations of $\{G^2, M\}$ generate a (not necessarily proper) subalgebra of D.

Let us decompose V, say, as a direct sum of G^2 modules, as follows:

Put

$$v_i \in \otimes^N V_Q$$

an element of the configuration space basis, so

$$G^2\, v_i = v_{f(i)}$$

for some permutation $f(i)$. Suppose we define a set of integers m_i by

$$G^{2p} v_i = v_i \qquad for \;\; p = m_i$$

and

$$G^{2p} v_i \neq v_i \qquad for \;\; p < m_i$$

then the fourier transform

$$\sum_{k=0}^{m_i - 1} \left(e^{2\pi i j/m_i} G^2 \right)^k v_i = w_j^{(i)}$$

is an indecomposable G^2 module. If P is any minimal set of configurations i such that repeated application of f to P hits all configurations then V has basis

$$\{ w_j^{(i)} \; | i \in P \; , \; 0 \leq j < m_i \}$$

in which G^2 is diagonal. Thus V decomposes into a direct sum of $T_{(N)}(x_1, x_2)$ modules indexed by j:

$$T_{(N)}(x_1, x_2) \;\; w_j^{(i)} = \sum_{k \in P} F_j^{ik}(x_1, x_2) \; w_j^{(k)}$$

where F_j^{ik} is a scalar. The dimension of the $j = 0$ subspace, for example, is $\dim(V/G^2 = 1) = |P|$.

Alternatively, writing

$$B = (1 + x_1 U_1)(1 + x_2 U_2)$$

then the stairway transfer matrix may be written

$$T'_{(N)} = (BG^2)^N \; G^{-2N}$$

where the last factor is central (and typically equal to 1). The spectrum can thus be obtained from that of BG^2. This is not translation invariant, but it is relatively easy to write down (the temperature dependent part is almost trivial).

The transfer matrix itself has entries which are polynomials in $x = e^\beta$ with positive coefficients, and so any x independent similarity transformations will preserve at least the polynomial property. Those blocks which can keep the positive property, such as the $j = 0$ block above, will also satisfy the conditions for Perron's theorem in the region x positive.

As an example, consider the periodic 2 site wide strip isotropic Q-state Potts model ($Q > 1$). Using the variable $z = (\exp(\beta) - 1)/Q$ the block relevant to the Potts model free energy (the *principal* block) may be written

$$\begin{pmatrix} z + \sqrt{Q} & z(z + \sqrt{Q}) \\ 1 & z(2 + z\sqrt{Q}) \end{pmatrix}.$$

There are many such examples of finite width systems with two dimensional principal blocks. Of course the result, in this form, is not very revealing, but note that it is now easy to write down the characteristic polynomial, which is a quadratic equation for the free energy. The eigenvalues will take the form

$$\lambda_{\pm} = A(x) \pm \sqrt{B(x)}$$

where $A(x)$ and $B(x)$ are polynomials in $\exp(\beta)$. The larger of the eigenvalues on the positive real axis is λ_+ (it must be exclusively one or the other on the whole positive axis, from Perron's theorem). The free energy, with suitable boundary conditions, is then

$$\lim_{N\to\infty} \ln(Z)/N = \lim_{N\to\infty} (1/N)\ln(\lambda_+^N + \lambda_-^N) = \ln(\lambda_+).$$

What has happened to our zeros?

Consider the following useful identity for any pair of scalars C, D:

$$C^N + D^N = \prod_n (C + \exp(2\pi i n/N)D) \tag{11.3}$$

where the product is from $n = 1, 2, .., N$ if N is odd; and from $n = 1/2, 3/2, .., N - 1/2$ for N even. Then

$$\ln(\lambda_+) = (1/2\pi)\int_0^\infty \ln(2[AA + B] + 2\cos(y)[AA - B])dy,$$

which may be thought of as the log of an infinite polynomial with lines of zeros on the loci $|\lambda_+| = |\lambda_-|$. These loci terminate at the zeros of B (where $\lambda_+ = \lambda_-$). Notice that the zeros of the partition function for finite N lie on the *same* loci, by equation 11.3.

For our example a direct computation reveals that the zeros of B lie at $z = e^{i\theta}$, where

$$\theta = \cos^{-1}(-1/2\sqrt{Q} \pm \sqrt{Q-1}/Q),$$

so the zeros of the partition function lie on the unit circle between these points. We will have some more examples shortly.

For a *pre* block diagonalised version of the transfer matrix it is always possible to choose boundary conditions such that the zeros still lie on the same loci. Even with arbitrary boundary conditions the zeros converge to these loci as $N \to \infty$, since

$$Z = K_+ \lambda_+^N + K_- \lambda_-^N + \sum_n K_n \lambda_n^N = 0$$

(where $K_+ = K_-$ since Z is polynomial) when

$$\left(\frac{\lambda_+}{\lambda_-}\right)^N = -1 - \sum_n \frac{K_n}{K_+} \left(\frac{\lambda_n}{\lambda_+}\right)^N .$$

The latter term tends to zero in the limit, and the identity is then satisfied when

$$\frac{\lambda_-}{\lambda_+} = \exp(i\pi(2n+1)/N)$$

(integer n).

In this sense finite lattice results can be used to give an image of a semi-infinite limiting distribution, and hence a full thermodynamic limit. Note, however, that line *density* information can only be pertinent if the dimensions of the lattice in the lateral and transverse directions are kept roughly similar (since otherwise it will be swamped by the branch points of the semi-infinite system, which we have already established do not reach the critical point). One way round this is to rescale the transverse couplings, as if approaching a continuum time formulation. In our examples we keep to globally square lattices for the sake of simplicity.

Note that if C, D can be written as polynomials in equation 11.3 then the logarithm of the corresponding infinite polynomial is

$$\frac{1}{2\pi} \int_0^\infty \ln(2[CC + DD] + 2\cos(y)[CD])dy = \begin{cases} \ln(C) & \text{where } |C| > |D| \\ \ln(D) & \text{where } |D| > |C| \end{cases}$$

and the zeros of the infinite polynomial form the *closed* boundary between these two regions.

Similar arguments to those given above hold for irreducible blocks of higher dimension, although there is not, in general, such a convenient integral representation. The basic picture is one of lines of zeros mapping out the loci of equal magnitude of the largest two or more eigenvalues, which terminate at the zeros of the algebraic determinant. The only complication is that the largest eigenvalues will not always be the same two as we move around the complex plane.

11.3 Finite lattice results

There are many results on zeros of the partition function in the literature. With respect to the Potts model these include results for various lattice shapes, and various dimensions (see, for example, Pearson 1982, Baxter 1987, Martin and Maillard 1986, Martin 1983 and references therein). Exact results about finite systems are, in the original framework of statistical mechanics, perhaps more meaningful than results which only hold in the thermodynamic limit. On the other hand, the lattice size N is typically of the order of 10^2 sites, compared to useful physical systems of 10^{24} atoms or so. The ideal solution is one in which N appears as a parameter. There are a few of these, as we see in chapter 4, but not many. In their stead, currency can be gained for the $N \sim 10^2$ finite results by noting that they already seem quite stable in *some* respects under changes in lattice size (provided we only consider regions of coupling parameter space in which boundary conditions are not specifically emphasising the finiteness of the lattice).

As far as the isotropic 2 dimensional Potts model is concerned, finite lattice results have produced overwhelming evidence to support the idea that the distribution of zeros close to the ferromagnetic phase transition point lies on an arc of a circle (c.f. Hintermann, Kunz and Wu 1978), but is not universally confined to such a circle. This circle is the locus of points for which the duality transformation corresponds to complex conjugation:

$$e^{\beta} = \frac{1-Q}{1+\sqrt{Q}e^{i\theta}} \qquad for \quad 0 \leq \theta \leq 2\pi.$$

Figures 11.1 and 11.2 dramatically illustrate the point for $Q = 3$. The scale in these and subsequent figures in this chapter is given by unit length of the positive real axis. The zeros are obtained by a Newton-Raphson based technique. Considerable care (i.e. very high precision arithmetic) is required with such high order polynomials, and even then the immediate vicinity of the point -1 should be treated with some scepticism. Fortunately this is well away from the physical region. We can see from equation 12.1, for example, that the behaviour of the specific heat for real β is dominated by the effect of the zeros closest to the positive real axis.

In figure 11.1 the zeros on the right lie exactly on the duality=complex conjugation circle (which cuts the positive real axis at the exact critical point). The line separation of zeros approaching x_c is consistent over a sequence of similar lattice sizes with a limiting line density corresponding to $\alpha = 1/3$. The first significant deviation from the circle distribution, even in figure 11.2, where the boundary conditions violate the duality symmetry, appears at the braid point. Recall (from chapter 2) that this is a point where *all* eigenvalues of the transfer matrix have degenerate magnitude. The picture close to the ferromagnetic region is similar for other Q values (see e.g. Martin 1987), but the line density of zeros varies.

On other shapes of lattice, and in other dimensions, there is no duality circle, but a one dimensional distribution of zeros is seen repeatedly in models with two phase ferromagnetic regions. Why is this, and what about models with more complicated phase structures....?

11.4 Energy, entropy and combinatorics

We are now in a good position to relate the mathematical picture of phase transitions to a physical one. To this end it is useful to be able to call upon models which exhibit a richer phase structure than the ferromagnetic Potts models. Let us consider other Z_Q symmetric models (see chapter 10). The configuration space is the same, but instead of a delta function interaction we allow an interaction Hamiltonian which depends on the spins as if they corresponded to dipole orientations *in the plane*. The Hamiltonian then depends on the relative orientations:

$$H(\{\beta_r\}) = \beta \sum_{<ij>} \chi_{\{\beta_r\}}(s_i - s_j)$$

where, for example,

$$\chi_{\{\beta_r\}}(s) = \sum_{r=1}^{[Q/2]} \beta_r \cos(2\pi r s/Q)$$

and, as usual, $s_i \in \{1, 2, .., Q\}$.

We restrict to integer valued Hamiltonians, for ease of interpretation (and without significant loss of generality, for our purpose). It is then most convenient to define each model by the list of integer energies associated to the various possible relative orientations $(s_i - s_j)$ in order of increasing angle:

$$(\chi(0), \chi(1), \chi(2), ..).$$

The Potts model is defined in this notation by $\chi = (1, 0, 0, ..)$; the Z_6 *vector* model (c.f. Savit 1980) is defined by (4,3,1,0), and so on. For models with chiral symmetry (that is $\chi(n) = \chi(-n)$) a maximum of $(Q/2) + 1$ integers are required.

It follows that the cases $Q = 2, 3$ are coverred by the Potts models and that, in fact, $Q > 4$ are the interesting cases. Here we will also restrict attention to non-negative energies.

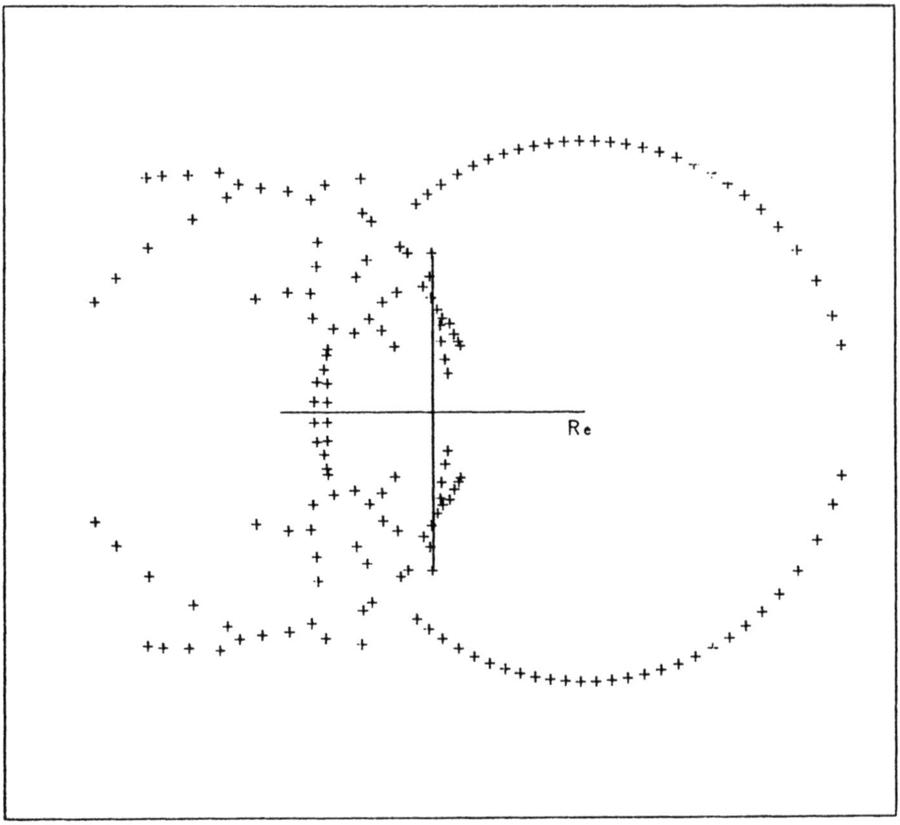

Figure 11.1: Zeros of the partition function for an 8×10 square lattice $Q = 3$-state Potts model with self-dual boundary conditions.

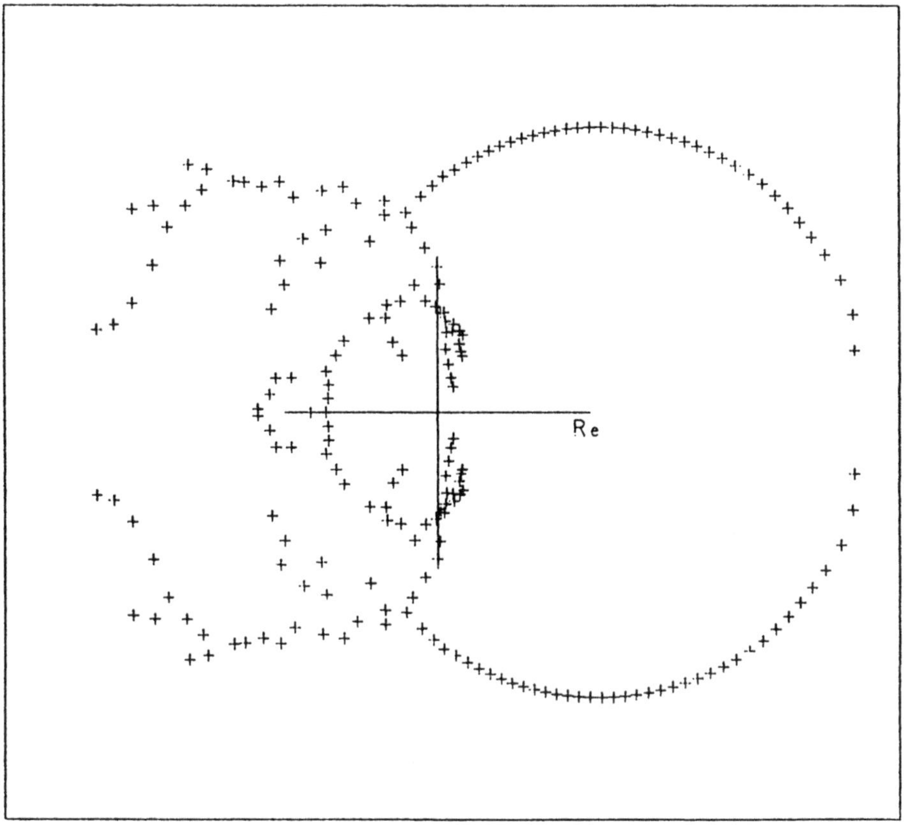

Figure 11.2: Zeros for a 10×12 lattice $Q = 3$-state model with duality breaking boundary conditions.

Independently of this energetic assignment χ, configurations of the Q valued spins may be represented by the boundaries of areas of aligned spins. We already considered such a picture when describing the entropy of the Potts model in chapter 1. Now, to cope with the various possible interactions, we need a bit more detail. On each boundary segment (on the dual lattice bond between the two spins in question, i and j, say) we will put $|s_i - s_j|$ bits of directed string or directed *flux lines*, with the direction depending on the sign of the difference. A maximum of $Q/2$ lines can appear on a bond. Clearly some of the oriented strings will be connectable together into closed loops (not necessarily in a unique way), and the remainder will form lines which terminate in sources (sinks) for Q strings. Following Einhorn *et al* (1980) we will call these vortices (antivortices).

Remember that this description of the configurations of the models is independent of the energy assignment. When we know the energy assignment we can determine the *similarity criteria* for entropy (from chapter 1) and unravel the phase structure of the model. In any case, at low temperatures (high β) the Boltzmann factors for configurations with few strings are largest, outweighing the relative rarity of such configurations. As the temperature rises the energy penalty for configurations with strings becomes less significant. We suppose that eventually their entropic (combinatorial) advantage makes them, collectively, a significant factor in the partition function. Since they correspond to faults in the long distance ordering, we anticipate that they drive an order/disorder phase transition.

Let us write

$$\exp(\beta) = r\exp(i\psi)$$

for $\exp(\beta)$ in the complex plane. Then, for large r and small ψ, terms in Z coming from configurations having low string density have large magnitude Boltzmann factors. Let us assume that such configurations have typical energy NE, where N is the total number of sites in the system (clearly the typical energy grows linearly with N). They are so large that other configurations can be neglected, despite their great multiplicity. However, since ψ is small these low density contributions add more or less coherently, so there cannot be any zeros of the partition function. We are assuming that the typical energy NE is well focused, but of course for sufficiently large r it is focused exactly at the ground state energy, so this is reasonable.

Let us assume further that the configurations corresponding to high string density have a typical energy $N(E - E')$, say. Again it is reasonable that a typical energy, if it exists, will depend linearly on N, but here the energy possibilities are unlikely to be well focused. We assume that they can at least be *represented* by the typical figure. This assumption is less easy to justify, but can be tested 'experimentally', as we will see. The high string configurations then add fairly coherently within themselves, but provided

$$\psi \sim ((2n+1)\pi/NE')$$

(n integer) they tend to cancel with the low string contributions. Of course at low temperatures they are a negligable effect. But as the temperature is increased their great entropy may cause their contribution to become comparable with the low string terms. At this stage $Z = 0$ for all integer n. Note that as $N \to \infty$ the zeros approach the real axis, and finite line density.

In the (4,3,1,0) model, for example (see table 11.1), overlapping strings are energetically penalised (the energy cost for two strings on the same bond is greater than that for two on different bonds). It is thus plausible that single string loops would become significant first, and drive this transition on their own. As the temperature is increased further, vortex/antivortex pairs would become significant, and the burgeoning contribution, through very high entropy, of pairs with divergent pair separation would eventually drive another transition to an even more disordered state. The zeros argument is the same as before (but in a different energetic regime).

Contrastingly, in the Potts models for instance, the energy penalty for multi-string loops is the same as for single string loops, so these high entropy objects would become significant at the first transition, and completely swamp the effect of the arrival of unbound vortices. That is, with energy *and* entropy in their favour, the multistring loops would provide a massive coherent contribution to the partition function which could not be cancelled by any less energetically favourable phenomenon.

Note, this is not only a physical argument for two phase transitions in some models. It simultaneously and inextricably suggests a richer structure of zeros of the partition function approaching the real axis than that exhibited by the Potts models (above). This prediction can be tested directly.

It also gives qualitative information about the sensitivity to energy assignment. The significance of vortices in this picture depends on the relationship between these energies. For example, the Z_5 (3,1,0) model (see table 11.1) requires less energy to have two strings together than separated. Thus multi-string loops are favoured over vortices, suggesting a Potts like structure. The borderline between two and three phase models should occur in models requiring roughly equal amounts of energy for separate or overlapping strings (e.g. (2,1,0) - see table 11.1). This can also be tested by examining the zero distributions.

In the following figures we plot the zero distributions in $\exp(-\beta)$ for a few key cases. The models involved are given in table 11.1. The exact computation of the partition functions is again made possible only by making

figure	Q	interaction energies	lattice size	transitions expected
11.3	5	(1,0,0)	7x9	1
11.4	5	(3,1,0)	6x7	1
11.5	5	(2,1,0)	7x9	1 or 2
11.6	5	(3,2,0)	6x7	2
11.7	6	(2,1,1,0)	6x7	1
11.8	6	(3,2,1,0)	6x7	1 or 2
11.9	6	(4,3,1,0)	6x7	2

Table 11.1: Collated figure details.

full use of the spatial symmetries and irreducible algebraic content of the models.

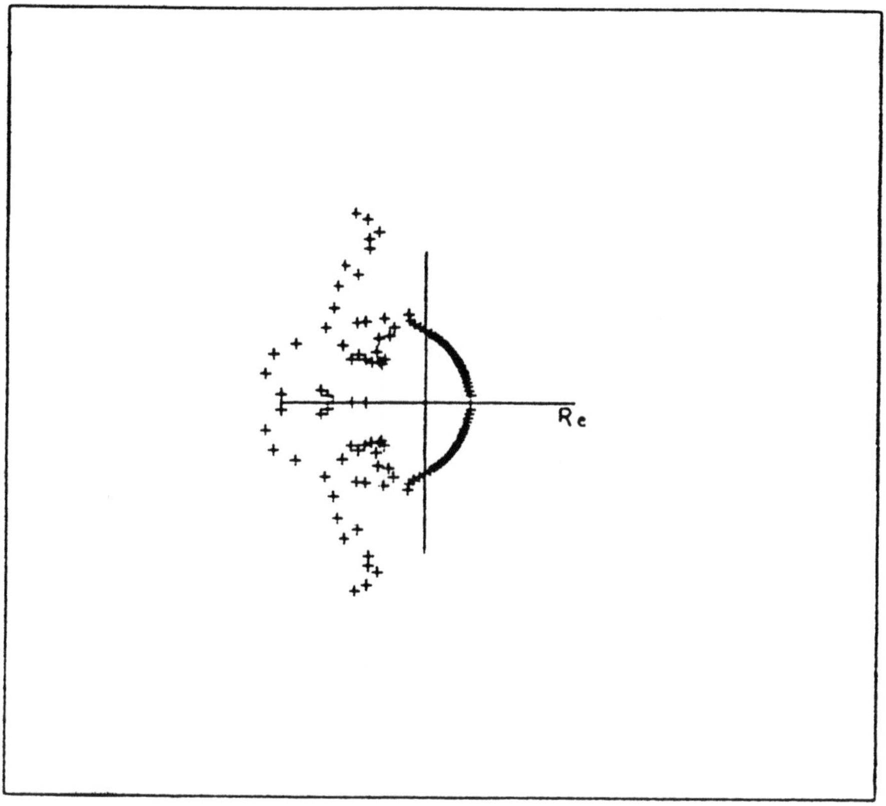

Figure 11.3: Zeros in $e^{-\beta}$ for the 7×9 lattice $\chi = (1,0,0)$ 5-state model.

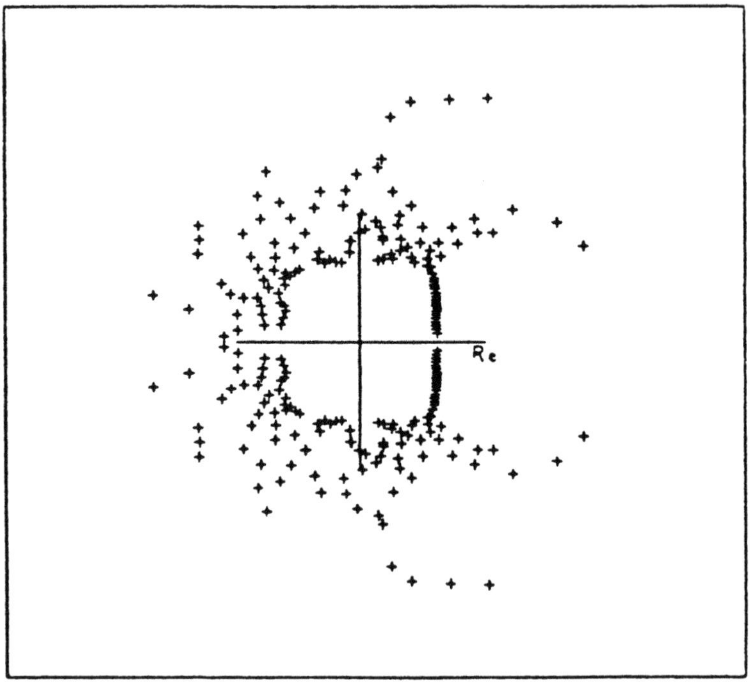

Figure 11.4: Zeros in $e^{-\beta}$ for the 6×7 lattice $\chi = (3, 1, 0)$ 5-state model.

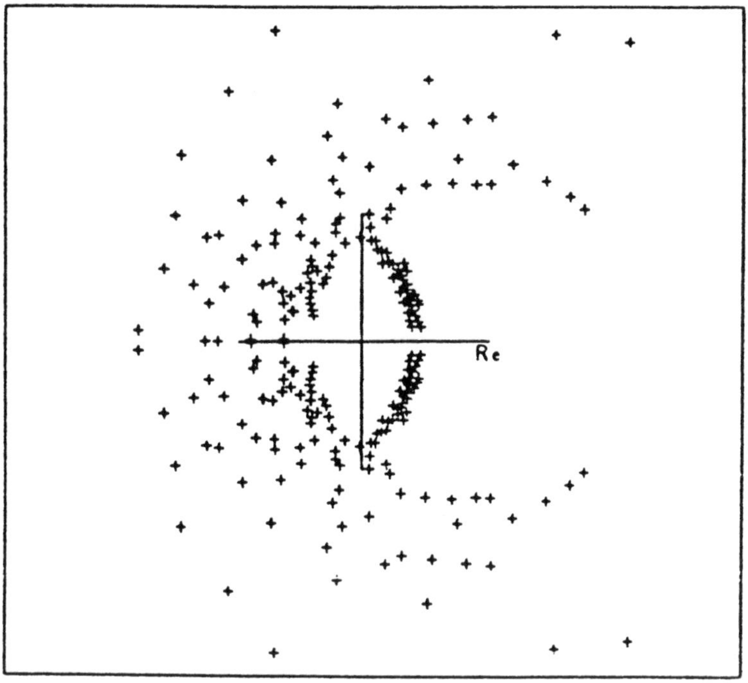

Figure 11.5: Zeros in $e^{-\beta}$ for the 6×7 lattice $\chi = (2, 1, 0)$ 5-state model.

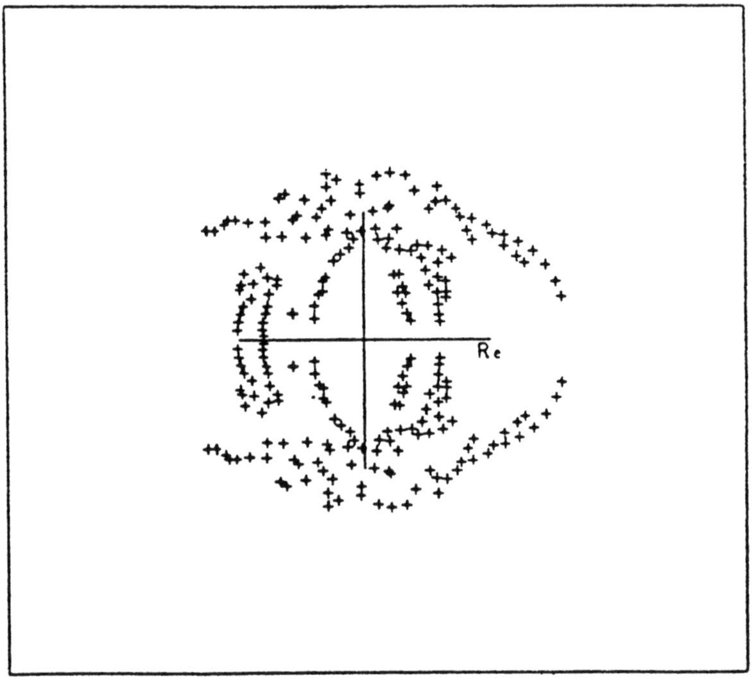

Figure 11.6: Zeros in $e^{-\beta}$ for the 6×7 lattice $\chi = (3, 2, 0)$ 5-state model.

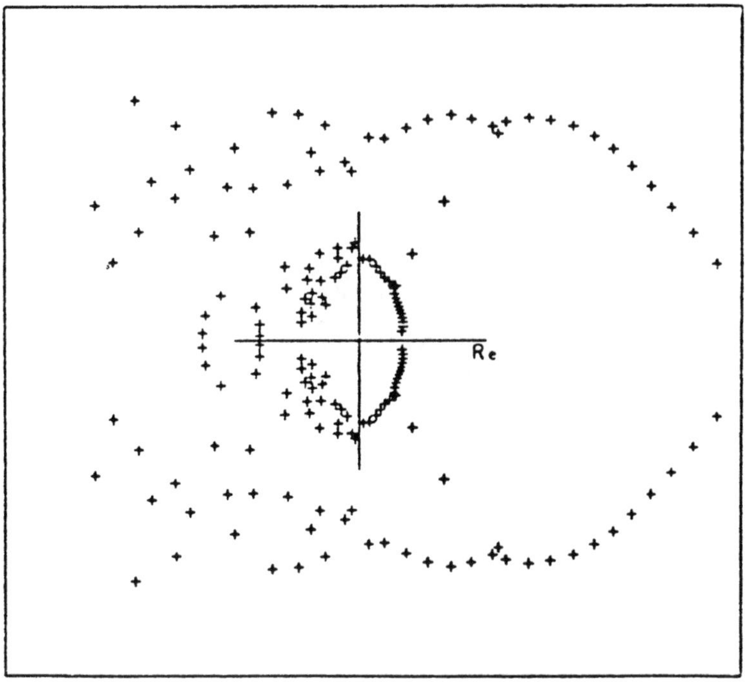

Figure 11.7: Zeros in $e^{-\beta}$ for the 6×7 lattice $\chi = (2,1,1,0)$ 6-state model.

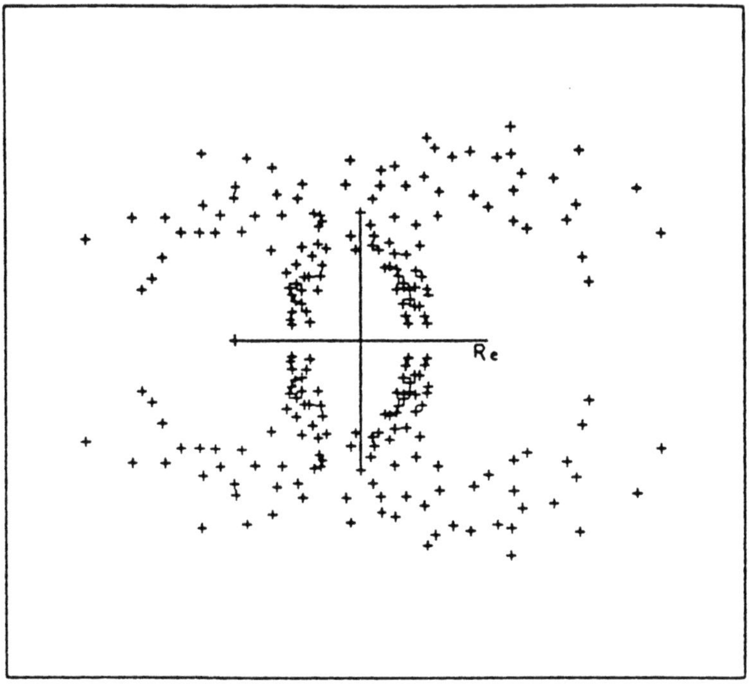

Figure 11.8: Zeros in $e^{-\beta}$ for the 6×7 lattice $\chi = (3, 2, 1, 0)$ 6-state model.

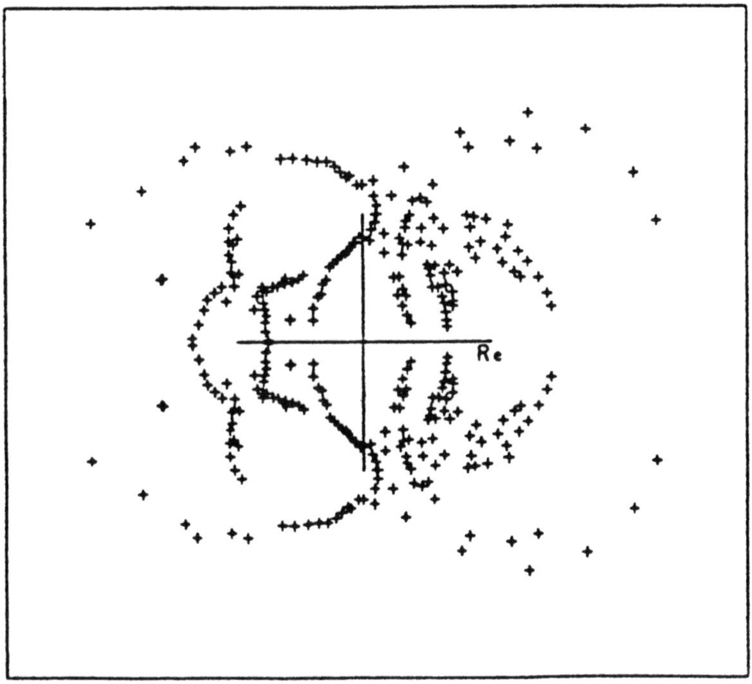

Figure 11.9: Zeros in $e^{-\beta}$ for the 6×7 lattice $\chi = (4,3,1,0)$ 6-state model.

Chapter 12

Vertex models and related algebras

In this chapter we look at the class of statistical mechanical models whose variables reside on the bonds of a square lattice, and whose interactions depend on the 4 such variables incident at a site.

Within this framework exist a rich variety of models and associated algebras. The vertex models are, in particular, the statistical mechanical source of the R-matrix solutions to the Yang-Baxter equations. Using them we will be able to show how to determine the order of Potts phase transitions, and to generalise the algebras discussed so far in various intriguing directions.

One facet of the algebraic formulation of statistical mechanics is that we can turn it on its head and derive a model from an algebra. The key step is finding a representation of the algebra whose basis vectors may be interpreted as configuration states for the model...

12.1 Homogeneous 6 vertex model

12.1.1 Vertex algebras

For γ a complex number, $x = e^{\gamma}$, and k a positive integer consider the algebra $V_k(e^{\gamma})$ defined by the generators $\{t_i; i = 1, 2, .., k\}$ and relations

$$(t_i - 1)(t_i + 1)(t_i - e^{\gamma}) = 0 \qquad (12.1)$$

$$t_i t_{i+1} t_i = t_{i+1} t_i t_{i+1} \qquad (12.2)$$

$$t_i t_{i+j} = t_{i+j} t_i \quad j \neq 1. \qquad (12.3)$$

As a quotient algebra of the braid group algebra this is a simple generalisation of the Hecke algebra $H_k(e^{\gamma/2})$. There are surjective homomorphisms from $V_k(e^\gamma)$ both to $H_k(e^{\gamma/2})$, by the quotient relation

$$(t_i + 1)(t_i - e^\gamma) = 0,$$

and to the symmetric group algebra S_{k+1}, by the quotient relation

$$(t_i - 1)(t_i + 1) = 0.$$

Note from equation 12.1 that when $x = 1$ these become the same *isomorphism*.

For all $x \neq 1$ the algebra $V_k(x)$ has a subalgebra generated by elements

$$U_i = \frac{(t_i + 1)(t_i - x)}{1 - x}$$

which obey the usual $H_k(1)$ relations.

We have the following reducible representations of $V_k(x)$. For each positive integer N define a representation on $\otimes^{k+1} V_N$ by the action of t_i on an arbitrary basis vector

$$v = v_1 \otimes v_2 \otimes \ldots \otimes v_{k+1} \tag{12.4}$$

as follows:

$$t_i\ v = x\ v \quad if \quad v_i = v_{i+1}$$

and

$$t_i\ v = v_1 \otimes v_2 \otimes \ldots \otimes v_{i+1} \otimes v_i \otimes \ldots \otimes v_{k+1} \quad if \quad v_i \neq v_{i+1}.$$

Now $2V_k(x)$ is the quotient algebra of $V_k(x)$ obtained by adding in the relations

$$\begin{aligned} t_i t_{i+1}^2 + t_{i+1}^2 t_i &= (x^2+1)t_i + x(t_i^2 - 1) - 2t_i t_{i+1} t_i \\ &\quad + 2x(t_i t_{i+1} + t_{i+1} t_i - x(t_i + t_{i+1}) + t_{i+1}^2) \end{aligned}$$

and similarly with $i \leftrightarrow i + 1$.

Note that in the case $x = 1$, with the first relation (12.1) being replaced by $t_i^2 = 1$, the other relations reduce to those defining the Temperley-Lieb algebra $T_k(1)$, so $2V_k(1) \cong T_k(1)$.

For $x \neq 1$ the algebra $2V_k(x)$ has a subalgebra generated by elements

$$U_i = \frac{(t_i + 1)(t_i - x)}{1 - x}$$

which obey the usual $T_k(1)$ relations (c.f. a specialisation of the Birman-Wenzl algebra, Birman and Wenzl 1988).

As in the Temperley-Lieb case the algebra $2V_k(x)$ has an exceptional structure for certain values of x, the most striking of which is the case $x = 1$, as will be further illustrated by the representation below. The generic algebra $2V_2(x)$ is a 10 dimensional multimatrix algebra with structure $M_3(C) \oplus C$.

The primitive idempotents of $2V_2(x)$ (for $x \neq \pm 1$) are

$$\begin{aligned} e_0 &= 1 - f_1 - f_2 + f_1 f_2 \\ e_1 &= f_1 - f_1 f_2 \\ e_2 &= f_2 - f_1 f_2 \\ e_3 &= f_1 f_2 \end{aligned}$$

where

$$f_i = \frac{t_i^2 - x^2}{1 - x^2},$$

and e_0 is central.

By taking $N = 2$ in equation 12.4 we find that the algebra $2V_k(x)$ has the following reducible representation:

$$t_i = 1_2 \otimes 1_2 \otimes \ldots \otimes \begin{pmatrix} x & 0 & 0 & 0 \\ 0 & 0 & 1 & 0 \\ 0 & 1 & 0 & 0 \\ 0 & 0 & 0 & x \end{pmatrix} \otimes \ldots \otimes 1_2,$$

where $t_i \in End(\otimes^{k+1} V_2)$, and the *R-matrix* is in the ith position in the tensor product. The generic irreducible content of this representation in the case $k = 2$ is $2S_1 \oplus 2S_3$. In the case $2V_2(1)$ this breaks up further, with S_3 replaced by $S_2 \oplus S_1$.

12.1.2 Deriving a model

The algebra $2V_k(x)$ provides a solution to the star-triangle relations

$$R_i(u)R_{i+1}(u+u')R_i(u') = R_{i+1}(u')R_i(u+u')R_{i+1}(u)$$

of the form

$$R_i(u) = a(u)1 + b(u)t_i + c(u)t_i^2$$

where a, b, c are *scalar* functions of the parameter u. Specifically, the relations are satisfied provided

$$\begin{aligned} c(u) &= \frac{\sinh(u+g)-\sinh(u)x-\sinh(g)}{x^2-1} \\ a(u) &= \sinh(g)-c(u) \\ b(u) &= \sinh(u) \end{aligned}$$

where g is to be regarded as a constant. Note that these 'weights' can only be made well defined at $x = \pm 1$ if $g = \gamma$.

Incidentally, these special cases of x which keep croping up can be used to characterise the boundaries of the analytically disjoint sectors of the free energy, and hence the phase structure, in the resultant statistical mechanical model. This connection between algebraic structure and phase structure, also observed in the Potts models, seems in fact to run through statistical mechanics (see also Baxter 1982).

As usual, the solutions to the fixed point equation $u = u' = u + u'$, which are $u = 0$ and, formally, $u = \infty$, correspond to the two special limits of the R-matrix, i.e. the trivial limit $R_i(0) \propto 1$ and, in the case $g = \gamma$, the braid limit

$$\lim_{u\to\infty} \frac{R_i(u)}{\sinh(u)} \to t_i.$$

Let us define weights

$$W = \sinh(u+g)$$

$$X = \sinh(g)$$

and

$$Y = \sinh(u).$$

For notational convenience in this section, if $F = F(u,g)$ (e.g. $F \in \{W, X, Y\}$) then we write $F' = F(u', g)$ and $F'' = F(u-u', g)$ and $F_- = F(u, -g)$.

The single interaction partition vector, arranged as a transfer matrix, obtained by the *adjoint* construction (see chapter 3) from the representation of $2V_k(x)$ above, is then

$$R_{(ik,lj)}(u) = \begin{pmatrix} W & 0 & 0 & 0 \\ 0 & X & Y & 0 \\ 0 & Y & X & 0 \\ 0 & 0 & 0 & W \end{pmatrix}.$$

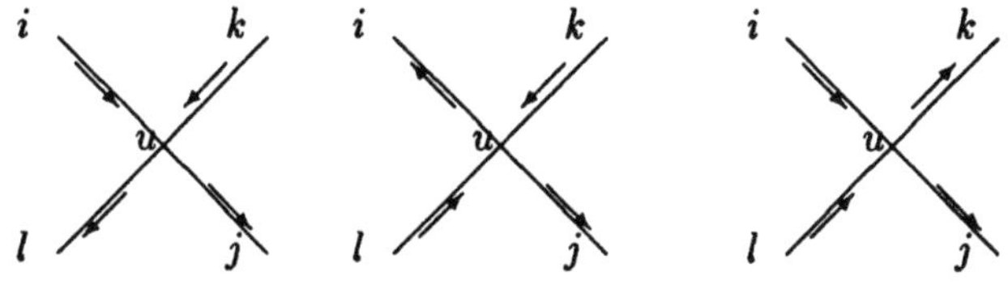

Figure 12.1: The single interaction partition vector $R_{(ik,lj)}(u)$.

Since the index pairs ik and lj each correspond to basis states in $V_2 \otimes V_2$, each pair may be thought of as configurations of two adjacent bond variables, each variable taking values from V_2.

It is traditional to interpret the 16 possible configurations at the interaction as arrow coverings of the 4 bonds incident at a vertex of the square lattice (see figure 12.1). The non-vanishing weights are then associated with the configurations in which the number of outgoing and incoming arrows is equal. Since only 6 interaction weights are non-vanishing we call this a 6 vertex model. The examples shown in the figure are $R_{(\downarrow\downarrow,\downarrow\downarrow)}(u) = W$, $R_{(\uparrow\downarrow,\uparrow\downarrow)}(u) = X$ and $R_{(\downarrow\uparrow,\uparrow\downarrow)}(u) = Y$.

Note that the weight to configuration assignments are orientation dependent. By convention the orientation is given by consistent positioning of the parameter in the diagram.

Physically the resultant model has some disturbing features for certain values of the parameters (see Baxter 1982), but we will persevere with it, as it still provides a medium for illustrating some useful points.

12.1.3 On solving the model

This model facilitates the illustration of an approach to the star-triangle relation different to that pursued in chapter 3. We will consider an N site row to row transfer matrix, with periodic boundary conditions at the ends of the layer, but we build this directly out of transfer matrices with given boundary conditions at the ends of the layer.

This differs from the approach in chapter 2, where we could only keep boundary information at the ends of the layer by introducing redundancy. In order to do so here we will reorient our local transfer matrices in the layer,

$$(t_{ij}(u))_{kl} = R_{(ik,lj)}(u) \tag{12.5}$$

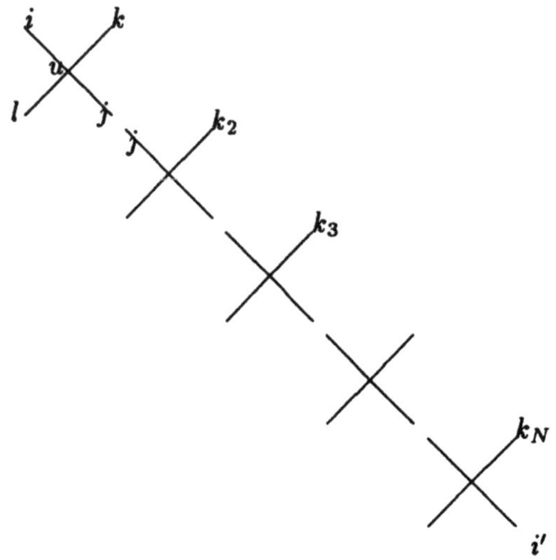

Figure 12.2: The single layer partition vector $T_{ii'}(u)$.

so that i, j relate to bond variables *within* the layer and only k, l are kept as conventional incoming and outgoing transfer matrix boundary information. We then have the layer transfer matrix

$$(T_{ii'}(u))_{k_1k_2..k_N,l_1l_2..l_N}$$

$$= \sum_{j,m,n,..,p=1,2} (t_{ij}(u))_{k_1l_1}(t_{jm}(u))_{k_2l_2}(t_{mn}(u))_{k_3l_3}...(t_{pi'}(u))_{k_Nl_N}$$

- see figure 12.2. We can then write a periodic transfer matrix,

$$T = \sum_{i\in V_2} T_{ii}(u) \quad \in End(\otimes^N V_2)$$

without the complications introduced in chapter 2. The price we pay for this here is that the rearrangement 12.5 obfuscates the beautiful algebraic structure. The advantage of it is that $T(u)$ and $T(u')$ commute, so the eigenvectors of $T(u)$ cannot depend on u.

To show how to determine the eigenvectors let us consider the transfer matrix in the case where the bonds at the *ends* of the layer (labelled i and i' above) are explicit boundary information. We write it in the form

$$T_{ii'}(u) = \begin{pmatrix} A & B \\ C & D \end{pmatrix} \tag{12.6}$$

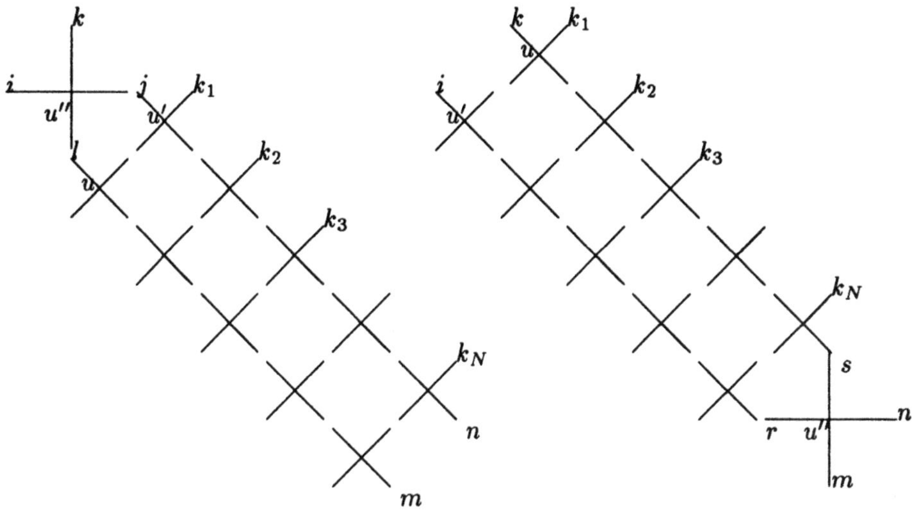

Figure 12.3: Commuting transfer matrices. Repeated use of the star-triangle relations takes the left hand diagram to the right hand diagram. Note that $u'' = u - u'$.

where the four sub-blocks are the transfer matrices with given end configurations, i.e. with i not necessarily equal to i'. It follows from the star-triangle relations that we have

$$R_{(ik,lj)}(u-u')T_{lm}(u)T_{jn}(u') = T_{ir}(u')T_{ks}(u)R_{(rs,mn)}(u-u')$$

as illustrated by figure 12.3.

This in turn implies some partial commutation relations among the sub-blocks of equation 12.6. Most usefully we have

$$AB' = (W''/Y'')B'A + (X/Y'')BA',$$

$$BB' = B'B$$

and

$$DB' = (W''_-/Y'')B'D - (X/Y'')BD'.$$

These relations can be used to determine the eigenvectors, and hence the free energy of the model in the thermodynamic limit, i.e. the largest eigenvalue of $T = A + D$, and other eigenvalues of the transfer matrix.

To see this, we introduce the basis states

$$v_0 = \otimes^N \begin{pmatrix} 1 \\ 0 \end{pmatrix}$$

and

$$v_i = \left[\otimes^{i-1} \begin{pmatrix} 1 \\ 0 \end{pmatrix}\right] \otimes \begin{pmatrix} 0 \\ 1 \end{pmatrix} \otimes \left[\otimes^{N-i} \begin{pmatrix} 1 \\ 0 \end{pmatrix}\right].$$

Note that

$$Av_0 = W^N v_0$$

$$Cv_0 = 0$$

and

$$Dv_0 = Y^N v_0.$$

Now for integer r define $w = \exp(2\pi i r/N)$ and

$$u' = \frac{1}{2}\ln(e^{-g} - we^{g} - w).$$

so that

$$\sinh(u' + g) = w \sinh(u'). \tag{12.7}$$

We then find that

$$B' v_0 = B(u')v_0$$

is also an eigenvector of $A + D$ for any integer r, since

$$\begin{aligned}(A+D)B'v_0 \;&=\; (W^N(W''/Y'') + Y^N(W''_-/Y''))B'v_0 \\ &\quad -[(X''/Y'')((W')^N - (Y')^N)]Bv_0\end{aligned}$$

where the factor in square brackets vanishes by equation 12.7.

This result is a manifestation of translation invariance. The action of B' on v_0 is to generate running wave states, that is superpositions of states with one spin flipped (and 'momentum' r). This is because $B = T_{12}$, and every summand in T_{12} has one more (direct product) factor of the form t_{12} than of the form t_{21}. Any factor t_{21} anihilates the 'vacuum' state v_0, so the surviving terms have exactly one t_{12} each. This is, up to an overall factor, a spin flip operator on the appropriate sub-basis vector. Altogether we have

$$B'v_0 = X'(Y')^N(W')^{-1}\sum_{i=1}^{N}\left(\frac{W'}{Y'}\right)^i v_i$$

The largest of the eigenvalues in this sector comes from the translation invariant eigenvector, i.e. $(W'/Y') = w = 1$, $r = 0$.

Similarly, there exist vectors $B_1B_2v_0 = B(u_1)B(u_2)v_0$, with u_1, u_2 given by similar conditions, which are eigenvectors with

$$(A+D)B_1B_2v_0 = \left(W^N \prod_{i=1}^{2}(W_i''/Y_i'') + Y^N \prod_{i=1}^{2}(W_{-i}''/Y_i'')\right) B_1B_2v_0.$$

Eventually, making repeated use of the partial commutation relations above, it is possible to determine various choices for n-tuples of complex numbers $u_1, u_2, ..u_n$ such that

$$v(u_1, u_2, .., u_n) = B(u_1)B(u_2)..B(u_n)v_0$$

is an eigenvector of $A+D$ (for another way of motivating this *Bethe ansatz* approach, and for further details, see Baxter 1982 or De Vega 1989).

On dual models

Duality transformation of a model can be split up into two operations: One in which the variables and interaction simplices are dualised in a geometrical sense, and another in which the interactions themselves are transformed. Conventional statistical mechanical duality determines both these steps (Savit 1980). As a rule, little can be learned from the dual model that is not implicit in the original one. Sometimes, however, we find that the configuration space of a dual model lends itself to a new representation of the associated algebra which reveals new information. The ABF model is a case in point (Andrews, Baxter and Forrester 1984). The dual model for the 6 vertex model is a similar face model.

12.2 Asymmetric 6 vertex models

The boundary conditions on a 6 vertex model are said to be charge free if the number of vertices which are sinks for horizontal arrows (and hence sources for vertical ones) is equal to the number which are sources for horizontal arrows. In the case of non-skew torroidal boundary conditions this conservation law holds on every horizontal and vertical line. This means that the partition function and transfer matrix are invariant under a transformation of the weights X on these two types of vertex which preserves their geometric mean. Let us write X_5 and X_6 for the distinguished horizontal source and sink weights respectively, then the partition function is not changed by changing these weights, provided $X_5X_6 = XX$.

Recall that the original weights are $W, X, Y = \sinh(u+g), \sinh(g), \sinh(u)$.

The new R-matrix is

$$R_{(ik,lj)}(u) = \begin{pmatrix} W & & & \\ & X_5 & Y & \\ & Y & X_6 & \\ & & & W \end{pmatrix}$$

This is certainly not a similarity transformation, since it has changed the trace. We must check the star-triangle relations from scratch. Let us consider a more general question, namely, supposing the existence of a braid limit R-matrix of the form

$$R(\infty) = \begin{pmatrix} x & & & \\ & \beta & 1 & \\ & 1 & \alpha & \\ & & & x \end{pmatrix}$$

what conditions does the braid relation place on α, β? The answer is readily obtained by direct calculation: either both α and β vanish (as in section 12.1) or one vanishes and the other is $x - x^{-1}$. In the latter case the braid generator obeys a quadratic local relation and the algebra generated is the Temperley-Lieb algebra. In fact the representation of the Temperley-Lieb algebra we get is the *defining* representation discussed in chapter 9.

The former case corresponds to the homogeneous 6 vertex model, the latter to an *asymmetric* homogeneous model in which the weights X_5 and X_6 obey the relation

$$\frac{X_5 - X_6}{Y} = (x - x^{-1}).$$

Now note that putting $g = \gamma$ then

$$(WW + YY - X_5 X_6)/2WY = -\cosh(g).$$

Combining these results and parameterising in terms of W, Y and x instead of u, Y and g we find that $X_5 = W + xY$ and $X_6 = W + x^{-1}Y$. Using these weights instead of the original ones we change the braid limit algebra, but the star-triangle relation, the transfer matrix, and the partition function are preserved (in the absence of boundary charges).

Perhaps the most useful calculation which may be performed by applying these properties is to compute the spontaneous staggered polarization, which is the difference between the probability of a single site being in configuration 5 and being in configuration 6. In fact this is a complicated calculation, fraught with technical dangers and requiring more space than is available here. We refer the reader to Baxter's original (1973) paper.

For our purposes it will be quite useful to have the result, however. The spontaneous staggered polarization is given up to sign by

$$p(5) - p(6) = \pm \prod_{n=1}^{\infty} (\frac{1 - q^{2n}}{1 + q^{2n}})^2 \tag{12.8}$$

for $Q > 4$; and is zero for $Q \leq 4$.

12.2.1 Inhomogeneous 6 vertex models

The inhomogeneous 6 vertex model is obtained by assigning different values to the parameters W and Y depending on whether a vertex is on the odd or even sublattice.

Note that renormalising all the weights on the even (or odd) sublattice by the same factor just produces an overall factor in the transfer matrix or the partition function. For example, renormalising on the odd sublattice by $(1/Y_o)$, and the even by $(1/W_e)$ we obtain

$$odd : W_o/Y_o, 1, W_o/Y_o + x^{\pm 1}$$

$$even : 1, Y_e/W_e, 1 + x^{\pm 1} Y_e/W_e.$$

If $W_o/Y_o = W_e/Y_e$ these weights are the same up to an overall factor, and we effectively have a homogeneous model again.

12.2.2 Equivalence with the Potts model

In general, as we have seen, the algebra associated to the asymmetric models is the Temperley-Lieb algebra. Furthermore, the transfer matrix has the same operator expression as that of the Potts model (the Potts coupling parameter $v = \exp(\beta) - 1$ is replaced by $v = \sqrt{Q}(W/Y)$). Thus, in as far as the *representations* of the algebra obtained from the transfer matrices in each case both contain a given irreducible representation, the eigenvalue spectra overlap.

We discussed this point in chapter 9. The vertex model representation contains every possible irreducible representation, and hence its spectrum contains that of the Potts model (ignoring degeneracy).

It further follows from the discussion in chapter 6 and the definitions of the representations that, for a realistic set of boundary conditions the free energies of the two models are the same. Here, and in general, the key question is one of interpretation, that is to say: to what physical observable does each subsequent spectral gap correspond ?

The question is somewhat academic, given that neither model has been solved. However, the homogeneous limit of the vertex model, which corresponds to the critical Potts model, does provide some vital information. We will show how it can be used to demonstrate that the Potts phase transition is first order for $Q > 4$ and second order for $Q = 2, 3, 4$.

The Potts phase transitions

The free energy of the critical Potts model can be given as a function of Q and the coupling parameter in a single direction, since the parameter in the perpendicular direction is determined by the condition that the model is critical. The result in itself is not very revealing. Except in as much as the form is different for $0 < Q < 4$ and $Q > 4$, the information content is limited by the fact that we cannot examine the response to small perturbations *away* from the critical line. The result is given in Baxter 1982.

Much more usefully, we can calculate the internal energy at the critical temperature. In a model with a first order phase transition this result must be ambiguous, since it depends on the direction from which the critical point was approached. At a second order transition it must be unique. Thus the order of transition can be determined even without knowing anything *exact* about off critical values. The first question is: how is the Potts internal energy information stored in the vertex model picture?

Here, for the sake of simplicity, we will only consider the isotropic case (and introduce an overall coupling variable $x = v_o/\sqrt{Q} = v_e/\sqrt{Q}$). The internal energy per site for an N site Potts model may then be written

$$U = (1/N)\frac{\partial x}{\partial \beta}\frac{\partial}{\partial x}\ln(Z_N).$$

Recall that $v_i = \exp(\beta_i) - 1$ and take $\beta_o = \beta_e = \beta$.

We can write the partition function $Z_N(x)$ in its vertex model form as

$$Z_N = Q^{N/2} \sum_{\{bond\}} x^{a'+b} q^{-d+c} (x+q)^{c'+d} (x+1/q)^{c+d'}$$

where the sum is over vertex model bond variable configurations; a' is the *total* number of vertices in configurations 1 and 2 on the even sublattice; b is the *total* number of vertices in configurations 3 and 4 on the odd sublattice; c (respectively c') is the total number in configuration 5 on the odd (even) sublattice; and d (d') the total number in 6 on the odd (even) sublattice. Note that we have neglected any boundary details, so this result will only be meaningful once we have contrived a thermodynamic limit.

We hence obtain the internal energy per site

$$U = (1/N)(x + Q^{-1/2})(\frac{< a' + b >}{x} + \frac{< c + d' >}{x + 1/q} + \frac{< d + c' >}{q + x}).$$

Now we can take the thermodynamic limit by removing the factor $1/N$ and reinterpreting each expectation value as that of a single vertex being in one of the given configurations. The problem is that, in general, we cannot evaluate the expectation values in the limit. An exceptional case is $x = 1$ where homogeneity of the vertex model version is restored. We will now consider this case.

Note that, although the symmetries of the homogeneous vertex model under interchange of sublattices and under global bond variable reversal may be spontaneously broken, the composite of the the two transformations *does* leave the model ground state unchanged. This means that we can express all the even lattice expectation values in terms of juxtaposed odd lattice expectation values. We can then rewrite the $x = 1$ result as

$$U = (1 + Q^{-1/2})(< a + b + c + d > + < c - d > (q - 1)/(q + 1))$$

where $< a+b+c+d >=< 1 >= 1$ and $< c-d >$ has a sign ambiguity when it is non-zero. This quantity is the spontaneous staggered polarization of the vertex model. So, the question of the Potts phase transition does indeed come down to a question about the homogeneous vertex model! Applying equation 12.8 (with $< c - d >= p(5) - p(6)$) we see that the $Q > 4$ result is finite, and $Q \leq 4$ gives zero. This latter result ensures a second order phase transition for $Q = 2, 3$. Note that the internal energy is unique in any case for $Q = 4$ since this corresponds to $q = 1$.

Now let us consider $Q > 4$, and the situation in which x is slightly larger than one. By comparing the product of (inhomogeneous) weights for an odd vertex in configuration 5 and an even vertex in configuration 6 with the product of weights for a complimentary pair we find that, in this situation, the former pair is prefered. If x is slightly less then one then the preference is reversed. Thus the inhomogeneous vertex model expectation value $< c - d >$ is positive when approaching the critical temperature from above and negative from below. This establishes the claimed result, $Q > 4$ has a first order phase transition.

12.3 Homogeneous Z_N vertex models

We can generalise the homogeneous 6 vertex model by allowing each bond variable to take $N \in \mathbf{Z}_+$ different values. The N^2 dimensional matrix of interactions can then be filled in in various ways consistent with the star triangle relations. We will mention a couple of the more interesting ones, although they do not appear to have any direct relevance to the Potts models.

Vertex algebra cases

Here three different weights W, X, Y are assigned, depending on whether all bonds at a vertex are the same, i.e. $R_{(bb,bb)} = W$; or different, $R_{(ba,ba)} = X$; $R_{(ba,ab)} = Y$ $(a \neq b \ : \ a, b \in 1, 2, .., N)$, with all other weights zero. These are $2N^2 - N$ vertex models.

For each N the corresponding algebra $NV_k(x)$ (i.e. the algebra with generators defined by the braid limit of this model) obeys the same local, braid and commutation relations as $V_k(x)$, but there are additional relations. The algebra associated with $N = N_1$ is a quotient of the algebra associated with $N = N_2$ if $N_2 > N_1$. In each case the $x = 1$ limit reduces to the corresponding quotient of the $q = 1$ Hecke algebra $H_k(1)$, i.e. the group algebra of the symmetric group S_{k+1} restricted to tableau with at most N rows.

These models are all amenable to generalisations of the Star-triangle/Bethe ansatz technique (see De Vega 1989) and most other remarks in the preceding sections of this chapter have an echo here. A notable exception is the corresponding generalisation of critical Potts models.

Asymmetric Z_N vertex models are based on the Hecke algebra (see chapter 9, Date *et al* 1987 and references therein).

Higher spin cases

In the so called higher spin cases the bond variables take values from, for example, $-N$ to N in integer steps, where N may be integer of half integer ($N = 1/2$ is the 6 vertex model case). The interactions associated with solutions of the Yang-Baxter equations require many different weights. They may be calculated directly by a *cabling* procedure (so called because of its interpretation at the level of braids, see chapter 13 and e.g. Wadati, Yamada and Deguchi 1989 and references therein).

Chapter 13

Braids and cables

In this chapter we show how to determine the structure of algebras obtained by cabling quotients of the braid group algebra. In addition to providing solvable models (see e.g. Date *et al* 1988), the cabling morphisms may be seen as the algebraic incarnation of block spin transformations.

13.1 Introduction

Recall that the braid group on n strings B_n may be defined by the generators $\{g_i, g_i^{-1}; i = 1, 2, .., n-1\}$ and relations

$$g_i g_i^{-1} = 1$$

$$g_i g_{i+1} g_i = g_{i+1} g_i g_{i+1}$$

$$g_i g_{i+j} = g_{i+j} g_i \qquad (j \neq \pm 1).$$

We also use B_n to denote the associated group algebra.

From chapter 5 we have

$$B_m \subset B_{m+n}$$

realised by

$$g_i \mapsto g_i^{(j)}$$

for any $0 \leq j \leq n$, and

$$B_n \otimes B_m \subset B_{n+m}$$

realized, for example, by

$$g_i \otimes 1 \mapsto g_i$$

$$1 \otimes g_i \mapsto g_i^{(n)}.$$

Suppose that m consecutive strings (as they occur along the top of a braid) may be removed from the plane either from over or underneath the box containing the braid, without cutting any string. Suppose that they are also consecutive at the bottom. Then these strings define an element of B_m in a natural way. We may highlight such an occurence of $b \in B_m$ as a removable subdiagram of $a \in B_{n+m}$ (say) by replacing the subdiagram with a labelled 'ribbon' covering the appropriate positions

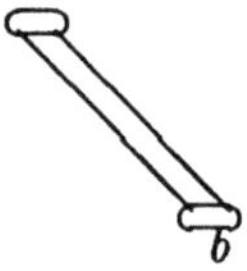

For example, for $g_i \in B_n$ and $b \in B_m$ define $g_i.b \in B_{nm}$ by

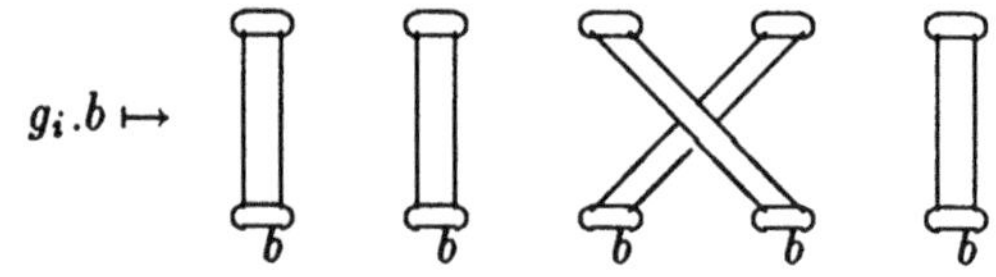

We then define an injection

$$B_n \otimes B_m \subset B_{nm}$$

by

$$g_i \otimes b \mapsto g_i.b.$$

For $m \in Z_+$ and e an idempotent in B_m we have the *cabling* morphism

$$B_n \xrightarrow{e} B_{nm}$$

defined by

$$e : g_i \mapsto g_i.e,$$

that is

13.1.1 Cabling quotients of B_n

Let e be an idempotent in B_m and A_{nm-1} be a quotient algebra of B_{nm} with quotient relations $\sim *$

$$B_{nm} \xrightarrow{\sim *} A_{nm-1}$$

so that

$$g_i \mapsto g_i$$

and

$$e \mapsto \epsilon$$

(for example, $B_{nm} \rightarrow H_{nm-1}(q)$ - by convention the index on quotient algebras refers to the number of generators, not the number of strings). Then for $\sim$ some equivalence relation on B_n, such that quotienting

$$B_n \xrightarrow{\sim} eA_{n-1}$$

$$g_i \mapsto g_i$$

we have the morphism of quotient algebras from

$$\begin{array}{ccc} B_n & \xrightarrow{e} & B_{nm} \\ \sim\downarrow & & \downarrow\sim * \\ eA_{n-1} & \xrightarrow{e} & A_{nm-1} \end{array}$$

defined by

$$g_i \mapsto g_i.\epsilon.$$

The question here is, given $\sim *$ and e, what is the equivalence relation $\sim$? In other words, what is the algebra eA_{n-1}?

This is quite an important question since, particularly if $eA_{n-1} \cong A_{n-1}$, this morphism can be closely linked with the idea of scale invariance at the critical point of the associated statistical mechanical models (see chapters 1 and 3).

13.2 Cabling Temperley-Lieb algebras

Here we only have space to look at an illustrative example (see also Murakami 1989, Wadati *et al* 1989 and references therein). We will take the case relevant to the Potts model, i.e. cabling Temperley-Lieb algebras, $A_{nm-1} = T_{nm-1}(q)$. To proceed the reader will need to have first covered chapter 6 (particularly boundary diagrams). A considerable degree of generalisation of this example will be straightforward for the interested reader, using chapter 9.

Let us start by considering a specific morphism of the type indicated above. In the case $m = 2$ we have

$$g_1.1 = t_2t_1t_3t_2 = (1 - qU_2)(1 - qU_1)(1 - qU_3)(1 - qU_2),$$

and two choices for ϵ, that is

$$\epsilon_1 = (1 - U_i/\sqrt{Q}); \quad \epsilon_2 = U_i/\sqrt{Q}.$$

Let us consider the former. We see by direct computation that with $I = 1.\epsilon_1$ then

$$g_1.\epsilon_1 = I - (q + q^3)IU_2I + q^4IU_2U_1U_3U_2I$$

since all the other terms in the expansion are killed off by I.

The question is: what is the algebra $e_1T_{n-1}(q)$ generated by $g.\epsilon_1$ and $G^n.\epsilon_1$ (and inverses)? Since it is automatically a quotient algebra of the group algebra of the braid group, the simplest questions to answer are: what is the local relation and what is the structure of the algebra?

Let us consider the generically isomorphic algebra generated by

$$I, IU_2I, IU_2U_3U_1U_2I$$

and translations. We can easily figure out the structure of this algebra by looking at the boundary diagram representation of the Temperley-Lieb algebra, and the effect of the above operators on the concommitant bases, introduced in chapter 6.

A convenient basis for $T_k(q)$

Let us briefly review these ideas. Words in $T_k(q)$ are in one-to-one correspondence with the set of possible connections of 2 parallel rows of $k+1$ nodes ('top' and 'bottom') by $k+1$ non-intersecting lines. The idea of composition by abutting such diagrams and discarding the identified nodes leads, given a factor of $\sqrt{Q}$ for each closed loop, to an algebra isomorphism (rather as in the braid diagrams, but with no crossing rather than no doubling back). At the same time, such diagrams are in one-to-one correspondence with pairs of sequences of $k+1$ numbers, obtained by numbering each line in the diagram distinctly from 1 to $k+1$ so that the first occurence of $i < j$ preceeds the first j on reading the nodes from left to right top, then left to right bottom.

Thus a generator $U_i \in T_k(q)$ may be represented by the pair

$$U_i \mapsto (1234 \ldots\, i\; i\; i+1 \ldots k,\; 1234 \ldots i'\; i'\; i+1 \ldots k)$$

	1						
11		12					
	112		123				
	122						
1122		1123		1234			
1221		1223					
		1233					
	11223		11234		12345		
	12213		12234				
	11233		12334				
	12233		12344				
	12332						
112233		112234		112345		123456	
122133		122134		122345			
112332		112334		123345			
122331		122334		123445			
123321		123324		123455			
		112344					
		122344					
		123344					
		123443					
	1122334		1122345		1123456		1234567
	1221334		1221345		1223456		
	1123324		1123345		1233456		
	1223314		1223345		1234456		
	1233214		1233245		1234556		
	1122344		1123445		1234566		
	1221344		1223445				
	1123344		1233445				
	1223344		1234435				
	1233244		1123455				
	1123443		1223455				
	1223443		1233455				
	1233442		1234455				
	1234432		1234554				

Table 13.1: Bratelli/basis diagram for $T_k(q)$ $(k = 0, 1, .., 6)$.

where $i' = k + 1$. Composition of generators $U_i U_j$ is, for example

$$(12334, 12554)(11234, 55234) = (11223, 44553)$$

It follows that the set of sequences for the top line of nodes with given maximum number occuring form a basis for a left $T_k(q)$ module (with the action implied above), on quotienting by all lower maximum number cases.

The basis states are the set of all left halves of pairs. Accordingly, the action of a generator on a basis state is, for example

$$(12334, 12554)(11223) = \sqrt{Q}(11223)$$

and, because of the quotient

$$(12334, 12554)(11234) = (11223) = 0.$$

The basis vectors for the first few cases are given, in the boundary line numbering notation (which is also discussed in chapter 6), in table 13.1. Each subcolumn of sequences is the complete list of basis vectors for a given module.

Let us define two types of moves for changing a boundary line numbering sequence of k numbers into a sequence of $k+1$ numbers. Move A is to add the next lowest positive integer which does not appear in the sequence; Move B is to add another copy of the highest number which occurs exactly once in the sequence, or to return $\emptyset$ if there is no such number.

Let us define M_i as the set of states in the ith row, and M_{ij} as the subcolumn of states in the ith row and j^{th} column of the table (note that each row has alternate columns empty). We write AM_{ij} for the subset of M_{i+1} obtained by acting with move A on every element of M_{ij}.

It follows from the definitions that the list of basis vectors for a given module is obtained by first applying move A to the list of states from the immediately above left position, and then applying move B to the list of states from the immediately above right position:

$$M_{i+1\,j} = AM_{ij-1} + BM_{ij+1}.$$

Note that this procedure involves no duplication of states, and that each possible state occurs.

A basis for the subalgebra

In general, let us denote the subalgebra obtained by using an idempotent associated with the module in the $(m+2-j)^{th}$ column of the Bratelli diagram, at the $T_{m-1}(q)$ level, by

$$e_j^m T_{k-1}(q) \subset T_{mk-1}(q).$$

	12			
1221	1223	1234		
122331	122134	122345	123456	
	122334	123445		
	123443			
12213443	12233145	12213456	12234567	12345678
12233441	12213445	12233456	12344567	
12344321	12233445	12344356	12345667	
	12344325	12234556		
	12234554	12344556		
	12344553	12345665		

Table 13.2: Bratelli/basis diagram for $e_1^2 T_{k-1}(q) \subset T_{2k-1}(q)$.

Now with $m = 2$ the subalgebra we want is only defined for odd numbers of generators in the original algebra. Using the central primitive idempotent ($j = 1$), then for $k = 1$ it is generated by the identity; for $k = 2$ it is generated by elements of the form IWI where $W \in \{1, U_2, U_2U_3U_1U_2\}$; for $k = 3$ it is generated by IWI with

$$W \in \{1, U_2, U_4, U_2U_3U_1U_2, U_4U_3U_5U_4\};$$

and so on.

Suppose V is the left $T_{2k-1}(q)$ module associated to the states in M_{kj}. Then IV is a left $e_1^2 T_{k-1}(q)$ module. On the other hand, the subset of $T_{2k-1}(q)$ basis vectors obtained by taking the top halves of the pairs for all words generated by the words W is the set which have no adjacent pair of numbers from an odd postion to an even position the same. These are precisely the states which are not sent to zero by the action of I. These states may thus be used to form bases for (generically simple) representations of the subalgebra (where the presence of the I is to be understood in determining the action). The first few of these states are given in table 13.2.

In the partition basis notation of chapter 8 these states correspond to the set which have no singleton in the first team.

The iterative procedure for generating a new row of basis vectors here is as follows:

First take the set of states from the immediate above left position and modify each one by adding another copy of the highest once occuring number (if there is one), this step is then repeated; next take the set of states from the immediate above position and modify each one by adding the highest once occuring number again, and then the the lowest possible new number; then take the set of states from the immediate above right position and modify each one by adding the lowest new number, and then repeating this step.

In our notation this is first

$$M_1 = M_{12} = \{(12)\},$$

then

$$M_{i+1\,j} = AAM_{ij-1} + ABM_{ij} + BBM_{ij+1} \qquad j > 0.$$

Note from chapter 5 that this implies that the dimensions of the vector space bases coincide with the multiplicities of irreducible representations in the decomposition of $\otimes^k l_{m-1}$ where l_{m-1} is the irreducible representation of $SU(2)$ on V_m.

This correspondence immediately generalises to all m (this was spotted by Westbury). Let us introduce $p = 1, 2$ for $m = even, odd$ and $x = (m+p+1)/2$. Then the general form for the iteration is

$$M_1 = M_{1x} = \{(12...m)\}$$

and

$$M_{i+1\,j} = \sum_{J=-m/2}^{m/2} A^{-J+m/2} B^{J+m/2} M_{i\,j+pJ} \qquad j > 0$$

where the sum is incremented in integer steps as usual.

Explicitly, the generators for $m = 3, k = 2$, for example, are IWI with $W \in \{U_3, U_3U_2U_4U_3, U_3U_2U_4U_1U_3U_5U_2U_4U_3\}$, and $G^3.\epsilon$. The generalisation is then obvious.

Since

$$T_k(q) = End_{U_qSU(2)}(\otimes^{k+1} V_2)$$

then central idempotents in $T_k(q)$ also give idempotents in the appropriate quotient of $U_qSU(2)$, and vice versa. In $T_k(q)$ we constructed a primitive central idempotent (out of primitive idempotents) for each simple module (chapter 6). By a dimension counting argument we can identify each of the primitive idempotents as a projector for a corresponding $U_qSU(2)$ submodule of $\otimes^{k+1}V_2$. In particular, the central primitive idempotent of $T_{m-1}(q)$, call it ϵ_1 (that is, E_{m+1} from chapter 6) acts by

$$\epsilon_1 \ (\otimes^m V_2) = V_{m+1}$$

as far as the $U_qSU(2)$ modules are concerned. That is, it picks out the highest dimensional irreducible in the decomposition (which happens to be non-degenerate). Altogether, then, this means that $I = 1.\epsilon_1 \in T_{mn-1}(q)$ acts by

$$1.\epsilon_1 \ (\otimes^{mn} V_2) = \otimes^n V_{m+1}.$$

In other words

$$e_1^m T_{mk-1}(q) = End_{U_qSU(2)}(\otimes^k V_{m+1}).$$

We now see that our iteration rule matches the decomposition rule for tensor products in $SU(2) \cong U_qSU(2)$ for q not a root of unity.

We also see immediately that the generic structure of $\epsilon T_{k-1}(q)$ is not the same as $T_{k-1}(q)$. On the other hand, we know that the non-generic cases will be much richer....

Before looking at this more closely, note that if we take an arbitrary primitive idempotent instead of ϵ_1, then the story is much the same. The action of each idempotent associated with a given simple module S_j (there are $\dim(S_j)$ of them) is more or less equivalent here. Each projects onto one of the $\dim(S_j)$ isomorphic simple $U_qSU(2)$ modules in $\otimes^m V_2$:

$$\epsilon_j \ (\otimes^m V_2) = V_{m+2-j}.$$

All the algebras which can appear by varying the idempotent are thus already covered by the central primitive idempotent for some lower m.

13.2.1 The local relation

To illustrate the special circumstances of the non-generic cases, consider the calculation of the local relation or, equivalently, of the eigenvalues of $g_i.\epsilon$.

It is sufficient to work with $T_{2m-1}(q)$. Regarding q as an indeterminate we note from the basis iteration program that this will give m inequivalent irreducible representations.

Let us increase the sophistication of our boundary diagram notation. We will represent the central primitive idempotent of $T_{m-1}(q)$ by (for example, with $m = 5$)

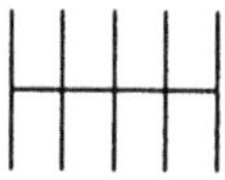

Of course this idempotent is well defined in larger algebras, so that for example

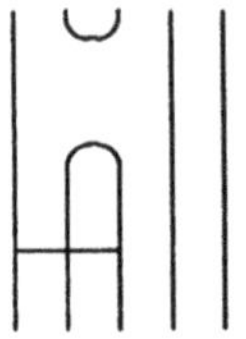

$$= 0$$

Note also that in $T_k(q)$ we have

 $-q$  $=$ 

Then the $m+1$ 1-dimensional $e_1^m T_1(q)$ modules are spanned, in the case $m = 4$ for example, by the basis vectors

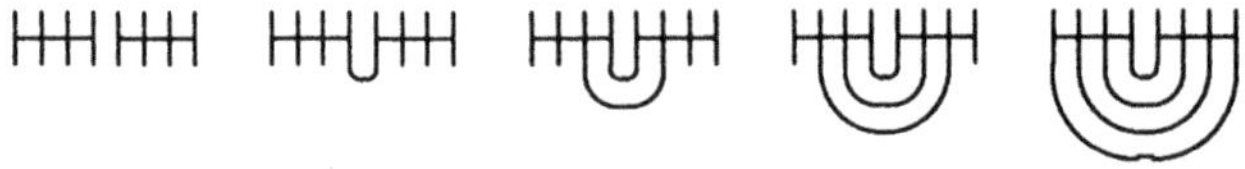

where as usual we have thrown away the (mirror symmetric) bottom half of each IWI diagram, as it remains unchanged; and where a quotient by vectors to the right is to be understood, also as in conventional Temperley-Lieb.

We must act with $g_1.\epsilon_1$ on the top of each of these and see what happens! To steamline things still further, we will represent any number of strings which move parallel to each other by a single line with an index giving the multiplicity. Every 1-dimensional space is then spanned by an object of the form

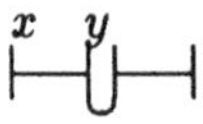

where $y + x = m$. We want to compute the first coefficient in the expansion

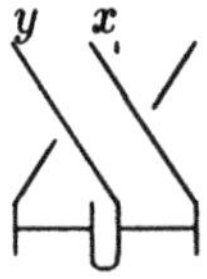 $= \sum_{z=y}^{m} C_z^m$ 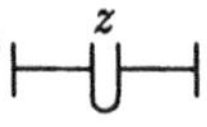

Note firstly that $g_i U_i = -q^2 U_i$, that is

 $= -q^2$ 

and hence $C_1^1 = -q^2$. Then, for example, with $x = 0, j = m - 1$, we can write

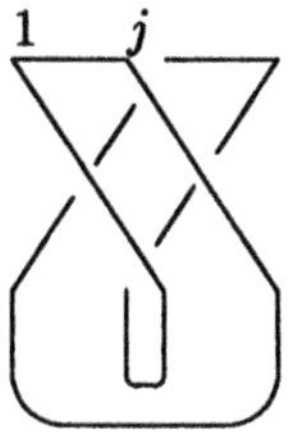
$= -q^2$
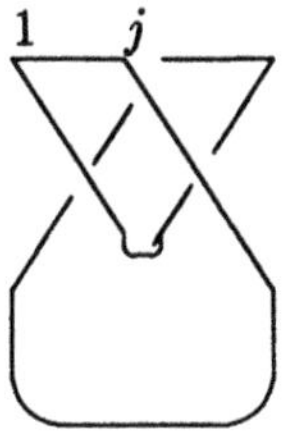
$= -q^2$
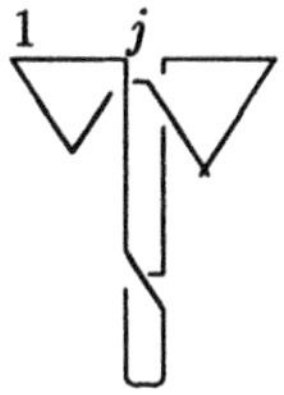

When the j line crosses the 1 line in the last picture the identity term in the expansion of the braid allows a U_i to reach, and hence kill, the idempotent ϵ. The only possibility is to take the $-qU_i$ term in the expansion of the braid. Obviously this happens j times on each side, so altogether we have

$= -q^2(-t^j)(-t^j)$
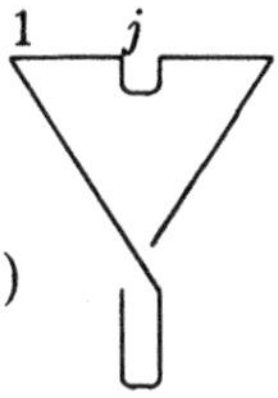
$= -q^{2(j+1)}C_j^j$
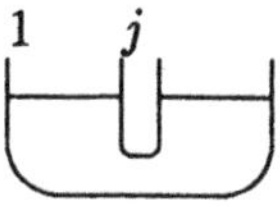

so that (using $C_1^1 = -q^2$) we have

$$C_m^m = (-1)^m q^{m(m+1)}.$$

Similarly, for the general case

$= C_y^y$
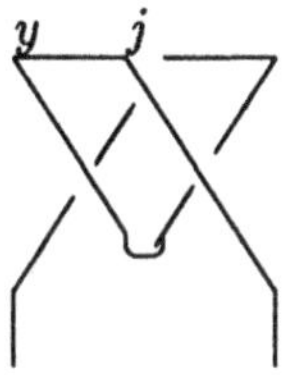
$= C_y^y$
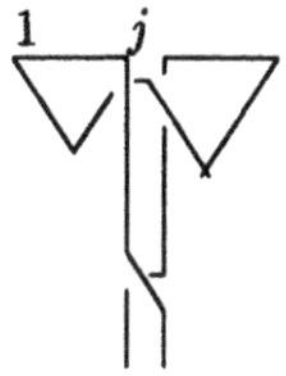

$= C_y^y(-q^{yj})(-q^{yj})$
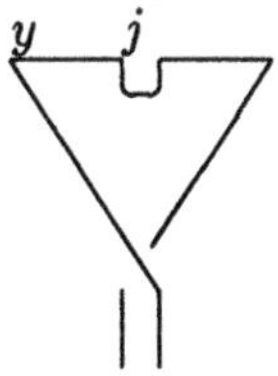
$= -q^{2(jy)}C_y^y$
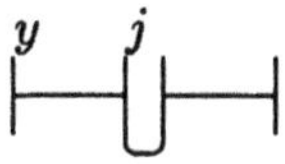

(where at the last stage we have used the quotient as well as the orthogonality property of the idempotent) so that

$$C_y^{j+y} = (-1)^y q^{y(y+1)+2yj}.$$

Overall, we see that for q indeterminate the local relation is never quadratic again, confirming that $\epsilon T_{k-1}(q) \not\cong T_{k-1}(q)$. On the other hand, we see that when q is a root of unity then a lower order local relation may occur, in which case the structure would also be different from the generic case. A quadratic local relation in particular would restore the possibility of $\epsilon T_{k-1}(q) \cong T_{k-1}(q)$.

Before pursuing this, what about other choices for ϵ...?

13.2.2 Other idempotents

What happens to the local relation if we take I to be some other idempotent?

The only possibility for $m = 2$ is $I' = 1.\epsilon_2 = \prod_{i\ odd}(U_i/\sqrt{Q})$. In this case we note that I' is a primitive idempotent for the whole algebra $T_{mk-1}(q)$, so

$$g(2) = (1 - 2(q+q^3)/\sqrt{Q} + q^4 - 2(q+q^3)\sqrt{Q} + (5+Q)q^2)I' = I'.$$

More generally, the idempotents are given, in the Temperley-Lieb case, in chapter 6. We note that each equivalence class may be represented by an idempotent of the form

$1/Q$ 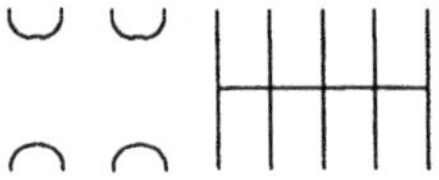

The subalgebra is thus generated by objects of the form

Q^{-4} 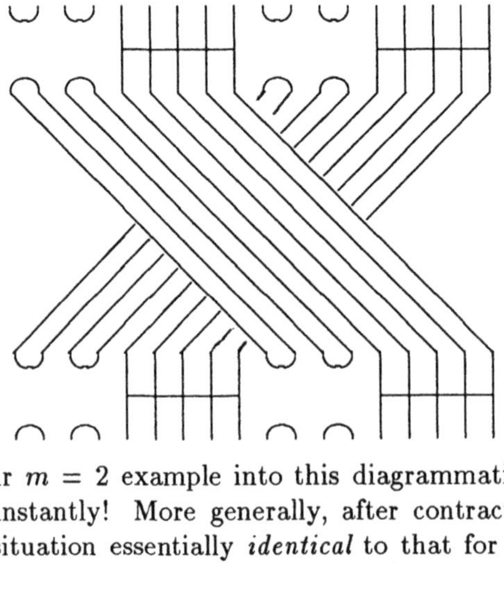

Plugging our $m = 2$ example into this diagrammatic scheme the answer comes out instantly! More generally, after contracting the closed loops we have a situation essentially *identical* to that for the central primitive idempotent at lower m obtained by ignoring the fixed cups and caps. The basis states for these algebras are also isomorphic to those for the case of the central primitive idempotent at lower m, noting that the apparent difference in the action of the generators simply produces a factor of $\sqrt{Q}$ for each cup, which then cancels with the normalisation on the idempotent.

Altogether then,

$$e_j^m T_{k-1}(q) \cong e_i^m T_{k-1}(q)$$

provided that the ϵ_1-like part of each idempotent has the same length.

Beraha q values

What happens when we apply this procedure to the Potts representation of the Temperley-Lieb algebra in the Beraha (non-generic) cases? In the Ising model, for example, we know that the idempotent E_5 vanishes (as it does in every unitarisable representation). This means we are not at liberty to use I as a projector except in the case $m = 2$. More generally, there are various tricks, in principle, for figuring out the structures of the special cases using the diagrammatic techniques described above. This programme had not been completed at time of going to press! The key question is, when does the cabled version of an algebra simply correspond to a cruder resolution in the same physical hierarchy of models?

Other source algebras

We give the list of primitive idempotents in the other quotients of the Hecke algebra, $NH_k(q)$, in chapter 9. From these it is just a question of turning the handle, controlling the calculation via the appropriate centraliser algebra, which is $U_q Sl(N)$ acting on its k-fold tensored fundamental representation (c.f. chapters 5 and 9).

Bibliography

Abe R 1967 Prog. Theor. Phys. 38,322.

Abe Y and A Katsura 1970 Prog. Theor. Phys. 43,1402

Abramowitz M and I A Stegun eds. 1965 Handbook of mathematical functions (New York: Dover).

Ahlfors L V 1979 Complex Analysis (McGraw Hill: New York).

Akutsu Y, A Kuniba and M Wadati 1986 J. Phys. Soc. Japan 55,1880, and references therein.

Anderson F W and K R Fuller 1973 Rings and Categories of Modules (New York: Springer Verlag).

Andrews G E, R J Baxter and P J Forrester 1984 J. Stat. Phys. 35,193.

Birman J S 1974 Ann. Math. Stud. 82,1.

Birman J S and H Wenzl 1987 Braids, link polynomials and a new algebra (unpublished).

Baxter R J 1973 J. Stat. Phys. 9,145.

\- 1982a Exactly Solved Models in Statistical Mechanics (New York: Academic)

\- 1982b Proc. R. Soc. A383,43.

\- 1985 in Integrable Systems in Statistical Mechanics (Singapore: World Scientific) p5.

\- 1986 J. Phys. A19,2821.

\- 1987 J. Phys A20,5241.

Baxter R J, S B Kelland and F Y Wu 1976 J.Phys. A9,397.

Baxter R J, H N V Temperley and S E Ashley 1978 Proc. R. Soc. A358,535.

Belavin A A, A M Polyakov and A B Zamolodchikov 1984 Nucl. Phys. B241,333.

Bellman R 1960 Introduction to Matrix Analysis (New York: McGraw Hill) p278.

Bjorken J D and S D Drell 1964 Relativistic Quantum Mechanics/ Relativistic Quantum Fields (New York: McGraw-Hill).

Blote H W J and M P Nightingale 1982 Physica 112A,405.

Cardy J 1987 in Phase Transitions and Critical Phenomena, eds. C Domb and J L Lebowitz, vol.11.

Cherednik I V 1986 Funct. Annal. and Appl. 20 N1,87-89.

Cohn P M 1989 Algebra (New York: Wiley).

Date E, M Jimbo, T Miwa and M Okado 1987 Nucl. Phys. B290 [FS20],231.

\- 1988 Adv. Stud. Pure Math. 16,17.

Deguchi T, M Wadati and Y Akutsu 1988 J. Phys. Soc. Japan 57,2921-2935.

De Vega H J 1989 Int. J. Mod. Phys. A4,2371.

Dipper R and G James 1987 Proc. London Math. Soc. 54,57.

Drinfeld V G 1986 in ICM proceedings, Berkeley, 798-820.

Einhorn M B, Savit R and Rabinovici E 1980 Nucl. Phys. B170 [FS1],16.

Elitzur S, R B Pearson and J Shigemitsu 1979 Phys. Rev. D19,3698.

Fisher M E 1964 in Lectures in Theoretical Physics vol.7c, ed. W E Brittin (Boulder: University of Colorado Press) p1.

Fraleigh J B 1976 A First Course in Abstract Algebra (Addison-Wesley: London).

Friedan D, Z Qiu and S Shenker 1984 Phys. Rev. Lett. 52,1575.

Goddard P, W Nahm and D Olive 1985 Phys. Lett. 160B,111.

Gutmann A J and Enting I G 1988 J. Phys. A21,L165, and private communication.

Gyoja A 1986 Osaka J. Math. 23,841.

Hamermesh M 1962 Group Theory (Oxford: Pergamon).

Hancock H 1958 Theory of Elliptic Functions (New York: Dover).

Hintermann A, H Kunz and F Y Wu 1978 J. Stat. Phys. 19,623.

Hoefsmit P N 1974 PhD thesis University of British Columbia.

Huang K 1963 Statistical Mechanics (New York: Wiley).

Huse D A 1984 Phys. Rev. B30,3908.

Ising E 1925 Z. Physik 31,253.

Jimbo M 1985 Lett. Math. Phys. 10,63-69.

Jimbo M, T Miwa and M Okado 1988 Commun. Math. Phys. 116,507-525.

Jones V F R 1983 Invent. Math. 72,1.

- 1985 Bull. Am. Math. Soc. 12,103.

Kac M and J C Ward 1952 Phys. Rev. 88,1332.

Kauffman L H c.1987 various unpublished works.

Kaufman B 1949 Phys Rev. 76,1232.

Kogut J B 1979 Rev. Mod. Phys. 51,659.

Kulish P P, N Yu Reshetikhin and E K Sklyanin 1981 I. Lett. Math. Phys. 5,393.

Kuniba A, Akutsu Y and M Wadati 1986 J. Phys. Soc. Japan 55,3285.

Landau L D and E M Lifshitz 1980 Statistical Physics (Lifshitz and L P Pitaevskii) (Oxford: Pergamon).

Lee T D and C N Yang Phys. Rev. 87,410.

Leech J W and D J Newman 1970 How to use Groups (London: Methuen).

Lickorish W B R 1988 Bull. London Math. Soc. 20,558-588.

Lusztig G 1989 unpublished (see also 1988 Advances in Math. 70,237).

Luther A and Peschel I 1975 Phys. Rev. B12,3908.

MacLane S and G Birkoff 1979 Algebra (London: Collier Macmillan).

Magnus W, A Karrass and D Solitar 1975 Combinatorial Group Theory (New York: Dover).

Maillard J M and Rammal R 1983 J. Phys. A16,353.

Martin P P 1983 Nucl. Phys. B225[FS9],497.

- 1985 in Integrable Systems in Statistical Mechanics, ed.s G M D'Ariano, A Montorsi and M G Rasetti (Singapore: World Scientific) p129.

- 1986a J. Phys. A19,3267

- 1986b J.Phys. A19,L1117.

- 1988 J. Phys. A21,577.

- 1988 J. Phys. A21,4415.

Martin P P and J M Maillard 1986 J. Phys. A19,L547.

McCoy B M and Wu T T 1973 The Two Dimensional Ising Model (Cambridge, MA: Harvard University Press).

Murakami J 1989 Osaka J. Math. 26,1.

Niemeyer Th, and J M J Van Leeuwen 1973 Phys. Rev. Lett. 34,1411; 1974 Physica 71,17-40.

Nienhuis B 1987 in Phase Transitions and Critical Phenomena, eds. C Domb and J L Lebowitz, vol.11.

Nilsen T S and P C Hemmer 1967 J. Chem. Phys. 46,2640.

Onsager L 1944 Phys. Rev. 54,117.

Pasquier V 1988 Nucl. Phys. B295 [FS21],491.

Pearson R B 1982 Phys. Rev. B26,6285.

Potts R B 1952 Proc. Camb. Phil. Soc. 48,106.

Ramond P 1989 Field Theory: A Modern Primer (New York: Addison Wesley).

Ravndal F 1976 Scaling and Renormalisation Groups (Copenhagen: Nordita).

Robinson G de B 1961 Representation Theory of the Symmetric Group (Toronto: University of Toronto Press).

Rosso M 1988 Commun. Math. Phys. 117,581.

Savit R 1980 Rev. Mod. Phys. 52,453.

Schultz T D, D C Mattis and E H Lieb 1964 Rev. Mod. Phys. 36,856.

Stanley H E 1971 Phase Transitions and Critical Phenomena (Oxford: Clarendon) p265.

Suzuki M and Fisher M E 1971 J. Math. Phys. A9,397.

Temperley H N V and E H Lieb 1971 Proc. R. Soc. A322,251.

Thacker H B and H Itoyama 1988 Nucl. Phys. B 5A,9.

Tsuchiya A and Y Kanie 1986 Publ. RIMS Kyoto Univ. 22,259; Duke Math. J. 53,1013.

Vdovichenko N V 1965 Soviet Physics JETP 21,350-2.

Wadati M, Y Yamada and T Deguchi 1989 J. Phys. Soc. Japan 58,1153.

Wenzl H 1988 Invent. Math. 92,349.

Whitney H 1932 Ann. Math. (N.Y.) 33,688.

Wood D W 1985 J. Phys. A18,L917.

Wu F Y 1982 Rev. Mod. Phys. 54,235.

Index

www.ingramcontent.com/pod-product-compliance
Ingram Content Group UK Ltd.
Pitfield, Milton Keynes, MK11 3LW, UK
UKHW020400150926
8624IPUK00004B/259

9 789810 200756